New Insights into Cardiovascular and Exercise Physiology

New Insights into Cardiovascular and Exercise Physiology

Guest Editors

Helena Lenasi
Ines Drenjančević

MDPI

Basel • Beijing • Wuhan • Barcelona • Belgrade • Novi Sad • Cluj • Manchester

Guest Editors
Helena Lenasi
Institute of Physiology
Faculty of Medicine
University of Ljubljana
Ljubljana
Slovenia

Ines Drenjančević
Department of Physiology
and Immunology
Faculty of Medicine Osijek
Josip Juraj Strossmayer
University of Osijek
Osijek
Croatia

Editorial Office
MDPI AG
Grosspeteranlage 5
4052 Basel, Switzerland

This is a reprint of the Special Issue, published open access by the journal *Life* (ISSN 2075-1729), freely accessible at: https://www.mdpi.com/journal/life/special_issues/82M07R46ZF.

For citation purposes, cite each article independently as indicated on the article page online and as indicated below:

Lastname, A.A.; Lastname, B.B. Article Title. *Journal Name* **Year**, *Volume Number*, Page Range.

ISBN 978-3-7258-3393-1 (Hbk)
ISBN 978-3-7258-3394-8 (PDF)
https://doi.org/10.3390/books978-3-7258-3394-8

Contents

About the Editors . . . . **vii**

Helena Lenasi and Ines Drenjančević
New Insights into Cardiovascular and Exercise Physiology: A Compendium of the Special Issue
Reprinted from: *Life* **2025**, *15*, 252, https://doi.org/10.3390/life15020252 1

Jernej Šorli and Helena Lenasi
The Effect of Acute Hyperglycaemia Induced by Oral Glucose Load on Heart Rate Variability and Skin Microvascular Reactivity in Young Adults
Reprinted from: *Life* **2024**, *14*, 56, https://doi.org/10.3390/life14010056 7

Luka Kolar, Petar Šušnjara, Marko Stupin, Ana Stupin, Ivana Jukić, Zrinka Mihaljević, et al.
Enhanced Microvascular Adaptation to Acute Physical Stress and Reduced Oxidative Stress in Male Athletes Who Consumed Chicken Eggs Enriched with n-3 Polyunsaturated Fatty Acids and Antioxidants—Randomized Clinical Trial
Reprinted from: *Life* **2023**, *13*, 2140, https://doi.org/10.3390/life13112140 23

Pedro Garcia Brás, António Valentim Gonçalves, João Ferreira Reis, Rita Ilhão Moreira, Tiago Pereira-da-Silva, Pedro Rio, et al.
Cardiopulmonary Exercise Testing in Patients with Heart Failure: Impact of Gender in Predictive Value for Heart Transplantation Listing
Reprinted from: *Life* **2023**, *13*, 1985, https://doi.org/10.3390/life13101985 39

Helena Lenasi, Eva Rihar, Jerneja Filipič, Matjaž Klemenc and Petja Fister
The Effect of Caffeine on Heart Rate Variability in Newborns: A Pilot Study
Reprinted from: *Life* **2023**, *13*, 1459, https://doi.org/10.3390/life13071459 56

Vassiliki Michou, Vassilios Liakopoulos, Stefanos Roumeliotis, Athanasios Roumeliotis, Maria Anifanti, Georgios Tsamos, et al.
Effects of Home-Based Exercise Training on Cardiac Autonomic Neuropathy and Metabolic Profile in Diabetic Hemodialysis Patients
Reprinted from: *Life* **2023**, *13*, 232, https://doi.org/10.3390/life13010232 66

Zdravko Anicic, Danica Janicijevic, Olivera M. Knezevic, Amador Garcia-Ramos, Milos R. Petrovic, Dimitrije Cabarkapa and Dragan M. Mirkov
Assessment of Countermovement Jump: What Should We Report?
Reprinted from: *Life* **2023**, *13*, 190, https://doi.org/10.3390/life13010190 83

Iva Jurov, Janez Toplišek and Marta Cvijić
Prediction of Maximal Oxygen Consumption in Cycle Ergometry in Competitive Cyclists
Reprinted from: *Life* **2023**, *13*, 160, https://doi.org/10.3390/life13010160 97

Paul Zimmermann, Nico Müller, Volker Schöffl, Benedikt Ehrlich, Othmar Moser and Isabelle Schöffl
The Energetic Costs of Uphill Locomotion in Trail Running: Physiological Consequences Due to Uphill Locomotion Pattern—A Feasibility Study
Reprinted from: *Life* **2022**, *12*, 2070, https://doi.org/10.3390/life12122070 . . . . **108**

Mojca Škafar, Jana Ambrožič, Janez Toplišek and Marta Cvijić
Role of Exercise Stress Echocardiography in Pulmonary Hypertension
Reprinted from: *Life* **2023**, *13*, 1385, https://doi.org/10.3390/life13061385 . . . . **121**

Iva Jurov, Jure Demšar and Thomas McCurdy
A Meta-Analysis of Sampled Maximal Aerobic Capacity Data for Boys Aged 11 Years Old or Less Obtained by Cycle Ergometry
Reprinted from: *Life* **2023**, *13*, 276, https://doi.org/10.3390/life13020276 . . . . **134**

Anton Ušaj, Jon Lihteneger Vidmajer and Sonja Lojen
Recovery after Running an "Everesting" Mountain Ultramarathon
Reprinted from: *Life* **2023**, *13*, 1946, https://doi.org/10.3390/life13101946 . . . . **152**

Ilija Srdanović, Dragana Dabović, Vladimir Ivanović, Milenko Čanković, Teodora Pantić, Maja Stefanović, et al.
Takotsubo Cardiomyopathy Occurring Simultaneously with Acute Myocardial Infarction
Reprinted from: *Life* **2023**, *13*, 1770, https://doi.org/10.3390/life13081770 . . . . **167**

About the Editors

Helena Lenasi

Helena Lenasi, MD, PhD is an Associate Professor of Physiology at the Faculty of Medicine at the University of Ljubljana. She completed her medical studies at the Faculty of Medicine in Ljubljana, Slovenia, in 1996, and defended her Master of Science degree at the same faculty in 2001 and her doctorate in 2003.

She mainly teaches physiology to medical and dental medicine students as well as students of pharmacy and the sports faculty. She serves as book and journal editor and revisor and has published several scientific papers. Her main scientific interests include cardiovascular and sports physiology, with a particular focus on human microcirculation and its (patho)physiological background.

Ines Drenjančević

Ines Drenjančević, MD, PhD is a Full-Time Professor at the Faculty of Medicine Osijek, Institute and Dept of Physiology and Immunology. She completed her medical studies at the Faculty of Medicine in Zagreb, Croatia, in 1993. She defended her Master of Science degree at the Faculty of Medicine in Zagreb, Croatia, in 1997 and received her doctorate in 2004 from the Medical College of Wisconsin, WI, USA. Prof Drenjančević is an award-winning nationally and internationally recognized scientist. She has written an abundance of scientific papers, and she stands out in particular for the number and quality of scientific and infrastructure projects that she has led at her home institution. She has been the Head of the National Scientific Center of Excellence for Personalized Health Care since 2015, and she has received numerous awards for her scientific work (Medical Faculty Osijek, Croatian Physiological Society, Republic of Croatia—Annual State Award for Science 2017; international awards (Ukraine, the Microcirculatory Society, USA).

Scientific interests: microcirculation; diet; oxidative stress; mechanisms of endothelium-mediated vascular reactivity; blood pressure regulation; cardiovascular physiology.

life

Editorial

New Insights into Cardiovascular and Exercise Physiology: A Compendium of the Special Issue

Helena Lenasi [1,*] and Ines Drenjančević [2,3,*]

1 Institute of Physiology, Faculty of Medicine, University of Ljubljana, 1000 Ljubljana, Slovenia
2 Department of Physiology and Immunology, Faculty of Medicine Osijek, University Josip Juraj Strossmayer, 31000 Osijek, Croatia
3 Scientific Centre of Excellence for Personalized Health Care University Josip Juraj Strossmayer, 31000 Osijek, Croatia
* Correspondence: helena.lenasi.ml@mf.uni-lj.si (H.L.); ines.drenjancevic@mefos.hr (I.D.)

1. Introduction

The capacity of the cardiovascular system to adjust to varying needs is immense. This Special Issue aims to present recent knowledge in cardiovascular and exercise physiology, and address some clinical implications.

Considering that the incidence of cardiovascular diseases is increasing, and that cardiovascular diseases present a major cause of death worldwide, the need to accomplish treatment strategies is also increasing. The knowledge of the physiological processes behind them remains a prerequisite for upgraded clinical studies.

Although many new mechanisms on the micro- and macro-scales have been revealed in recent years, which is also due to the evolution of modern techniques and new concepts that are emerging almost daily, there are still many controversies and enigmas. The never-ending challenge remains how to translate the findings obtained in cell cultures and studies using animal models into human physiology. In this respect, in vivo studies on humans are encouraged, as they reflect the real situation, and could strongly support clinical work.

Physical exercise has been increasingly regarded as one of the potentially beneficial measures to improve cardiovascular health, interfering with numerous elements of the cardiovascular system and implying a multitude of potential mechanisms [1]. Its long-term beneficial effects are well known; nevertheless, many questions remain unresolved. Where is the line between benefit and harm? How do different types of exercise affect the cardiovascular system in health and disease? What is the most appropriate measure regarding duration, repetition, and recovery, and whether and how exercise regimes can be individually adjusted? In addition, a proper evaluation of the physical parameters and cardio-pulmonary and training status are crucial, both in the light of the athlete's achievements and improvements, and from the clinical point of view.

In this Special Issue, scientists from various fields of cardiovascular and/or sports medicine have presented their new findings: all studies were performed in humans, from investigations obtained in healthy individuals, ranging throughout the lifespan, from newborns to elderly, as well as in patients. The compilation encompasses eight original articles, two review papers, and two case reports, which are presented below, first presenting papers dealing with healthy persons, and later focusing on patients.

Received: 22 January 2025
Revised: 3 February 2025
Accepted: 5 February 2025
Published: 7 February 2025

Citation: Lenasi, H.; Drenjančević, I. New Insights into Cardiovascular and Exercise Physiology: A Compendium of the Special Issue. *Life* **2025**, *15*, 252. https://doi.org/10.3390/life15020252

2. An Overview of the Published Articles

To start with the study involving the youngest participants, Lenasi et al. investigated the effects of caffeine in preterm newborns [2]. Caffeine, a central nervous system (CNS) stimulant of the methylxanthine class, exerts many complex and pleiotropic physiological effects on various organ systems. In newborns, it has empirically been used to treat apnea [3]. However, the study by Lenasi et al. aimed to assess whether it affects the maturation of the autonomic nervous system (ANS) in preterm newborns by analyzing their heart rate variability (HRV). Although caffeine increased breathing frequency (BF), it did not affect arterial oxygen saturation, body temperature, or the heart rate during sleeping in a supine position or any other HRV parameters. It appears that the maintenance dose of caffeine is too low to affect the heart rate and HRV. This study challenges further research to establish the exact mechanisms of caffeine's multiple beneficial effects. Moreover, it exposes HRV as a useful tool for assessing the activity of the ANS in human studies.

Another study that also applied HRV as a measure of ANS activity was the study by Šorli and Lenasi [4], which elucidated the potential impacts of acute hyperglycemia induced by a standard oral glucose loading test on the microvascular reactivity at the time point of the peak plasma glucose concentration in young, healthy participants. Acute effects of hyperglycemia on microcirculation remain controversial in healthy individuals; decreased endothelial functions have been implicated, likely due to increased oxidative stress, decreased endothelial nitric oxide synthase (eNOS) activity, and the reduced bioavailability of vasodilator nitric oxide (NO) [5]. As insulin was also shown to interfere with endothelial signaling pathways, how these (to some extent) contradictory mechanisms interact at the vascular level in vivo is debatable. The study showed significant correlations between some post-occlusive reactive hyperemia (PORH) parameters, HRV parameters, and the plasma glucose concentration after glucose load, implying diminished vascular reactivity evoked by hyperglycemia in healthy subjects with lower glucose tolerance. Accordingly, this study exposes the need to develop stronger tools/markers for the timely detection of predisposed individuals.

Microvascular reactivity is mediated by endothelium-dependent mechanisms, which could be modified by exercise, but also by various external influences, such as a diet, related to a decrease in oxidative stress and an increase in antioxidative mechanisms. In athletes, although acute strenuous exercise leads to acute endothelial dysfunction, exercise increases their microvascular responsiveness, or leads to better utilization of vasodilatory capacity, consistent with the hormesis hypothesis. Dietary intake of n-3 polyunsaturated fatty acids [6] and antioxidant-enriched food has a beneficial effect on vasodilation. The results of the study of Kolar et al. provide evidence that athletes should enrich their diet with natural sources of micronutrients in the form of functional foods [7].

The next studies addressed the cardiopulmonary exercise testing (CPET). The study by Jurov et al. [8] sought an optimal model to estimate maximal oxygen consumption (VO_{2max}) in trained cyclists. The general prediction equations applicable in nontrained individuals may underestimate the real VO_{2max} in trained persons. This study, conducted on 496 male and 84 female competitive cyclists, included six predictors: power output, body weight, body height, fat mass, fat-free mass, and age, and compared its accuracy to the traditional American College of Sports Medicine (ACSM) equation [9]. Three new equations were finally derived; power output and body weight were shown to be the most important parameters for predicting VO_{2max}. This study demonstrates that the traditional ACSM equation strongly underestimates VO_{2max}, exposing the need to evaluate prediction models for other athletes with a special focus on disciplines that demand high aerobic capacity.

Another valuable test used for monitoring sports performance and addressing the effectiveness of different training programs and neuromuscular fatigue is the countermovement jump (CMJ) with force platform. This device is considered the gold standard testing device, yielding different variable outputs [10]. The reliability of the CMJ-derived variables is greatly affected by the complexity of the computational methods: more computation renders less reliability. The study by Aničić et al. [11] aimed to assemble the list of necessary, highly reliable metrics derived from CMJ, and showed that only 24 out of 45 CMJ-derived variables had an acceptable reliability. The most reliable variables were performance variables, followed by kinetic variables, and finally kinematic variables. These variables were included in the principal component analysis, and loaded a total of four factors, explaining 91% of the CMJ variance: the performance component (variables responsible for overall jump performance), eccentric component (variables related to the breaking phase), concentric component (variables related to the upward phase), and jump strategy component (variables influencing the jumping style). This study revealed important implications for sports scientists and practitioners regarding the CMJ-derived metrics.

The study by Zimmermann et al. [12] aimed to define the physiological differences between two uphill locomotion patterns, namely uphill running versus uphill walking, in trail running (TR) athletes. The obtained data of the TR athletes were compared for anthropometric data and CPET parameters, such as maximal ventilation ($V^{\cdot}E_{max}$), VO_{2max}, maximal BF (BF_{max}), peak oxygen pulse, and the energy cost of running (C_r). All TR athletes showed excellent performance data, whereby across both different uphill locomotion strategies, significant differences were revealed solely for $V^{\cdot}E_{max}$ and time to reach the mountain peak. These results provide new insights, and might contribute to a comprehensive understanding of cardiorespiratory consequences of short uphill locomotion strategy in TR athletes.

The next study, by Jurov et al. [13], is a meta-analysis investigating associations between VO_{2max} and body mass, year of the study, and country of origin of healthy prepubertal boys (mean age under 11 years old). Study aimed to extract new reference values for cardiorespiratory fitness, considering 95 study samples. The results show that the absolute VO_{2max} ($Lmin^{-1}$) is higher in more recent studies, whereas the mean relative VO_{2max} is lower, stressing the important impact of body mass. Aerobic capacity normalized to body weight does not change with age. Cardiorespiratory fitness (CRF) in prepubertal boys is declining, associated with an increasing body mass over the last few decades. The study stresses the importance of distinguishing between the VO_{2max} and peak oxygen consumption (VO_{2peak}) indicators. Given the lack of age- and gender-specific CRF reference values in prepubertal children, the study finally exposes a need to develop observed distributions of VO_{2max} based on criterion methods rather than estimated or regression-based predicted values and to compare the VO_{2max} between prepubertal boys and girls.

In addition to healthy individuals, exercise testing is a key in the risk stratification of patients with heart failure (HF). The study of Bras et al. [14] aimed to assess the predictive value of the heart transplantation (HTx) thresholds in HF in women and men by performing a prospective evaluation of HF patients who underwent CPET from 2009 to 2018 for the composite endpoint of cardiovascular mortality and urgent HTx. A total of 458 patients underwent CPET, with a composite endpoint frequency of 10.5% in females vs. 16.0% in males in the 36-month follow-up. According to the 2016 International Society for Heart Lung Transplantation (ISHLT) listing criteria for HTx, the recommended thresholds for pVO_2 ($\leq$12 mL/kg/min, or $\leq$14 mL/kg/min if intolerant to β-blockers), minute ventilation–CO_2 production relationship (VE/VCO_2 slope > 35), and percent of predicted $VO_{2peak} \leq 55\%$, showed a significantly higher overall diagnostic effectiveness in women compared to men [15]. Moreover, specific VO_{2peak}, VE/VCO_2 slope, and percent of predicted VO_{2peak} cut-offs in each sex group presented a higher prognostic power than the recommended thresholds. The authors concluded that individualized sex-specific thresholds may improve patient selection for HTx.

The narrative review by Škafar et al. [16] evaluated exercise stress echocardiography (ESE), a non-invasive, inexpensive, and widely available imaging modality for estimating systolic pulmonary arterial pressure (sPAP) as a method for diagnosis of pulmonary hypertension (PH), which is associated with numerous respiratory and/or cardiovascular diseases, and diagnosed by right heart catheterization. The newest ESC Guidelines for the diagnosis and treatment of PH reintroduced the diagnosis of the exercise PH [17]. ESE can distinguish between noncardiac and cardiac causes of unexplained dyspnea, and is useful in patients with connective tissue disease. However, there is a lack of validated criteria and prospective clinical studies. The paper reviews pulmonary hemodynamic responses to exercise, briefly describes the modalities for assessing pulmonary hemodynamics, and discusses contemporary key clinical applications of ESE in patients with PH.

The study by Michou et al. [18] also addressed patients. Increased physical activity is highly recommended in patients with diabetes mellitus and one of its complications, namely chronic kidney disease. Cardiac autonomic neuropathy is a complication of diabetes, and may contribute to increased risks. Since many diabetic patients with chronic kidney disease require hemodialysis, maintaining an exercise routine might be demanding for them. Thus, home-based exercise programs may help improve their overall fitness and cardiac functioning. The randomized controlled study assessed the effects of a 6-month home-based exercise training program consisting of three combined (aerobic and strengthening) exercise sessions per week on the non-dialysis days. The results demonstrated increased levels of HDL and a decrease in HbA1c, improved cardiovascular fitness indices, including decreased heart rate and systolic blood pressure, and favorable effects on sympathetic and vagal nerve activity and the sympathovagal balance.

The two last studies, by Ušaj et al. and Srdanović et al., respectively, are case reports.

Ušaj et al. [19] traced a single runner with a VO_{2peak} 74 $mL \cdot min^{-1} \cdot kg^{-1}$ over a preparatory period of two months (Everesting) and the subsequent one-month recovery period, which is a known period of reduced performance. During the first phase of the recovery, enhanced peak creatine kinase (800%) and C-reactive protein (44%) levels explained the decreased performance. In contrast, decreased performance during the second, longer phase was associated with a decreased lactate threshold and VO_2 (21% and 17%, respectively), as well as an increased energetic cost of running (15%) and higher endogenous carbohydrate oxidation rates (87%), lactate concentrations (170%), and respiratory muscle fatigue sensations that remained elevated for up to one month. These alterations may represent the characteristics of the second phase of the recovery process after Everesting.

Takotsubo Cardiomyopathy (TCM) is a reversible cardiomyopathy characterized by transient regional systolic dysfunction of the left ventricle. It can clinically mimic myocardial infarction without obstruction of coronary arteries; however, it presents as a distinct condition. Interestingly, TCM can be concomitant to acute myocardial infarction with ST elevation (STEMI), which is the case in 5–8% of women with STEMI. Srdanović et al. [20] presented a 78-year-old woman who experienced simultaneous TCM and STEMI, which was successfully resolved with intensive medical care. This case showed that concomitant TCM and STEMI can lead to cardiogenic shock, which makes treatment challenging.

3. Concluding Remarks

This compilation of articles encompasses a diverse range of research, aspects, and approaches to exercise, with the common denominator being that they were all performed in humans, including healthy participants with differences in cardiopulmonary status and patients. This Special Issue stresses the importance of exercise as a daily routine, also appropriate as home-based exercises which might postpone, or at least ameliorate, disease progression. Overall, the findings reveal important implications for sports scientists, practitioners, and clinicians, stressing the importance of adjusted and proper exercise testing in healthy and in patients. Many manuscripts support the need for gender stratifications when adjusting either exercise testing or exercise regimes or evaluating a cohort of healthy participants and patients. In addition, the articles expose a need for standardization of various diagnostic tools in sports medicine and standardization of the evaluated parameters. In the future, special emphasis should be given to establishing the cut-off/predicted/recommended values of VO_{2max} in children, athletes, and patients, adjusted to the particular diseases. Since cardiovascular physiology is a huge field, encompassing many complex players and mechanisms, this Special Issue also exposes some unresolved questions, and encourages further research in this vast and complex field.

We believe this Special Issue would attract readers, with each manuscript included in this Special Issue uniquely contributing a puzzle to the whole mosaic and opening challenges and questions for future research.

Conflicts of Interest: The authors declare no conflicts of interest.

References

1. Isath, A.; Koziol, K.J.; Garber, C.E.; Martinez, M.N.; Emery, M.S.; Baggish, A.L.; Naidu, S.S.; Lavie, C.J.; Arena, R.; Krittanawong, C. Exercise and cardiovascular health: A state-of-the-art review. *Prog. Cardiovasc. Dis.* **2023**, *79*, 44–52. [CrossRef] [PubMed]
2. Lenasi, H.; Rihar, E.; Filipič, J.; Klemenc, M.; Fister, P. The Effect of Caffeine on Heart Rate Variability in Newborns: A Pilot Study. *Life* **2023**, *13*, 1459. [CrossRef] [PubMed]
3. Long, J.-Y.; Guo, H.-L.; He, X.; Hu, Y.-H.; Xia, Y.; Cheng, R.; Ding, X.-S.; Chen, F.; Xu, J. Caffeine for the Pharmacological Treatment of Apnea of Prematurity in the NICU: Dose Selection Conundrum, Therapeutic Drug Monitoring and Genetic Factors. *Front. Pharmacol.* **2021**, *12*, 681842. [CrossRef] [PubMed]
4. Šorli, J.; Lenasi, H. The Effect of Acute Hyperglycaemia Induced by Oral Glucose Load on Heart Rate Variability and Skin Microvascular Reactivity in Young Adults. *Life* **2024**, *14*, 56. [CrossRef] [PubMed]
5. Mapanga, R.F.; Essop, M.F. Damaging Effects of Hyperglycemia on Cardiovascular Function: Spotlight on Glucose Metabolic Pathways. *Am. J. Physiol. Heart Circ. Physiol.* **2016**, *310*, H153–H173. [CrossRef]
6. Oppedisano, F.; Macrì, R.; Gliozzi, M.; Musolino, V.; Carresi, C.; Maiuolo, J.; Bosco, F.; Nucera, S.; Caterina Zito, M.; Guarnieri, L.; et al. The Anti-Inflammatory and Antioxidant Properties of n-3 PUFAs: Their Role in Cardiovascular Protection. *Biomedicines* **2020**, *8*, 306. [CrossRef] [PubMed]
7. Kolar, L.; Šušnjara, P.; Stupin, M.; Stupin, A.; Jukić, I.; Mihaljević, Z.; Kolobarić, N.; Bebek, I.; Nejašmić, D.; Lovrić, M.; et al. Enhanced Microvascular Adaptation to Acute Physical Stress and Reduced Oxidative Stress in Male Athletes Who Consumed Chicken Eggs Enriched with n-3 Polyunsaturated Fatty Acids and Antioxidants—Randomized Clinical Trial. *Life* **2023**, *13*, 2140. [CrossRef] [PubMed]
8. Jurov, I.; Toplišek, J.; Cvijić, M. Prediction of Maximal Oxygen Consumption in Cycle Ergometry in Competitive Cyclists. *Life* **2023**, *13*, 160. [CrossRef] [PubMed]
9. Balady, G.J.; Arena, R.; Sietsema, K.; Myers, J.; Coke, L.; Fletcher, G.F.; Forman, D.; Franklin, B.; Guazzi, M.; Gulati, M.; et al. Clinician's Guide to Cardiopulmonary Exercise Testing in Adults: A Scientific Statement from the American Heart Association. *Circulation* **2010**, *122*, 191–225. [CrossRef] [PubMed]
10. Claudino, J.G.; Cronin, J.; Mezêncio, B.; McMaster, D.T.; McGuigan, M.; Tricoli, V.; Amadio, A.C.; Serrão, J.C. The countermovement jump to monitor neuromuscular status: A meta-analysis. *J. Sci. Med. Sport* **2017**, *20*, 397–402. [CrossRef]
11. Anicic, Z.; Janicijevic, D.; Knezevic, O.M.; Garcia-Ramos, A.; Petrovic, M.R.; Cabarkapa, D.; Mirkov, D.M. Assessment of Countermovement Jump: What Should We Report? *Life* **2023**, *13*, 190. [CrossRef]
12. Zimmermann, P.; Müller, N.; Schöffl, V.; Ehrlich, B.; Moser, O.; Schöffl, I. The Energetic Costs of Uphill Locomotion in Trail Running: Physiological Consequences Due to Uphill Locomotion Pattern—A Feasibility Study. *Life* **2022**, *12*, 2070. [CrossRef]
13. Jurov, I.; Demšar, J.; McCurdy, T. A Meta-Analysis of Sampled Maximal Aerobic Capacity Data for Boys Aged 11 Years Old or Less Obtained by Cycle Ergometry. *Life* **2023**, *13*, 276. [CrossRef]
14. Brás, P.G.; Gonçalves, A.V.; Reis, J.F.; Moreira, R.I.; Pereira-da-Silva, T.; Rio, P.; Timóteo, A.T.; Silva, S.; Soares, R.M.; Cruz Ferreira, R. Cardiopulmonary Exercise Testing in Patients with Heart Failure: Impact of Gender in Predictive Value for Heart Transplantation Listing. *Life* **2023**, *13*, 1985. [CrossRef] [PubMed]
15. Mehra, M.R.; Canter, C.E.; Hannan, M.M.; Semigran, M.J.; Uber, P.A.; Baran, D.A.; Danziger-Isakov, L.; Kirklin, J.K.; Kirk, R.; Kushwaha, S.S.; et al. The 2016 International Society for Heart Lung Transplantation listing criteria for heart transplantation: A 10-year update. *J. Heart Lung Transpl.* **2016**, *35*, 1–23. [CrossRef]

16. Škafar, M.; Ambrožič, J.; Toplišek, J.; Cvijić, M. Role of Exercise Stress Echocardiography in Pulmonary Hypertension. *Life* **2023**, *13*, 1385. [CrossRef]
17. Humbert, M.; Kovacs, G.; Hoeper, M.M.; Badagliacca, R.; Berger, R.M.F.; Brida, M.; Carlsen, J.; Coats, A.J.S.; Escribano-Subias, P.; Ferrari, P.; et al. 2022 ESC/ERS Guidelines for the diagnosis and treatment of pulmonary hypertension. *Eur. Heart J.* **2022**, *43*, 3618–3731. [CrossRef] [PubMed]
18. Michou, V.; Liakopoulos, V.; Roumeliotis, S.; Roumeliotis, A.; Anifanti, M.; Tsamos, G.; Papagianni, A.; Zempekakis, P.; Deligiannis, A.; Kouidi, E. Effects of Home-Based Exercise Training on Cardiac Autonomic Neuropathy and Metabolic Profile in Diabetic Hemodialysis Patients. *Life* **2023**, *13*, 232. [CrossRef] [PubMed]
19. Ušaj, A.; Lihteneger Vidmajer, J.; Lojen, S. Recovery after Running an "Everesting" Mountain Ultramarathon. *Life* **2023**, *13*, 1946. [CrossRef] [PubMed]
20. Srdanović, I.; Dabović, D.; Ivanović, V.; Čanković, M.; Pantić, T.; Stefanović, M.; Dimić, S.; Crnomarković, B.; Bjelobrk, M.; Govedarica, M.; et al. Takotsubo Cardiomyopathy Occurring Simultaneously with Acute Myocardial Infarction. *Life* **2023**, *13*, 1770. [CrossRef] [PubMed]

life

MDPI

Article

The Effect of Acute Hyperglycaemia Induced by Oral Glucose Load on Heart Rate Variability and Skin Microvascular Reactivity in Young Adults

Jernej Šorli [1] and Helena Lenasi [2,*]

[1] General Hospital Dr. Franc Derganc, 5290 Šempeter pri Gorici, Slovenia
[2] Institute of Physiology, Faculty of Medicine, University of Ljubljana, Zaloška 4, 1000 Ljubljana, Slovenia
* Correspondence: helena.lenasi.ml@mf.uni-lj.si

Abstract: We aimed to elucidate the effects of acute hyperglycaemia, induced by an oral glucose tolerance test (OGTT), on the autonomic nervous system (ANS) and skin microvascular reactivity at the time point of peak plasma glucose concentration (c_{glc}) in 20 young, healthy participants. We assessed their heart rate variability (HRV) as a measure of the ANS activity and the parameters of post-occlusive reactive hyperaemia (PORH) to estimate skin microvascular reactivity as measured by laser Doppler (LD) fluxmetry. The tests were repeated 30 min after a standard OGTT (75 g glucose dissolved in 250 mL water) and, in a separate control experiment, after drinking the same amount of water. Participants had their c_{glc} and serum insulin measured at three consecutive time-points according to the testing protocol. The low-frequency (LF) spectral power, the LF to high-frequency (LF/HF) ratio, and the diastolic blood pressure increased significantly more after water than after OGTT, and there was a trend of the peak LD flux of PORH decreasing more after OGTT than after water. Significant correlations between some PORH and all the HRV parameters and c_{glc} increase after OGTT were found, implying diminished vascular reactivity evoked by hyperglycaemia in healthy subjects with lower glucose tolerance.

Keywords: hyperglycaemia; oral glucose tolerance test; heart rate variability; sympathetic nervous system; insulin; laser Doppler fluxmetry; microcirculation; endothelium; post-occlusive reactive hyperaemia; glucose tolerance

Citation: Šorli, J.; Lenasi, H. The Effect of Acute Hyperglycaemia Induced by Oral Glucose Load on Heart Rate Variability and Skin Microvascular Reactivity in Young Adults. *Life* **2024**, *14*, 56. https://doi.org/10.3390/life14010056

Academic Editor: Panagiotis Georgianos

Received: 13 November 2023
Revised: 22 December 2023
Accepted: 26 December 2023
Published: 28 December 2023

1. Introduction

The long-term effects of chronic hyperglycaemia on vascular integrity are well known, with the principal pathogenic mechanism being (micro)vascular endothelial dysfunction [1,2]. In endothelial cells, hyperglycaemia has been shown to cause mitochondrial hyperpolarization and endothelial nitric oxide synthase (eNOS) uncoupling, thereby increasing reactive oxygen species (ROS) and decreasing endothelial vasodilators production as well as activating inflammatory pathways, which impairs endothelial vasodilatory capacity [1–3]. According to in vitro studies, it usually takes hours to days for ROS production to increase, with evidence suggesting alterations in RNA expression [3–5]. On the other hand, decreased vasodilator production was observed after only 20 min incubation in high-glucose solutions in some experimental settings; however, the effect was significant at unphysiologically high glucose concentrations (higher than 25 mM) and was absent at lower concentrations [5,6].

Interestingly, acute hyperglycaemia has been shown to impair microvascular dilation in humans in vivo on different timescales, ranging from 20 min to three hours, depending on experimental settings [7–14], although with several opposing findings [15–21]. In some studies, even an improvement of vascular function during exposure to acute hyperglycaemia was observed [18,20]. Several authors suggested that the endothelium-dependent vasodilation is impaired only in states of deranged insulin sensitivity and not in healthy subjects [14,16,19,22,23]. The discrepant findings imply that in vivo, the mechanisms are far more complex, and there might be other endothelium-dependent and -independent mechanisms altering the vasodilatory capacity during acute hyperglycaemia.

One possible mechanism affecting the vascular vasodilatory potential might be mediated by insulin, which was shown to increase endothelial vasodilator production in various settings in vitro [24–27]. The action of insulin on endothelial cells was observed in seconds following insulin administration, rendering the glucose- and insulin-counteracting hypothesis viable. Moreover, there is evidence for the in vivo effects of insulin on microcirculation: Iredahl et al. showed that transdermal application of insulin induced endothelium-dependent vasodilation in skin microcirculation in healthy subjects, presumably mediated by nitric oxide (NO) [28], while Rossi et al. implicated its effect on the vascular smooth muscle [29]. In addition, insulin was implicated in the increase in functional capillary density, which has been shown to directly correlate with insulin sensitivity in human skin microcirculation, reinforcing the idea that capillary recruitment plays an important role in enhancing insulin-mediated glucose uptake [30].

Another mechanism by which acute hyperglycaemia affects vascular reactivity might be mediated by the autonomic nervous system (ANS). Both glucose and insulin have been shown to alter the ANS activity in humans [18,31–39], although with some opposing findings [20,40–42].

It has not been thoroughly evaluated how these to some extent contradictory mechanisms integrate at the level of skin microcirculation, which has a complex anatomical and functional organization [43–45]. While microvasculature in all skin areas is innervated by the vasoconstrictor fibres of the sympathetic nervous system (SNS), the non-glabrous, non-acral skin areas also receive the sympathetic vasodilatory fibres. However, the noradrenergic vasoconstrictor response is the most powerful in glabrous acral parts with abundant arteriovenous anastomoses (AVAs), i.e., special anatomical structures that are not present in other skin sites [43–45]. Other control mechanisms include hypoxic metabolites, intrinsic myogenic activity, axon reflex mediated, and, most importantly, endothelium-dependent mechanisms, with different contributions thereof to the regulation of vascular tone in various sites [43,44]. The abundance of different mechanisms involved in the regulation of microvascular blood flow makes an evaluation of a particular mechanism separately from the others difficult. Moreover, due to large spatial and temporal variability, rather than measuring resting blood flow, vascular reactivity should usually be assessed [44,46].

In light of that mentioned above, our aim was to investigate the impact of acute hyperglycaemia induced by an OGTT on microvascular function in young, healthy volunteers at the time point of peak plasma glucose concentration (c_{glc}). To this end, we assessed skin microvascular reactivity by inducing post-occlusive reactive hyperaemia (PORH), a potential measure of endothelial function, and performed heart rate variability (HRV) analysis to assess ANS. Additionally, we measured plasma glucose and insulin concentration at particulate time points before and after OGTT. Based on animal model studies and some previous reports, we hypothesised that vascular reactivity would be diminished after OGTT, potentially reflecting impaired endothelium-dependent vasodilation and/or alterations in the activity of the ANS.

2. Materials and Methods

2.1. Participants

Twenty healthy volunteers (15 males, 5 females) aged 21.6 years ± 1.6 years, non-smokers, and all with a normal BMI were invited to undergo the study protocol (ingestion of high-glucose solution dissolved in water) and the control protocol (ingestion of water) on two separate occasions. Female participants were confirmed to have been in the follicular phase of the menstrual cycle. The participants were asked to attend the trial in a fasting state (after having consumed "normal" diet in the last week before the measurements), to ingest nothing but water in small amounts 12 h prior to arrival to the laboratory, and to avoid strenuous exercise 24 h before the trial; none of them had been taking any medications.

The study was approved by the National Medical Ethics Committee (No. 0120-175/2017/6) and conducted in accordance with the principles of the Declaration of Helsinki and the European Convention on Human Rights and Biomedicine. Written informed consent was obtained from the participants before starting the experiment.

2.2. Skin Perfusion and Reactive Hyperaemia Assessment

Skin perfusion was assessed by means of laser Doppler (LD) fluximetry, using Periflux 4001 Master/4002 Satellite system (Perimed, Järfälla, Sweden). In short, the method is based on the Doppler shift of the laser light penetrating approximately 1 mm in depth and reflected from moving erythrocytes and measured by a probe attached to the skin [43]. The depth corresponds to the deep subpapillary microvascular layer consisting of arterioles, capillaries, venules, and arteriovenous anastomoses (AVAs); LD flux is linearly related to blood flow in skin microcirculation and expressed in arbitrary perfusion units (PU) [43]. After zero-calibration, LDF was measured in baseline, resting conditions; afterwards, a 3 min occlusion of the brachial artery was performed and the parameters of PORH assessed to test vascular reactivity. Although many mechanisms contribute to the phenomenon of PORH [43,47], PORH has been used in clinical practice to assess the endothelial function. As skin microvascular flow is controlled by different mechanisms partly depending on the measuring site, we measured skin blood flow in two corresponding sites: on the volar forearm (non-glabrous, non-acral site) and on the finger pulp, which is a glabrous acral site with abundant AVAs.

2.3. Heart Rate Variability Analysis

The contribution of the sympathetic and the parasympathetic component of the ANS to the regulation of the heart rate (HR) can be evaluated by heart rate variability (HRV) analysis [48–51]. We performed frequency spectral analysis of the R-R interval in 10 min ECG recordings in resting conditions using NeuroKit software package [52]. Low-frequency (LF) spectral power and high-frequency (HF) spectral power were determined from frequencies between 0.04 Hz–0.15 Hz and 0.15 Hz–0.40 Hz, respectively, which are believed to reflect baroreceptor-reflex-dependent (LF) and respiration-dependent (HF) cardiovagal and SNS-mediated HRV [48,53].

2.4. Protocol

All measurements were performed in the morning in a temperature-controlled room (23 °C). After arrival at the laboratory, participants underwent a 30 min acclimatization period in a supine position.

2.4.1. Plasma Glucose Measurement

The participants had their capillary blood samples taken from the right index finger to assess their c_{glc} spectrophotometrically using HemoCue® Glucose 201+ (HemoCue AB, Ängelholm, Sweden). Blood samples were obtained three times: upon arrival prior to any measurements, 30 min after OGTT/control trial (before the start of the haemodynamic measurements), and at the end of haemodynamic measurements. The time was chosen based on a previous pilot experiment designed to define the time when c_{glc} reached its peak value.

2.4.2. Measurement of the Haemodynamic Parameters

Throughout the protocol, a three-channel ECG was recorded for continuous HR measurement. A digital inflatable cuff was placed on the proximal phalanx of the right middle finger for continuous digital artery blood pressure (BP) measurement (Finapress™ BP monitor, Ohmeda 2300, Englewood, CO, USA). Double adhesive LD probes were placed on the volar site of the left forearm and on the left middle finger pulp. The participants were asked to remain in the same comfortable position throughout the experiment and avoid any movements. All haemodynamic variables were measured in the following manner:

- 10 min baseline period;
- 3 min supra-systolic occlusion of the brachial artery;
- 10 min after the release of the occlusion (normalisation period).

Following the above procedure, the participants underwent a standard OGTT: They were asked to drink 75 g glucose dissolved in 250 mL water in a five-minute period. Thirty minutes after glucose ingestion, c_{glc} was determined and the haemodynamic measurements performed as described.

The same protocol was repeated on a separate occasion under the same experimental conditions as a control trial, when participants ingested 250 mL water instead of glucose solution.

2.5. Glucose and Insulin Kinetics

A separate trial was performed to assess plasma glucose and insulin kinetics before (fasting state, time 0) and 30, 50 (corresponding to the start and the end of the haemodynamic measurements, respectively), and 90 min after OGTT. Each participant had an intravenous cannula inserted in a forearm peripheral vein and venous blood (VB) samples collected in pairs: the insulin sample (test tube for serum preparation) immediately after the glucose sample (anticoagulant-coated tube). All test tubes were centrifuged and placed on ice immediately after collection. Serum insulin concentration ($c_{ins\text{-}VB}$) was determined by the commercially available solid-phase, enzyme-labelled chemiluminescent immunometric assay, using the "Immulite2000 Insulin" kit and Immulite2000-XPI analyser (both Siemens Healthcare Diagnostics Products Ltd., Camberley, UK). For quantitative determination of plasma glucose ($c_{glc\text{-}VB}$), the enzymatic UV test (hexokinase method) was used (Olympus AU400 system; Mishima Olympus CO., Ltd., Tokyo, Japan; reagent: Beckman Coulter, Inc., Kildare, Ireland).

2.6. Data Acquisition and Statistical Analysis

Haemodynamic data (with a sampling rate of 500 Hz) were acquired and processed by Nevrokard software package (https://www.nevrokard.eu/contact.htm, accessed on 22 December 2023) (Nevrokard Kiauta, d.o.o., Izola, Slovenia).

Time-averaged systolic (SP), mean (MP), diastolic (DP) and pulse (PP) pressure, HR (expressed reciprocally as R-R interval—t_{RR}), resting LDF (LD_{rest}), LF, HF, and LF/HF ratio were determined from the initial 10 min resting interval preceding each trial. As for the PORH assessment, the peak post-occlusion LD—LD_{peak}, the time interval from the occlusion release to LD_{peak}—t_{peak}, the area under the curve of PORH—AUC, and the time-averaged LD flux of the normalization period—LD_{base} were determined from PORH curves for both skin sites.

For the systemic error control, all the initial values (pre-OGTT and pre-water) were compared by Mann–Whitney test. Variables were then compared between OGTT and water (control) as pre-post treatment changes (Δvariable = post-treatment − pre-treatment) using Wilcoxon test. PORH and HRV parameters changes following OGTT were divided into two groups (of comparable sample size) depending on Δc_{glc} 30 min after OGTT—"low glucose" (LG) group (LG, $\Delta c_{glc} < 2.3$ mM) and "high glucose" (HG) group (HG, $\Delta c_{glc} \geq 2.3$ mM). LG, HG, and post-water changes were compared to each other using ANOVA (post hoc Tukey). The haemodynamic variables (LDF, PORH, and HRV) were correlated with Δc_{glc} using univariate linear regression.

Consecutive c_{glc} and c_{ins} obtained in the second glucose and insulin kinetics trial were analysed in the corresponding time points after OGTT using repeated-measures ANOVA (post hoc Tukey). Maximal $c_{glc\text{-}VB}$ and $c_{ins\text{-}VB}$ were compared using linear regression.

Variables are presented as median and interquartile range (Q_1–Q_3) or mean ± SD where appropriate. Normality of distribution was checked using Shapiro–Wilk test. For all statistical tests, the significance level of 0.05 applies, and 95% confidence interval (CI) is given. Statistical analysis and graphical representation of data were carried out using Jamovi statistical package [54].

3. Results

Fasting c_{glc} (before any provocation) obtained from the capillary blood sample was 5.0 mM ± 0.4 mM pre-OGTT and 5.2 mM ± 0.4 mM pre-water; no significant difference was found between pre-OGTT and pre-water (CI: −0.4 mM–0.1 mM, $p = 0.123$, independent sample t-test). Further, 30 min after the provocation (OGTT and water, respectively), c_{glc} increased for 2.2 mM ± 0.5 mM after OGTT and 0.0 mM ± 0.5 mM after water (CI: 1.8 mM–2.58 mM, $p = 0.000$, paired sample t-test). The increase in c_{glc} at three consecutive time points throughout the OGTT protocol is presented in Figure 1. The data of c_{glc} after the control trial with water are not shown, as c_{glc} did not exhibit any significance with respect to the fasting value.

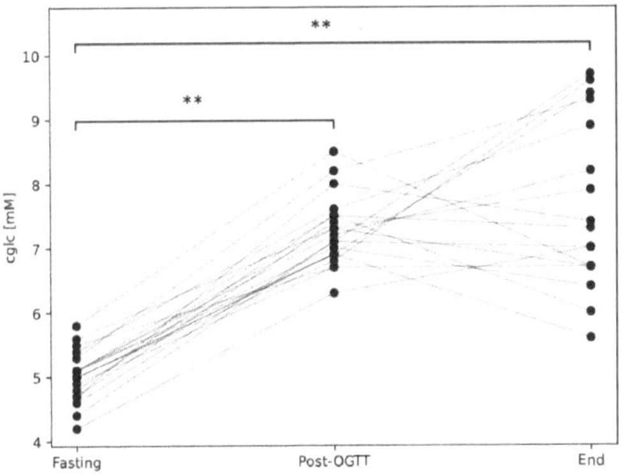

Figure 1. Plasma glucose concentration (c_{glc}) determined in capillary blood sample before and after OGTT. Fasting, before OGTT; Post-OGTT, 30 min after OGTT; End, at the end of the protocol (i.e., 50 min after OGTT). ** $p < 0.001$, paired t-test, with respect to fasting.

Baseline values of the haemodynamic variables before any provocation (either OGTT or water) are summarised in Table 1. It is worth mentioning that the haemodynamic variables did not differ pre-OGTT compared to pre-water justifying subsequent comparison of the changes evoked by treatment. Moreover, PORH had no effect on BP and HR.

Table 1. Baseline haemodynamic variables prior to OGTT or water protocol.

	Median (Q1–Q3)
Resting	
RR (ms)	968 (921–995)
SP (mmHg)	115.0 (108.0–122.0)
MP (mmHg)	81.5 (77.5–87.6)
DP (mmHg)	65.7 (60.1–69.4)
PP (mmHg)	51.6 (43.9–54.8)
$LD_{rest-fp}$ (PU)	244 (174–329)
$LD_{rest-vf}$ (PU)	5.9 (4.7–7.5)
LF (n.u.)	34.8 (21.3–54.9)
HF (n.u.)	54.0 (41.5–66.5)
LF/HF	0.57 (0.34–1.33)
PORH, volar forearm	
LD_{peak} (PU)	41 (30–50)
LD_{base} (PU)	6.3 (5.3–10.7)
t_{peak} (s)	10.6 (6.8–12.1)
AUC (PU^2)	941 (694–1138)

Table 1. *Cont.*

	Median (Q1–Q3)
PORH, finger pulp	
LD_{peak} (PU)	336 (281–454)
LD_{base} (PU)	196 (153–258)
t_{peak} (s)	25.8 (18.5–39.6)
AUC (PU^2)	4477 (2261–11,824)

Data are presented as median and interquartile range (Q1–Q3); N = 20. RR, the duration of the R-R interval of the ECG signal; DP, diastolic digital artery blood pressure; MP, mean blood pressure; SP, systolic blood pressure; PP, pulse pressure; LF, low-frequency spectral power; HF, high-frequency spectral power, LH/HF, ratio of the corresponding spectra; $LD_{rest\text{-}fp}$, resting laser Doppler flux on the finger pulp; $LD_{rest\text{-}vf}$, resting laser Doppler flux on the volar forearm; PU, perfusion units; n.u., normalised units. See Section 2.6 for further abbreviations of PORH (post-occlusion reactive hyperaemia).

3.1. Haemodynamic Changes after OGTT and Water

The changes of some haemodynamic variables were significantly different following OGTT compared to water and are presented in Table 2. Glucose and water induced an increase in systolic and diastolic BP and had no effect on the HR, yet only the diastolic BP increment was significantly larger after water compared to glucose. Also, LF and LF/HF increased significantly more after water than after OGTT, and there was a trend of LD_{peak} decreasing more after OGTT than after water (Table 2) in the finger pulp.

Table 2. Haemodynamic variables after OGTT and water protocol in resting conditions and after post-occlusive reactive hyperaemia (PORH).

	OGTT	Water	p	CI (95%)
Resting				
ΔRR (ms)	−16 (−50; 6)	26 (−26; 57)	0.120	−58; 8
ΔSP (mmHg)	10.2 (−0.5; 18.7) §	11.6 (6.9; 16.3) §	0.284	−9.6; 3.8
ΔMP (mmHg)	4.1 (−2.9; 9.1)	6.4 (4.3; 11.0)	0.145	−8.6; 1.5
ΔDP (mmHg)	0.7 (−3.6; 5.6)	6.2. (2.3; 7.2) §	0.045 *	−7.9; −0.2
ΔPP (mmHg)	7.3 (3.3; 11.7) §	5.9 (1.4; 9.2) §	0.890	−3.2; 5.0
$\Delta LD_{rest\text{-}fp}$ (PU)	−54 (−117; −11) §	−56 (−74; −17) §	0.734	−45; 40
$\Delta LD_{rest\text{-}vf}$ (PU)	0.7 (−0.5; 1.9)	0.5 (−0.3; 1.0)	0.304	−0.6; 3.1
ΔLF (n.u.)	2.4 (−3.1; 13.0)	7.2 (2.2; 14.2) §	0.022 *	−17.1; −1.0
ΔHF (n.u.)	−2.3 (−11.3; 3.5)	−8.3 (−13.3; −1.9) §	0.107	−1.3; 10.3
ΔLF/HF	0.05 (−0.17; 0.40)	0.205 (0.03; 0.62) §	0.022 *	−0.93; −0.04
PORH, volar forearm				
ΔLD_{peak} (PU)	−1 (−4; 6)	3 (−1; 6)	0.485	−8; 3
ΔLD_{base} (PU)	0.2 (−0.7; 1.2)	0.3 (−0.7; 1.2)	1.000	−1.1; 1.5
Δt_{peak} (s)	−0.2 (−2.2; 3.6)	−1.2 (−2.3; 0.4) §	0.119	−0.6; 4.9
ΔAUC (PU^2)	97 (−146; 243)	126 (−104; 246)	0.442	−171; 369
PORH, finger pulp				
ΔLD_{peak} (PU)	−32 (−70; −9) §	−18 (−37; 4)	0.071	−74; 3
ΔLD_{base} (PU)	−13.0 (−53.1; 3.4) §	−34.2 (−73.1; −22.1) §	0.468	−85.0; 39.8
Δt_{peak} (s)	1.2 (−5.7; 20.8)	5.4 (−7.2; 14.0)	0.734	−15.4; 23.2
ΔAUC (PU^2)	2170 (−813; 3932)	962 (−1786; 3528)	0.963	−3812; 3955

Data are presented as median and interquartile range; N = 20. Δvariable, post-treatment–pre-treatment (holds for all variables listed); LF, low-frequency spectral power; HF, high-frequency spectral power; LH/HF, ratio of the corresponding spectra; $LD_{rest\text{-}fp}$, resting laser Doppler flux measured on the finger pulp; $LD_{rest\text{-}vf}$, resting laser Doppler flux measured on the volar forearm; LD_{peak}, LD_{base}, and t_{peak}, AUC–PORH parameters (refer to Methods section); PU, perfusion units; n.u., normalised units; CI, confidence interval. § post-provocation vs. pre-provocation, Wilcoxon test; * $p < 0.05$ Wilcoxon test, OGTT vs. water change.

3.2. Dependence of Haemodynamic Changes on Plasma Glucose Concentration Increase after OGTT

We found significant correlations between some PORH and all the HRV parameters and Δc_{glc} (Figures 2 and 3), with the standardised regression coefficients and the corresponding p-values of Pearson's test presented in Table 3.

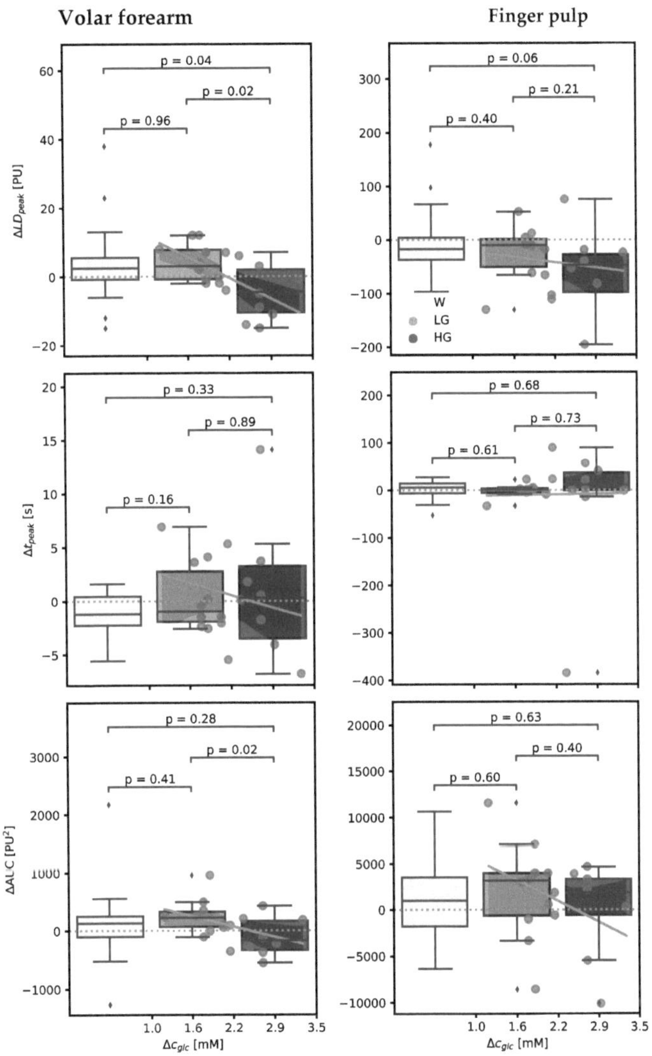

Figure 2. The post-occlusion reactive hyperaemia (PORH) parameters change (Δ) correlated to increase of glucose concentration assessed from capillary blood (Δc_{glc}) after oral glucose tolerance test (OGTT) (dark bars) and water load (W, control, the first—bright bar) on the forearm (**left** plots) and on the finger pulp (**right** plots), with the regression line. Middle bar represents the low-glucose (LG) group ($\Delta c_{glc} < 2.3$ mM); the bar on the right (the darkest colour) represents the high-glucose (HG) group ($\Delta c_{glc} \geq 2.3$ mM). ΔLD_{peak}, pre-post glucose/water load change in peak laser Doppler flux; Δt_{peak}, pre-post post glucose/water load change in time to LD_{peak}; ΔAUC, pre-post glucose/water load change in the area under the hyperaemic response curve; PU, perfusion unit; p, p-values of ANOVA, post hoc Tukey.

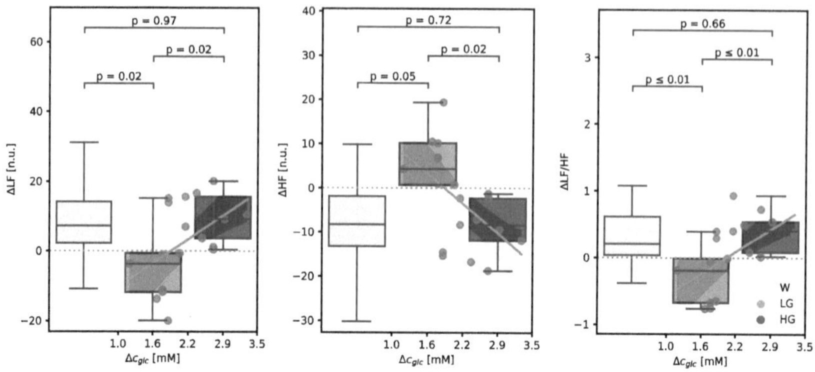

Figure 3. The heart rate variability (HRV) parameters change (Δ) relative to glucose concentration increase (Δc_{glc}) after oral glucose tolerance test (OGTT) and water load (W, control, left bar) with the regression line. Middle bar represents the low-glucose (LG) group (Δc_{glc} < 2.3 mM); the bar on the right (the darkest colour) represents the high-glucose (HG) group ($\Delta c_{glc} \geq$ 2.3 mM). ΔLF, pre-post glucose/water load change in low-frequency band of HRV; ΔHF, pre-post glucose/water change in high-frequency band of HRV; ΔLF/HF, pre-post glucose/water load change in low- to high-frequency band ratio of HRV; n.u., normalised unit; *p*, *p*-value of ANOVA, post hoc Tukey.

Table 3. Linear regression of post-occlusive reactive hyperaemia (PORH) and heart rate variability (HRV) parameters with respect to glucose concentration increase after OGTT.

	r	*p*
	HRV	
LF	0.488	0.040 *
HF	−0.528	0.024 *
LF/HF	0.542	0.020 *
	PORH, volar forearm	
ΔLD_{peak}	−0.619	0.004 *
Δt_{peak}	−0.196	0.408
ΔAUC	−0.408	0.074
	PORH, finger pulp	
ΔLD_{peak}	−0.126	0.597
Δt_{peak}	0.012	0.961
ΔAUC	−0.358	0.132

r, Pearson's correlation coefficient; N = 20; * $p < 0.05$, Pearson's test. See Section 2.6 for abbreviations description.

According to post-OGTT Δc_{glc}, we divided subjects into "low glucose" (LG) and "high glucose" (HG) groups, respectively (Δc_{glc} < 2.3 mM for LG and ≥2.3 for HG), and those changes were compared between both groups and to the changes following water load (Figures 2 and 3). Changes in PORH parameters on the volar forearm were greater in the HG group and were of opposite direction compared to the ones in the LG group and the ones after water load, whereas changes on the finger pulp exhibited the same directions in the HG and LG groups, respectively, and showed some dose dependence (Figure 2). Summarizing, on the volar forearm, ΔLDF_{peak} significantly negatively correlated with Δc_{glc}, whereas in the pulp, negative correlations between ΔLDF_{peak} and ΔAUC and Δc_{glc} did not reach statistical significance. HRV parameters showed consistent dynamics of changes between groups, with the HG group exhibiting similar changes to the ones after water load and opposite changes to the LG group (Figure 3). LF (similar as LF/HF ratio) was positively, and HF negatively correlated to Δc_{glc}. The division of the results into two groups regarding post-OGTT Δc_{glc} seems reasonable, as the last Δc_{glc} obtained after the end of haemodynamic measurements exhibited significant scattering, compromising a uniform interpretation of the data.

3.3. Glucose and Insulin Kinetics

Fasting $c_{glc\text{-}VB}$ and $c_{ins\text{-}VB}$ were 4.3 mM $\pm$ 0.3 mM and 2.6 μUml^{-1} $\pm$ 1.9 μUml^{-1}, respectively. Both $c_{glc\text{-}VB}$ and $c_{ins\text{-}VB}$ increased after OGTT, as shown in Table 4. Maximal $c_{glc\text{-}VB}$ and $c_{ins\text{-}VB}$ were positively correlated (Pearson regression coefficient r = 0.59; $p \leq 0.05$). Worth noting is that glucose concentration of the venous blood samples exhibited similar values to the ones obtained from the capillary blood.

Table 4. Plasma glucose and serum insulin concentration change with respect to fasting values following oral glucose load.

Time (min)	$\Delta c_{glc\text{-}VB}$ (mM)	p	$\Delta_{cins\text{-}VB}$ (μUml^{-1})	p
30	1.7 ± 1.3	0.017 *	20.2 ± 21.7	0.007 *
50	1.5 ± 1.7	0.034 *	28.1 ± 28.9	0.000 *
90	0.9 ± 1.2	0.560	32.6 ± 21.5	0.000 *

Data are presented as means ± SD (standard deviation). $\Delta c_{glc\text{-}VB}$, change in plasma glucose concentration, assessed in venous blood samples (VB) with respect to fasting value; $\Delta c_{ins\text{-}VB}$, change in serum insulin concentration with respect to fasting value; * $p < 0.05$, repeated-measures ANOVA, post hoc Tukey test.

4. Discussion

To the best of our knowledge, this is the first study simultaneously assessing vascular reactivity and SNS activity noninvasively while providing evidence that the measured changes occurred during peak c_{glc}, i.e., 30 min after an oral glucose load. Our results imply that OGTT affects SNS activity, as assessed by HRV analysis in healthy young humans and modestly impairs vascular reactivity assessed by inducing PORH in skin microcirculation, albeit less conclusive. Both SNS activity and vascular reactivity changes were found to significantly depend on the magnitude of Δc_{glc} after OGTT, a possible measure of glucose tolerance. The findings might help elucidate the physiological changes following orally administered glucose load on one hand and address the discrepancies between existing publications in the field on the other.

As for the effect of glucose and insulin on the activity of the SNS, several authors observed that both high c_{glc} or c_{ins} administered intravenously increased SNS activity in healthy humans [18,31,32,36–38,40] and altered the baroreflex sensitivity [41,55]. Based on our findings, we believe that after an oral administration of glucose, such as in OGTT, this effect might be counterbalanced by vagal reflexes due to carbohydrate content in the gastrointestinal tract and, possibly, anticipation [23,56]. However, some studies performed in healthy participants where glucose was administered orally also observed increased SNS activity [34,39,57] and partly attributed it to the effect of insulin, which contrasts with our study. Firstly, a possible explanation for the discrepancy might be that the participants in those studies were 10–80 years older than our participants, which could imply different glucose tolerance and/or different ANS response. The relationship between age and glucose tolerance has been well established as well as the relationship between glucose tolerance and ANS response to hyperglycaemia and hyperinsulinemia [40,41]. In the studies mentioned above, where glucose was administered orally, the c_{glc} after the load was significantly higher compared to our study, implying a higher glucose tolerance among our participants who were indeed much younger and very homogenous in age, which was not the case in many available studies. Positive correlations between LH (and LF/HF) and Δc_{glc} after OGTT in our study imply glucose's effect on the SNS. Additionally, we observed a decrease in SNS activity in the LG group, while there was no change in the HG group, supporting the glucose tolerance conjecture. If glucose tolerance was indeed higher in the LG group, as supported by the results correlating $c_{glc\text{-}VB}$ or $c_{ins\text{-}VB}$ in our study, it is possible that other studies failed to observe potential SNS attenuation because of lower glucose tolerance in older participants [34,39,57]. Moreover, SNS activity has been correlated to insulin resistance in many independent studies [27,38,39].

Secondly, our study is unique in terms of the time after OGTT when we assessed the parameters of HRV. According to glucose kinetics studies, plasma glucose and intercellular glucose concentration in skeletal muscles and subcutaneous tissue are balanced rapidly, with a maximum delay of 8 min [58,59]. Moreover, insulin-induced capillary recruitment increases glucose uptake by the tissues [30,60]. Based on our pilot experiment, we were able to determine the time interval of the peak c_{glc} and performed the subsequent haemodynamic measurements in this time frame (30 min after OGTT). In the above studies [34,39,57], the SNS activity was assessed at least one hour after OGTT. It is possible that at that time, the aforementioned vagal response diminishes, and the sympathetic dominance prevails. In one study performing OGTT and another using intravenous infusion of insulin to obtain physiological c_{ins}, they repeated the SNS activity measurement immediately after c_{glc} normalisation and showed that the SNS activity returned to normal by then [31,39]. On the other hand, using unphysiologically high c_{ins} (70–150 μU/mL) or keeping c_{ins} elevated for a longer period (2 h) induced a prolonged increase in SNS activity even after normalization of c_{glc} [31,32]. Therefore, physiological elevation of c_{glc} and c_{ins} in healthy humans, at least the ones with high glucose tolerance, should be expected to coincide with SNS activity changes. Moreover, when experimentally inducing hyperinsulinemia by a continuous infusion of insulin (with and without glucose clamping), Berkelaar et al. [40] as well as Schroeder et al. [61] were not able to unequivocally confirm an impact on the vagal tone, yet the mode of inducing hyperinsulinemia in their studies was quite different and indeed importantly differed from the physiological conditions simulated in our study. On the other hand, Horton et al. showed that potential detrimental effects of high glucose were blunted by insulin's vasoactive actions despite insulin effects on the SNS [62], yet their results could not be directly compared to our study, as they assessed microvascular response in the skeletal muscle microvasculature, which might behave differently from cutaneous microcirculation. Anderson et al. also showed that hyperinsulinemia induced by insulin infusion induced SNS activation and a concomitant increase in forearm blood flow as assessed by venous occlusion plethysmography [32]. Nevertheless, the effects of insulin on the ANS should be differently interpreted when considering the effects on macro- and microvasculature and the heart or on the central nervous system [37,55].

Moreover, different responses in various studies might be explained by the time lapse between the peak concentration of c_{glc} and c_{ins}, respectively. In this time frame, the contribution of glucose and insulin to vascular reactivity and the activity of the ANS might vary.

Aside from potential effects on the SNS affecting microcirculation, combined effects of acute hyperglycaemia and hyperinsulinemia on other, locally mediated mechanisms and the endothelial component regulating skin microcirculation should be considered when referring to microvascular reactivity, at least in the time frame in which we performed measurements, keeping in line with the known effects of glucose on endothelial function [1,3,7,27]. Interestingly, our results showed inconclusive evidence regarding vascular reactivity following OGTT in general, but an evident forearm vascular reactivity decrease in the HG group. Similarly, as for the impact of OGTT on the SNS activity elucidated above, we believe that the effect of acute hyperglycaemia on skin microcirculation depends on glucose tolerance, assumingly impairing vascular reactivity in subjects exhibiting lower glucose tolerance without an effect in those with high glucose tolerance, which fits well into the assumption of optimal glucose uptake in high insulin sensitivity [30]. Indeed, a positive correlation between both the LF spectrum and the LF/HF ratio (reflecting SNS activity) and Δc_{glc} and a negative correlation between the forearm peak LDF and AUC and Δc_{glc} after OGTT in our study imply deranged cutaneous microvascular control in subjects with lower glucose tolerance. In higher glucose tolerance, the integrated effect of OGTT on SNS and local and endothelium-derived factors remains unexplained. It may be that the attenuated SNS activity or a direct vasoactive insulin action balance potentially negative effects of

OGTT on local and endothelial factors. This may be the reason why in some well-controlled studies, no changes of vascular reactivity after OGTT were noticed [17,19,21]. Indeed, Natali et al. concluded that potentially negative effects of acute hyperglycaemia on endothelial function (as assessed by iontophoresis of acetylcholine) might be counterbalanced by direct vascular effects of hyperinsulinemia in healthy subjects [16]. Another possibility is that acute hyperglycaemia does not affect microvascular reactivity in subjects with higher glucose tolerance (i.e., higher insulin sensitivity).

As several studies found no vascular reactivity alterations either after OGTT or after intravascular glucose infusion of various durations [14–23], the included subjects might have had higher glucose tolerance/insulin sensitivity, which was not assessed in these studies. Nevertheless, the results of the studies are discrepant, as some did show different vascular reactivity response depending on glucose tolerance [14,16,22,23], while others even reported an impaired vascular reactivity in healthy subjects without addressing glucose tolerance [7–13,19]. Lower glucose tolerance in some studies could be due to included participants who were at least ten years older compared to our participants [8–10,35]. The only comparable study regarding the age of the included cohort as well as similar protocols to test microvascular reactivity did show an impairment of PORH but did not assess the parameters of HRV to test the ANS or correlate the glucose change to the variables assessed [7].

Moreover, many studies lacked a control experiment, which we believe is necessary, as the control trial with water in our study invoked some changes of haemodynamic and HRV parameters. In addition, the direct comparison between the studies is questionable, as the time elapsed between oral glucose load and vascular reactivity testing differed, with most studies assessing vascular reactivity one hour after glucose load or even longer. Moreover, many studies used different approaches to test vascular reactivity, which renders direct comparison questionable.

It might also be that potentially detrimental effects of glucose need a certain time frame for glucose to act on intracellular mechanisms, such as an increase in oxidative stress, increase in protein kinase C activity, and NOS uncoupling, all decreasing the endothelial vasodilator capacity [1–3,30] and implying a need to more systematically assess different times of high-glucose exposure on microvascular (endothelial) function.

Finally, it should be noted that aside from affecting the ANS, oral consumption of nutrients also challenges some other gastrointestinal hormones, which, besides metabolic and hormonal influences, have also been implicated in inducing indirect and potentially direct vasoactive effects [63,64]. Moreover, there are implications of different responses regarding glucose uptake and metabolic and vasodilatory effects of insulin depending on the mode of insulin action, i.e., a physiological increase after an OGTT or applied locally per micro-dialysis [60]. These observations additionally support the complex interplay of various players and hamper solid conclusions.

In the present study, we observed no differences in vascular reactivity between the LG group and the control. Therefore, we did not provide sufficient evidence for any detrimental effects on local and endothelial factors controlling skin microcirculation in subjects with higher glucose tolerance, whereas the effect in lower glucose tolerance was obvious, which exposes a need to establish surrogate markers to test glucose tolerance also in the healthy. Assessing serum insulin concentration after OGTT might be regarded as an additional clinical marker of glucose tolerance.

Study Strengths and Limitations

The strength of our study is an integrated approach that most studies lack, namely a simultaneous assessment of glucose and insulin kinetics as well as microvascular reactivity and HRV 30 min after OGTT. We acknowledge several limitations. One of the limitations is a rather small sample size, rendering direct evaluation of the potential impact of glucose tolerance questionable. Nevertheless, compared to other studies, our sample was even larger, as the majority included only ten subjects. Gender stratification and/or and sex hormone assessment would be desirable since female sex hormones significantly affect the microvascular response. Yet, by including only females who were in the early follicular phase, we could partially overcome this doubt.

Additional limitation is the use of tests for ANS activity and vascular endothelial function, which were rather indirect and thus did not provide a proper insight into the physiological mechanisms behind them.

It is important to note that HRV spectra as a measure of the ANS activity must be interpreted with caution, as the interpretation of the standard HRV parameters (LF and HF spectral power and the LF/HF ratio) has often been criticised under the claim that it does not directly reflect the ANS activity but rather the modulation of ANS by cardiovascular reflexes, i.e., baroreceptor reflex (LF) and respiratory sinus arrythmia (HF) [48–51,53]. Using other complementary methods, such as measuring muscle sympathetic neural activity [18,65] or assessing electrodermal activity (EDA), reflecting changes in cutaneous sympathetic activity related to cognitive or emotional stress [66] might increase the reliability of our results [18,65]. Nevertheless, in vivo assessment of ANS activity is difficult, and accordingly, HRV remains the gold standard for indirect assessment of the ANS activity also in the clinics.

Intended to be as non-invasive as possible, the control protocol with water load seemed to be a provocation itself, as it induced subtle alterations in some haemodynamic variables potentially attributable to volume load and elicited physiological reflexes, such as increased sympathetic vasoconstrictor activity and cardiac vagal tone [67]. This stresses the importance of the control trial on one hand but on the other may render some evidence less supportive. Water ingestion has been associated with increased SNS [65] similarly as shown in our study, which is believed to depend on the osmolality of ingested fluid [67]. To avoid the potential problem of hypo-osmolarity, performing a positive control experiment using mannitol solution instead of glucose would be desirable. Indeed, Hoffmann et al. reported increased forearm blood flow after injection of both dextrose and mannitol solutions of the same osmolality, whereas no effect was observed after saline injection [18]. Accordingly, the SNS changes measured in our study could be the result of decreased systemic vascular resistance due to increased osmolality (and not actively induced by glucose per se), although probably of a minimal order of magnitude (as our probands only drank 250 mL of solution). An improvement of our study would therefore be to compare the same effects following iso-osmolar mannitol ingestion, which was not feasible, as mannitol is known to induce osmotic diarrhoea [68]; furthermore, its absorption in the gastrointestinal tract is questionable. It could also be that the observed alterations in the ANS activity were the consequence of psychological stress, as laying supine still for one hour may be a stressor large enough to provoke changes in the ANS balance [66].

In our study, we simulated a postprandial state by performing OGTT. We could obtain a better insight into a corresponding physiological response of SNS and vascular reactivity to acute hyperglycaemia if we extended the study by using protein-rich fluid in addition to OGTT or simply by having subjects eat a well-controlled meal [69]. Including additional biochemical analyses such as, i.e., some markers of inflammation and oxidative stress, might strengthen the results [70]. Indeed, Parker et al. showed that a combined meal induces microvascular impairment in skeletal muscle despite an increase in central haemodynamic measures [64]. Since skin blood flow regulation differs, similar studies in skin microcirculation are warranted.

Taken together, one of the original findings of the present study is that even in healthy subjects with normal glycaemic control, glucose tolerance might be regarded as an important determinant of the SNS and vascular reactivity after OGTT. As the clinical standard for determining glucose tolerance is c_{glc} two hours after OGTT when c_{glc} has already normalised in healthy subjects, establishing a more sensitive measure of glucose tolerance in healthy subjects would be desirable [56,71]. Additional studies assessing similar haemodynamic parameters should be repeated in different time frames after OGTT with a simultaneous assessment of c_{glc} or c_{ins}.

5. Conclusions

In healthy humans, OGTT slightly impacts SNS activity as assessed by HRV analysis and modestly impairs microvascular reactivity in the volar forearm as measured 30 min after glucose load. Both effects are significantly associated with glucose tolerance. The SNS attenuation presumably reflects feeding-related vagal reflexes, which seem to be more pronounced in subjects with higher glucose tolerance. The impairment of vascular reactivity is only evident in lower glucose tolerance and might include endothelium-mediated and other local mechanisms. Even in the healthy, it is glucose tolerance that determines vascular reactivity shortly after OGTT, exposing the need to develop stronger tools/markers for a timely detection of predisposed individuals. We hope that our study will contribute to a better understanding of the underlying physiological changes following oral glucose load and, possibly, reveal pathological implications in diabetic vascular disease.

Author Contributions: Both authors have made a direct intellectual contribution to this work. Conceptualization, H.L.; methodology, H.L. and J.Š.; measurements, J.Š.; validation, J.Š. and H.L.; formal analysis, J.Š.; data curation, J.Š.; writing—original draft preparation, J.Š.; writing—review and editing, H.L.; supervision, H.L.; project administration, H.L.; funding acquisition, H.L. All authors have read and agreed to the published version of the manuscript.

Funding: The study was funded by the Slovenian Research Agency, Grant No. P3-0019.

Institutional Review Board Statement: The study was approved by the National Medical Ethics Committee of the Republic of Slovenia (No. 0120-175/2017/6, Date of approval: 23 September 2020).

Informed Consent Statement: Informed consent was obtained from all subjects involved in the study.

Data Availability Statement: The data presented in this study are available on request from the corresponding author.

Acknowledgments: We would like to thank all the volunteers who participated in the study. We also thank Viki Kotar for venous blood sampling, Alenka France Štiglic and co-workers from the Institute of Clinical Chemistry and biochemistry of the University Clinical Centre Ljubljana and Blaž Krhin, and his co-workers from the Laboratory of Clinical Radiochemistry at the Clinical University Centre for performing blood samples biochemical analyses for glucose and insulin determination.

Conflicts of Interest: The authors declare no conflicts of interest.

References

1. Gero, D. Hyperglycemia-Induced Endothelial Dysfunction. In *Endothelial Dysfunction—Old Concepts and New Challenges*; InTech: London, UK, 2018. [CrossRef]
2. Mapanga, R.F.; Essop, M.F. Damaging Effects of Hyperglycemia on Cardiovascular Function: Spotlight on Glucose Metabolic Pathways. *Am. J. Physiol. Heart Circ. Physiol.* **2016**, *310*, H153–H173. [CrossRef] [PubMed]
3. Zhou, Z.W.; Xie, X.L.; Zhou, S.F.; Li, C.G. Mechanism of Reversal of High Glucose-Induced Endothelial Nitric Oxide Synthase Uncoupling by Tanshinone IIA in Human Endothelial Cell Line EA.Hy926. *Eur. J. Pharmacol.* **2012**, *697*, 97–105. [CrossRef] [PubMed]
4. Ng, H.H.; Leo, C.H.; O'Sullivan, K.; Alexander, S.-A.; Davies, M.J.; Schiesser, C.H.; Parry, L.J. 1,4-Anhydro-4-Seleno-d-Talitol (SeTal) Protects Endothelial Function in the Mouse Aorta by Scavenging Superoxide Radicals under Conditions of Acute Oxidative Stress. *Biochem. Pharmacol.* **2017**, *128*, 34–45. [CrossRef] [PubMed]
5. Tsai, S.H.; Hein, T.W.; Kuo, L.; Yang, V.C. High Glucose Impairs EDHF-Mediated Dilation of Coronary Arterioles via Reduced Cytochrome P450 Activity. *Microvasc. Res.* **2011**, *82*, 356–363. [CrossRef] [PubMed]
6. Solymar, M.; Ivic, I.; Balasko, M.; Fulop, B.D.; Toth, G.; Tamas, A.; Reman, G.; Koller, A.; Reglodi, D. Pituitary Adenylate Cyclase-Activating Polypeptide Ameliorates Vascular Dysfunction Induced by Hyperglycaemia. *Diab Vasc. Dis. Res.* **2018**, *15*, 277–285. [CrossRef] [PubMed]
7. Varsamis, P.; Walther, G.; Share, B.; Taylor, F.; Stewart, S.; Lorenzen, C.; Loader, J. Transient Endothelial Dysfunction Induced by Sugar-Sweetened Beverage Consumption May Be Attenuated by a Single Bout of Aerobic Exercise. *Microvasc. Res.* **2018**, *115*, 8–11. [CrossRef] [PubMed]
8. Loader, J.; Meziat, C.; Watts, R.; Lorenzen, C.; Sigaudo-Roussel, D.; Stewart, S.; Reboul, C.; Meyer, G.; Walther, G. Effects of Sugar-Sweetened Beverage Consumption on Microvascular and Macrovascular Function in a Healthy Population. *Arter. Thromb. Vasc. Biol.* **2017**, *37*, 1250–1260. [CrossRef]

9. Akbari, C.M.; Saouaf, R.; Barnhill, D.F.; Newman, P.A.; LoGerfo, F.W.; Veves, A. Endothelium-Dependent Vasodilatation Is Impaired in Both Microcirculation and Macrocirculation during Acute Hyperglycemia. *J. Vasc. Surg.* **1998**, *28*, 687–694. [CrossRef]
10. Williams, S.B.; Goldfine, A.B.; Timimi, F.K.; Ting, H.H.; Roddy, M.-A.; Simonson, D.C.; Creager, M.A. Acute Hyperglycemia Attenuates Endothelium-Dependent Vasodilation in Humans In Vivo. *Circulation* **1998**, *97*, 1695–1701. [CrossRef]
11. De Marchi, S.; Prior, M.; Rigoni, A.; Zecchetto, S.; Rulfo, F.; Arosio, E. Ascorbic Acid Prevents Vascular Dysfunction Induced by Oral Glucose Load in Healthy Subjects. *Eur. J. Intern. Med.* **2012**, *23*, 54–57. [CrossRef]
12. Title, L.M.; Cummings, P.M.; Giddens, K.; Nassar, B.A. Oral Glucose Loading Acutely Attenuates Endothelium-Dependent Vasodilation in Healthy Adults without Diabetes: An Effect Prevented by Vitamins C and E. *J. Am. Coll. Cardiol.* **2000**, *36*, 2185–2191. [CrossRef] [PubMed]
13. Grassi, D.; Desideri, G.; Necozione, S.; Ruggieri, F.; Blumberg, J.B.; Stornello, M.; Ferri, C. Protective Effects of Flavanol-Rich Dark Chocolate on Endothelial Function and Wave Reflection During Acute Hyperglycemia. *Hypertension* **2012**, *60*, 827–832. [CrossRef] [PubMed]
14. Tomiyama, H.; Kimura, Y.; Okazaki, R.; Kushiro, T.; Abe, M.; Kuwabara, Y.; Yoshida, H.; Kuwata, S.; Kinouchi, T.; Doba, N. Close Relationship of Abnormal Glucose Tolerance With Endothelial Dysfunction in Hypertension. *Hypertension* **2000**, *36*, 245–249. [CrossRef] [PubMed]
15. Houben, A.J.H.M.; Schaper, N.C.; Kruseman, A.C.N. Acute Effects of Local Hyperglycaemia on Peripheral Blood Flow in Man. *Diabet. Med.* **1993**, *10*, 39–43. [CrossRef] [PubMed]
16. Natali, A.; Baldi, S.; Vittone, F.; Muscelli, E.; Casolaro, A.; Morgantini, C.; Palombo, C.; Ferrannini, E. Effects of Glucose Tolerance on the Changes Provoked by Glucose Ingestion in Microvascular Function. *Diabetologia* **2008**, *51*, 862–871. [CrossRef] [PubMed]
17. Major-Pedersen, A.; Ihlemann, N.; Hermann, T.S.; Christiansen, B.; Dominguez, H.; Kveiborg, B.; Nielsen, D.B.; Svendsen, O.L.; Køber, L.; Torp-Pedersen, C. Effects of Oral Glucose Load on Endothelial Function and on Insulin and Glucose Fluctuations in Healthy Individuals. *Exp. Diabetes Res.* **2008**, *2008*, 672021. [CrossRef] [PubMed]
18. Hoffman, R.P.; Hausberg, M.; Sinkey, C.A.; Anderson, E.A. Hyperglycemia Without Hyperinsulinemia Produces Both Sympathetic Neural Activation and Vasodilation in Normal Humans. *J. Diabetes Complicat.* **1999**, *13*, 17–22. [CrossRef] [PubMed]
19. Tomiyama, H.; Kushiro, T.; Okazaki, R.; Yoshida, H.; Doba, N.; Yamashina, A. Influences of Increased Oxidative Stress on Endothelial Function, Platelets Function, and Fibrinolysis in Hypertension Associated with Glucose Intolerance. *Hypertens. Res.* **2003**, *26*, 295–300. [CrossRef]
20. van Gurp, P.J.; Rongen, G.A.; Lenders, J.W.M.; Al Nabawy, A.K.M.; Timmers, H.J.L.M.; Tack, C.J. Sustained Hyperglycaemia Increases Muscle Blood Flow but Does Not Affect Sympathetic Activity in Resting Humans. *Eur. J. Appl. Physiol.* **2005**, *93*, 648–654. [CrossRef]
21. Horová, E.; Mazoch, J.; Hilgertová, J.; Kvasnička, J.; Škrha, J.; Šoupal, J.; Prázný, M. Acute Hyperglycemia Does Not Impair Microvascular Reactivity and Endothelial Function during Hyperinsulinemic Isoglycemic and Hyperglycemic Clamp in Type 1 Diabetic Patients. *Exp. Diabetes Res.* **2012**, *2012*, 851487. [CrossRef]
22. Houben, A.J.H.M.; Kruseman, A.C.N.; Bouhouch, E.; Slaaf, D.W.; Schaper, N.C. Peripheral Macro-and Microcirculation in Short-term Insulin-dependent Diabetes Mellitus: The Role of Prostaglandins in Early Haemodynamic Changes. *Eur. J. Clin. Investig.* **1993**, *23*, 662–667. [CrossRef] [PubMed]
23. Woerdeman, J.; Meijer, R.I.; Eringa, E.C.; Hoekstra, T.; Smulders, Y.M.; Serné, E.H. Insulin Sensitivity Determines Effects of Insulin and Meal Ingestion on Systemic Vascular Resistance in Healthy Subjects. *Microcirculation* **2016**, *23*, 62–68. [CrossRef] [PubMed]
24. Imaeda, K.; Okayama, N.; Okouchi, M.; Omi, H.; Kato, T.; Akao, M.; Imai, S.; Uranishi, H.; Takeuchi, Y.; Ohara, H.; et al. Effects of Insulin on the Acetylcholine-Induced Hyperpolarization in the Guinea Pig Mesenteric Arterioles. *J. Diabetes Complicat.* **2004**, *18*, 356–362. [CrossRef] [PubMed]
25. Huang, S.-S.; Lu, Y.-J.; Huang, J.-P.; Wu, Y.-T.; Day, Y.-J.; Hung, L.-M. The Essential Role of Endothelial Nitric Oxide Synthase Activation in Insulin-Mediated Neuroprotection against Ischemic Stroke in Diabetes. *J. Vasc. Surg.* **2014**, *59*, 483–491. [CrossRef]
26. Cardillo, C.; Nambi, S.S.; Kilcoyne, C.M.; Choucair, W.K.; Katz, A.; Quon, M.J.; Panza, J.A. Insulin Stimulates Both Endothelin and Nitric Oxide Activity in the Human Forearm. *Circulation* **1999**, *100*, 820–825. [CrossRef]
27. Potenza, M.A.; Addabbo, F.; Montagnani, M. Vascular Actions of Insulin with Implications for Endothelial Dysfunction. *Am. J. Physiol. Endocrinol. Metab.* **2009**, *297*, E568–E577. [CrossRef]
28. Iredahl, F.; Tesselaar, E.; Sarker, S.; Farnebo, S.; Sjöberg, F. The Microvascular Response to Transdermal Iontophoresis of Insulin Is Mediated by Nitric Oxide. *Microcirculation* **2013**, *20*, 717–723. [CrossRef]
29. Rossi, M.; Maurizio, S.; Carpi, A. Skin Blood Flowmotion Response to Insulin Iontophoresis in Normal Subjects. *Microvasc. Res.* **2005**, *70*, 17–22. [CrossRef]
30. Kolka, C.M.; Bergman, R.N. The Endothelium in Diabetes: Its Role in Insulin Access and Diabetic Complications. *Rev. Endocr. Metab. Disord.* **2013**, *14*, 13–19. [CrossRef]
31. Rowe, J.W.; Young, J.B.; Minaker, K.L.; Stevens, A.L.; Pallotta, J.; Landsberg, L. Effect of Insulin and Glucose Infusions on Sympathetic Nervous System Activity in Normal Man. *Diabetes* **1981**, *30*, 219–225. [CrossRef]
32. Anderson, E.A.; Hoffman, R.P.; Balon, T.W.; Sinkey, C.A.; Mark, A.L. Hyperinsulinemia Produces Both Sympathetic Neural Activation and Vasodilation in Normal Humans. *J. Clin. Investig.* **1991**, *87*, 2246–2252. [CrossRef] [PubMed]
33. Marfella, R.; De Angelis, L.; Siniscalchi, M.; Rossi, F.; Giugliano, D.; Nappo, F. The Effect of Acute Hyperglycaemia on QTc Duration in Healthy Man. *Diabetologia* **2000**, *43*, 571–575. [CrossRef] [PubMed]

34. Weissman, A.; Lowenstein, L.; Peleg, A.; Thaler, I.; Zimmer, E.Z. Power Spectral Analysis of Heart Rate Variability During the 100-g Oral Glucose Tolerance Test in Pregnant Women. *Diabetes Care* **2006**, *29*, 571–574. [CrossRef] [PubMed]
35. Takei, Y.; Tomiyama, H.; Tanaka, N.; Yamashina, A. Close Relationship Between Sympathetic Activation and Coronary Microvascular Dysfunction During Acute Hyperglycemia in Subjects With Atherosclerotic Risk Factors. *Circ. J.* **2007**, *71*, 202–206. [CrossRef] [PubMed]
36. Emdin, M.; Gastaldelli, A.; Muscelli, E.; Macerata, A.; Natali, A.; Camastra, S.; Ferrannini, E. Hyperinsulinemia and Autonomic Nervous System Dysfunction in Obesity. *Circulation* **2001**, *103*, 513–519. [CrossRef] [PubMed]
37. Van De Borne, P.; Hausberg, M.; Hoffman, R.P.; Mark, A.L.; Anderson, E.A. Hyperinsulinemia Produces Cardiac Vagal Withdrawal and Nonuniform Sympathetic Activation in Normal Subjects. *Am. J. Physiol. Regul. Integr. Comp. Physiol.* **1999**, *276*, R178–R183. [CrossRef]
38. Paolisso, G.; Manzella, D.; Tagliamonte, M.R.; Rizzo, M.R.; Gambardella, A.; Varricchio, M. Effects of Different Insulin Infusion Rates on Heart Rate Variability in Lean and Obese Subjects. *Metabolism* **1999**, *48*, 755–762. [CrossRef]
39. Paolisso, G.; Manzella, D.; Ferrara, N.; Gambardella, A.; Abete, P.; Tagliamonte, M.R.; De Lucia, D.; Furgi, G.; Picone, C.; Gentile, S.; et al. Glucose Ingestion Affects Cardiac ANS in Healthy Subjects with Different Amounts of Body Fat. *Am. J. Physiol. Endocrinol. Metab.* **1997**, *273*, E471. [CrossRef]
40. Berkelaar, M.; Eekhoff, E.M.W.; Simonis-Bik, A.M.C.; Boomsma, D.I.; Diamant, M.; Ijzerman, R.G.; Dekker, J.M.; 'T Hart, L.M.; De Geus, E.J.C. Effects of Induced Hyperinsulinaemia with and without Hyperglycaemia on Measures of Cardiac Vagal Control. *Diabetologia* **2013**, *56*, 1436–1443. [CrossRef]
41. Watkins, L.L.; Surwit, R.S.; Grossman, P.; Sherwood, A. Is There a Glycemic Threshold for Impaired Autonomic Control? *Diabetes Care* **2000**, *23*, 826–830. [CrossRef]
42. Mundinger, T.O.; Cooper, E.; Coleman, M.P.; Taborsky, G.J. Short-Term Diabetic Hyperglycemia Suppresses Celiac Ganglia Neurotransmission, Thereby Impairing Sympathetically Mediated Glucagon Responses. *Am. J. Physiol. Endocrinol. Metab.* **2015**, *309*, E246–E255. [CrossRef] [PubMed]
43. Lenasi, H. Assessment of Human Skin Microcirculation and Its Endothelial Function Using Laser Doppler Flowmetry. In *Medical Imaging*; InTech: London, UK, 2011. [CrossRef]
44. Kralj, L.; Lenasi, H. Wavelet Analysis of Laser Doppler Microcirculatory Signals: Current Applications and Limitations. *Front. Physiol.* **2023**, *13*, 1076445. [CrossRef] [PubMed]
45. Braverman, I.M. The Cutaneous Microcirculation: Ultrastructure and Microanatomical Organization. *Microcirculation* **1997**, *4*, 329–340. [CrossRef] [PubMed]
46. Tikhonova, I.V.; Tankanag, A.V.; Chemeris, N.K. Time–Amplitude Analysis of Skin Blood Flow Oscillations during the Post-Occlusive Reactive Hyperemia in Human. *Microvasc. Res.* **2010**, *80*, 58–64. [CrossRef] [PubMed]
47. Yvonne-Tee GB, R.A.H.A.W.A.R.A. Method Optimization on the Use of Postocclusive Hyperemia Model to Assess Microvascular Function. *Clin. Hemorheol. Microcirc.* **2008**, *38*, 119–133. [PubMed]
48. Grassi, G.; Esler, M. How to Assess Sympathetic Activity in Humans. *J. Hypertens.* **1999**, *17*, 719–734. [CrossRef] [PubMed]
49. Eckberg, D.L. Point:Counterpoint: Respiratory Sinus Arrhythmia Is Due to a Central Mechanism vs. Respiratory Sinus Arrhythmia Is Due to the Baroreflex Mechanism. *J. Appl. Physiol.* **2009**, *106*, 1740–1742. [CrossRef] [PubMed]
50. Reyes del Paso, G.A.; Langewitz, W.; Mulder, L.J.M.; van Roon, A.; Duschek, S. The Utility of Low Frequency Heart Rate Variability as an Index of Sympathetic Cardiac Tone: A Review with Emphasis on a Reanalysis of Previous Studies. *Psychophysiology* **2013**, *50*, 477–487. [CrossRef]
51. Lenasi, H.; Rihar, E.; Filipič, J.; Klemenc, M.; Fister, P. The Effect of Caffeine on Heart Rate Variability in Newborns: A Pilot Study. *Life* **2023**, *13*, 1459. [CrossRef]
52. Makowski, D.; Pham, T., Lau, Z.J.; Brammer, J.C.; Lespinasse, F.; Pham, H.; Schölzel, C.; Chen, S.H.A. NeuroKit2: A Python Toolbox for Neurophysiological Signal Processing. *Behav. Res. Methods* **2021**, *53*, 1689–1696. [CrossRef]
53. Goldstein, D.S.; Bentho, O.; Park, M.-Y.; Sharabi, Y. Low-Frequency Power of Heart Rate Variability Is Not a Measure of Cardiac Sympathetic Tone but May Be a Measure of Modulation of Cardiac Autonomic Outflows by Baroreflexes. *Exp. Physiol.* **2011**, *96*, 1255–1261. [CrossRef] [PubMed]
54. The Jamovi Project; 2019. Available online: https://www.jamovi.org/cloud.html (accessed on 22 December 2023).
55. Limberg, J.K.; Soares, R.N.; Padilla, J. Role of the Autonomic Nervous System in the Hemodynamic Response to Hyperinsulinemia—Implications for Obesity and Insulin Resistance. *Curr. Diab Rep.* **2022**, *22*, 169–175. [CrossRef] [PubMed]
56. Kramer, C.K.; Vuksan, V.; Choi, H.; Zinman, B.; Retnakaran, R. Emerging Parameters of the Insulin and Glucose Response on the Oral Glucose Tolerance Test: Reproducibility and Implications for Glucose Homeostasis in Individuals with and without Diabetes. *Diabetes Res. Clin. Pract.* **2014**, *105*, 88–95. [CrossRef] [PubMed]
57. Paolisso; Manzella; Rizzo; Barbieri; Gambardella. Varricchio Effects of Glucose Ingestion on Cardiac Autonomic Nervous System in Healthy Centenarians: Differences with Aged Subjects. *Eur. J. Clin. Investig.* **2000**, *30*, 277–284. [CrossRef]
58. Jansson, P.A.; Fowelin, J.; Smith, U.; Lonnroth, P. Characterization by Microdialysis of Intracellular Glucose Level in Subcutaneous Tissue in Humans. *Am. J. Physiol. Endocrinol. Metab.* **1988**, *255*, E218–E220. [CrossRef] [PubMed]

59. Gheysens, O.; Postnov, A.; Deroose, C.M.; Vandermeulen, C.; de Hoon, J.; Declercq, R.; Dennie, J.; Mixson, L.; De Lepeleire, I.; Van Laere, K.; et al. Quantification, Variability, and Reproducibility of Basal Skeletal Muscle Glucose Uptake in Healthy Humans Using ^{18} F-FDG PET/CT. *J. Nucl. Med.* **2015**, *56*, 1520–1526. [CrossRef] [PubMed]
60. Iredahl, F.; Högstedt, A.; Henricson, J.; Sjöberg, F.; Tesselaar, E.; Farnebo, S. Skin Glucose Metabolism and Microvascular Blood Flow during Local Insulin Delivery and after an Oral Glucose Load. *Microcirculation* **2016**, *23*, 597–605. [CrossRef] [PubMed]
61. Schroeder, E.B.; Chambless, L.E.; Liao, D.; Prineas, R.J.; Evans, G.W.; Rosamond, W.D.; Heiss, G. Diabetes, Glucose, Insulin, and Heart Rate Variability. *Diabetes Care* **2005**, *28*, 668–674. [CrossRef]
62. Horton, W.B.; Jahn, L.A.; Hartline, L.M.; Aylor, K.W.; Patrie, J.T.; Barrett, E.J. Hyperglycemia Does Not Inhibit Insulin's Effects on Microvascular Perfusion in Healthy Humans: A Randomized Crossover Study. *Am. J. Physiol. Endocrinol. Metab.* **2020**, *319*, E753–E762. [CrossRef]
63. Roberts-Thomson, K.M.; Parker, L.; Betik, A.C.; Wadley, G.D.; Della Gatta, P.A.; Marwick, T.H.; Keske, M.A. Oral and Intravenous Glucose Administration Elicit Opposing Microvascular Blood Flow Responses in Skeletal Muscle of Healthy People: Role of Incretins. *J. Physiol.* **2022**, *600*, 1667–1681. [CrossRef]
64. Parker, L.; Morrison, D.J.; Betik, A.C.; Roberts-Thomson, K.; Kaur, G.; Wadley, G.D.; Shaw, C.S.; Keske, M.A. High-Glucose Mixed-Nutrient Meal Ingestion Impairs Skeletal Muscle Microvascular Blood Flow in Healthy Young Men. *Am. J. Physiol. Endocrinol. Metab.* **2020**, *318*, E1014–E1021. [CrossRef] [PubMed]
65. Scott, E.M.; Greenwood, J.P.; Gilbey, S.G.; Stoker, J.B.; Mary, D. AWater Ingestion Increases Sympathetic Vasoconstrictor Discharge in Normal Human Subjects. *Clin. Sci.* **2001**, *100*, 335–342. [CrossRef]
66. Posada-Quintero, H.F.; Chon, K.H. Innovations in Electrodermal Activity Data Collection and Signal Processing: A Systematic Review. *Sensors* **2020**, *20*, 479. [CrossRef] [PubMed]
67. Brown, C.M.; Barberini, L.; Dulloo, A.G.; Montani, J.-P. Cardiovascular Responses to Water Drinking: Does Osmolality Play a Role? *Am. J. Physiol. Regul. Integr. Comp. Physiol.* **2005**, *289*, R1687–R1692. [CrossRef] [PubMed]
68. Ajaj, W.; Goehde, S.C.; Schneemann, H.; Ruehm, S.G.; Debatin, J.F.; Lauenstein, T.C. Dose Optimization of Mannitol Solution for Small Bowel Distension in MRI. *J. Magn. Reson. Imaging* **2004**, *20*, 648–653. [CrossRef]
69. Lages, M.; Barros, R.; Moreira, P.; Guarino, M.P. Metabolic Effects of an Oral Glucose Tolerance Test Compared to the Mixed Meal Tolerance Tests: A Narrative Review. *Nutrients* **2022**, *14*, 2032. [CrossRef]
70. Joy, N.G.; Perkins, J.M.; Mikeladze, M.; Younk, L.; Tate, D.B.; Davis, S.N. Comparative Effects of Acute Hypoglycemia and Hyperglycemia on Pro-Atherothrombotic Biomarkers and Endothelial Function in Non-Diabetic Humans. *J. Diabetes Complicat.* **2016**, *30*, 1275–1281. [CrossRef]
71. Morris, C.; O'Grada, C.; Ryan, M.; Roche, H.M.; Gibney, M.J.; Gibney, E.R.; Brennan, L. Identification of Differential Responses to an Oral Glucose Tolerance Test in Healthy Adults. *PLoS ONE* **2013**, *8*, e72890. [CrossRef]

life

MDPI

Article

Enhanced Microvascular Adaptation to Acute Physical Stress and Reduced Oxidative Stress in Male Athletes Who Consumed Chicken Eggs Enriched with n-3 Polyunsaturated Fatty Acids and Antioxidants—Randomized Clinical Trial

Luka Kolar [1,2,†], Petar Šušnjara [2,3,†], Marko Stupin [2,3,4,*], Ana Stupin [2,3], Ivana Jukić [2,3], Zrinka Mihaljević [2,3], Nikolina Kolobarić [2,3], Iva Bebek [5], Diana Nejašmić [5], Marija Lovrić [5] and Ines Drenjančević [2,3,*]

1 Department of Internal Medicine, National Memorial Hospital Vukovar, 32000 Vukovar, Croatia; lukakolar.vu@gmail.com
2 Scientific Center of Excellence for Personalized Health Care, Josip Juraj Strossmayer University of Osijek, 31000 Osijek, Croatia; psusnjara@mefos.hr (P.Š.); ana.stupin@mefos.hr (A.S.); ivana.jukic@mefos.hr (I.J.); zmihaljevic@mefos.hr (Z.M.); nikolina.bilic.dujmusic@gmail.com (N.K.)
3 Department of Physiology and Immunology, Faculty of Medicine Osijek, Josip Juraj Strossmayer University of Osijek, 31000 Osijek, Croatia
4 Department for Cardiovascular Disease, University Hospital Centre Osijek, 31000 Osijek, Croatia
5 BICRO BIOCENTAR d.o.o., 10000 Zagreb, Croatia; ivabebek@gmail.com (I.B.); diana.nejasmic@biocentre.hr (D.N.); marija.lovric@biocentre.hr (M.L.)
* Correspondence: marko.stupin@mefos.hr (M.S.); ines.drenjancevic@mefos.hr (I.D.)
† These authors contributed equally to this work.

Abstract: This randomized interventional study aimed to determine the effects of n-3 polyunsaturated fatty acids, selenium, vitamin E, and lutein supplementation in the form of enriched chicken egg consumption on microvascular endothelium-dependent vasodilation, oxidative stress, and microvascular response to an acute strenuous training session (ASTS) in competitive athletes. Thirty-one male athletes were assigned to a control (n = 17) or a Nutri4 group (n = 14) who consumed three regular or enriched chicken eggs per day, respectively, for 3 weeks. Significantly enhanced endothelium-dependent responses to vascular occlusion (PORH) and iontophoresis of acetylcholine (AChID) were observed in the Nutri4 group but not in the control group after egg consumption. Formation of peroxynitrite and hydrogen peroxide in peripheral blood mononuclear cells, as well as serum concentration of 8-iso prostaglandin F2α, decreased in the Nutri4 group while remaining unchanged in controls. PORH and AChID were reduced post-ASTS compared with pre-ASTS, both before and after the diets, in both groups. However, the range of PORH responsiveness to ASTS (ΔPORH) increased after consumption of enriched eggs. These results suggest that consumption of enriched chicken eggs has a beneficial effect on microvascular endothelium-dependent vasodilation and the reduction of oxidative stress levels in competitive athletes. Also, microvascular adaptation to the ASTS was improved after consumption of Nutri4 eggs.

Keywords: athletes; enriched eggs; functional food; microcirculation; oxidative stress

Citation: Kolar, L.; Šušnjara, P.; Stupin, M.; Stupin, A.; Jukić, I.; Mihaljević, Z.; Kolobarić, N.; Bebek, I.; Nejašmić, D.; Lovrić, M.; et al. Enhanced Microvascular Adaptation to Acute Physical Stress and Reduced Oxidative Stress in Male Athletes Who Consumed Chicken Eggs Enriched with n-3 Polyunsaturated Fatty Acids and Antioxidants—Randomized Clinical Trial. *Life* **2023**, *13*, 2140. https://doi.org/10.3390/life13112140

Academic Editor: Stefania Lamponi

Received: 11 September 2023
Revised: 28 October 2023
Accepted: 29 October 2023
Published: 31 October 2023

1. Introduction

Proper nutrition has a very important role in the performance of professional athletes, helping them to maintain ideal body weight and body composition specific to sports and faster recovery [1]. However, it is alarming that nutritional practices of elite athletes are often suboptimal, considering that the pursuit of better results and new records is forcing athletes to train longer and harder than ever. Analyses of dietary habits in various athlete groups found that a substantial proportion of the studied populations did not reach the dietary goals for many macro- and micronutrients, including n-3 polyunsaturated fatty acids (n-3 PUFAs), vitamins, and trace elements such as iron (Fe), zinc (Zn), and selenium

(Se) [2,3]. Proper intake of micronutrients, especially those with antioxidant properties, is important to maintain oxidative balance and physiological homeostasis in athletes since strenuous exercise and overtraining are known to increase acute oxidative stress [4,5]. Increased oxidative stress and reduced antioxidant capacity (measured as reduced ferric reducing ability of plasma (FRAP), increased 8-iso-prostaglandin F2α level (indicator of in vivo lipid peroxidation), increased total oxidant status [6], increased production capability of reactive oxygen species (ROS) in peripheral blood mononuclear cells (PBMCs) [7], etc.), have been repeatedly demonstrated following exercise training. Today, athletes are also very interested in nutrients that can improve athletic performance and recovery; they often use them to increase metabolic capacity, improve muscle growth, delay the onset of fatigue, and shorten recovery times [8]. However, nowadays, athletes avoid taking large amounts of supplements, especially in the form of pills [9]. Thus, to achieve better sports performance but also a healthier and more natural diet, consuming so-called functional foods could have a two-fold potential in the nutrition of elite athletes. The term "functional foods" refers to foods of natural origin that, in addition to an adequate nutritional effect, have a beneficial effect on one or more target functions in the body by improving health and general well-being or reducing the risk of disease [10].

Recently, a new functional food has been developed, namely, chicken eggs enriched with n-3 PUFAs, Se, lutein, and vitamin E [11], that could present a means to deliver nutrients in higher concentration per meal in a more convenient way. Chicken eggs are suitable for modification of the fatty acid profile since the content of n-3 PUFAs can be increased with the addition of different proportions of fish oil to the mixtures for laying hens. Eggs are also suitable for the deposition of higher concentrations of antioxidants like Se, vitamin E, and lutein by adding organic Se, marigold extract as a source of lutein, and a vitamin mixture to feed for laying hens, respectively [11]. Because of the anti-inflammatory and antioxidative proprieties of n-3 PUFAs [12], there is an increasing interest in the potential benefits of n-3 PUFA supplementation in athletes. One of the potential antioxidant effects of n-3 PUFAs is the reduction of intracellular oxidative stress, which was demonstrated by the potential of n-3 PUFAs (EPA and DHA) to reduce intracellular H_2O_2-induced DNA damage in human aortic endothelial cells [13]. Available data on n-3 PUFA supplementation indicate various improvements in skeletal muscle performance, energy metabolism and endurance performance in a population of athletes. For example, supplementation of n-3 PUFAs for 4 weeks may successfully attenuate minor aspects of exercise-induced muscle damage, e.g., soreness-associated exercise avoidance [14]. On the other hand, even though n-3 PUFA supplementation in healthy volunteers did not affect brachial artery diameter, conductance, and blood flow at rest, the increases in these measures were greater during contraction with the n-3 PUFA supplementation, suggesting the potential of n-3 PUFAs to improve vascular adaptation to exercise [15]. Similarly, a recent study by our research group showed that supplementation with n-3 PUFAs in the form of n-3 PUFA-enriched chicken eggs improved resting microvascular endothelial function and contributed to vascular adaptation to acute exhaustion in competitive athletes [16]. Besides n-3 PUFAs, micronutrients with antioxidant and anti-inflammatory properties have the potential to improve cardiovascular health, especially in conditions characterized by increased oxidative stress [17,18]. Preservation of endothelial vascular function has been shown as a significant protective effect of vitamin E supplementation in both healthy and cardiovascular disease populations, where it manifested as a reduction in carotid intima-media thickness and a reduction in the production of biomarkers of endothelial activation and endothelium–leukocyte interaction [19,20]. Exercise-induced oxidative damage in all endurance athletes and enhanced production of oxidized low-density lipoprotein (oxLDL) following strenuous endurance exercise may be prevented by vitamin E supplementation and by maintaining higher vitamin E status [21]. Because Se is an integral component of several antioxidant enzymes, most notably glutathione peroxidase (GPx), which exerts its antioxidant effects by reducing the concentration of peroxides such as hydrogen peroxide (H_2O_2), consumption of Se could play a central role in preventing and influencing the clinical course of various diseases (e.g., cancer, diabetes, cardiovascular disease, infections) [22,23] as well as being used to improve athletic performance and exercise recovery [24]. A recent systematic review reported that Se supplementation did not benefit athletic performance but prevented Se deficiency in athletes with high-intensity training. Optimal plasma Se levels have been shown to be important in minimizing chronic exercise-induced oxidative effects and modulating the training effect on mitochondrial changes [24].

A recent study by our group reported that the serum level of pro-inflammatory cytokine IL -17A decreases and neuronal nitric oxide synthase (nNOS) expression increases after consumption of chicken eggs enriched with n-3 PUFAs, Se, vitamin E, and lutein in young healthy subjects, suggesting that the combined effect of n-3 PUFAs and antioxidant micronutrients may play an important role in improving vascular endothelial properties (particularly those dependent on nitric oxide) and maintaining an inflammation-free environment under resting conditions [25]. Because of the paucity of data on the concomitant action of various micronutrients possessing anti-inflammatory and antioxidative properties, especially in the form of functional food, it would be prudent to investigate the effect of functional foods on vascular endothelial function and oxidative status in demanding populations such as competitive athletes. Since the earliest adaptive changes in physiological conditions, as well as the final organ damage in various pathological conditions, occur precisely at the microvascular level, it is essential to evaluate blood flow responses to standardized stimuli, specifically in the microcirculation. Our hypothesis is that competitive athletes may benefit from the consumption of chicken eggs enriched with n-3 PUFAs, Se, lutein, and vitamin E in terms of microvascular endothelial function and a more favorable oxidative environment. Thus, the present study aimed to (a) investigate the effects of intake of chicken eggs enriched in n-3 PUFAs, Se, lutein, and vitamin E (Nutri4 eggs) on skin microvascular endothelium-dependent and -independent vasodilation in athletes; (b) examine if functional food consumption modifies microvascular adaptation to acute exhausting training; and (c) determine the effects of functional food consumption on oxidative stress and antioxidant capacity in competitive athletes.

2. Materials and Methods

2.1. Study Population

Thirty-one young, healthy male competitive athletes participated in this study. They were recruited from local rowing and track and field clubs. All participating athletes had been training for at least 1 year, 5 and 12 times per week. In addition to being active competitive athletes, eligibility criteria included age between 18 and 30 years, normal body mass index (BMI, 18.5 to 24.9 kg/m^2), normal arterial blood pressure values (BP, <140/90 mmHg), and normal serum lipid levels (total cholesterol < 5.00 mmol/L, triglycerides < 1.70 mmol/L, HDL cholesterol > 1.00 mmol/L, LDL cholesterol < 3.00 mmol/L). Exclusion criteria were hypertension, coronary artery disease, diabetes, hyperlipidemia, renal insufficiency, cerebrovascular and peripheral artery disease, history of smoking, and use of drugs or substances that might affect the endothelium. Written informed consent was obtained from each subject. The study protocol and procedures met the standards of the latest revision of the Declaration of Helsinki and were approved by the Ethics Committee of the Faculty of Medicine, University of Osijek, Osijek, Croatia (Cl: 602-04/21-08/07; No: 2158-61-07-21-151). This study is part of a clinical research study investigating the effects of functionally enriched chicken eggs on cardiovascular function registered at ClinicalTrials.gov (accessed on 11 September 2023) (NCT04564690).

2.2. Production of n-3 PUFA-, Selenium-, Lutein- and Vitamin E-Enriched Chicken Eggs

Chicken eggs enriched with n-3 PUFAs, lutein, Se, and vitamin E were prepared according to the protocol of a research group from the Faculty of Agrobiotechnical Sciences, University of Osijek [11]. Rapeseed oil (1.5%) in the feed mixtures fed to laying hens was replaced by a mixture of fish oil (1.5%) and linseed oil (2%), with 0.43 mg/kg of selenium mixture and 100 mg/kg of mixture of vitamin E and lutein added. Such feed mixture for laying hens resulted in the production of Nutri4 eggs, the content of which is described in detail in the study of our collaborators from the Faculty of Agrobiotechnical Sciences Osijek [11].

2.3. Study Design

Subjects were instructed to eat three hard-boiled chicken eggs daily (63 eggs total) during the 21-day study protocol. Participants were divided into the experimental Nutri4 group (14 subjects), which consumed chicken eggs fortified with n-3 PUFAs, Se, lutein, and vitamin E (three per day; approximately 1056 mg n-3 PUFAs per day, 0.0573 mg Se per day, 3.29 mg vitamin E per day, and 1.85 mg lutein per day) and in the control group (17 subjects) who consumed normal chicken eggs from the same farm (three per day; about 438 mg n-3 PUFAs per day, 0.0549 mg Se per day, 1.785 mg vitamin E per day, and 0.33 mg lutein per day). Regular and Nutri4 chicken eggs were the same size (commercial size M), and neither the subject nor the researcher knew which group the subjects belonged to. For the purposes of the study, subjects were instructed not to consume any other foods rich in n-3 PUFAs, Se, lutein, and vitamin E, or any other form of these micronutrients during the study protocol, but only the eggs given to them for study purposes. The study was conducted in the Laboratory for Clinical and Sports Physiology, Department of Physiology and Immunology, Faculty of Medicine, University of Osijek. The study consisted of two visits, and all measurements described below were taken on the first day and the day immediately following the end of the protocol. All tests took place in the morning after an overnight fast; participants were instructed not to engage in strenuous activity in the 24 h before the tests.

2.4. Basic Anthropometric, Cardiovascular, and Biochemical Measurements

Height and body weight, to calculate subjects' BMI, were measured using a personal scale with a height meter (RADWAG, Radom, Poland). A tape meter was used for measurement of waist and hip circumference to calculate waist-to-hip ratio (WHR). Three consecutive measurements of arterial BP and heart rate (HR) were performed in the sitting position with an automatic oscillometer (OMRON M3, OMRON Healthcare Inc., Osaka, Japan). Venous blood samples were collected using a blood collection system and vacutainers (BD Becton, Dickinson and Company, Franklin Lakes, NJ, USA). Fasting venous blood samples were analyzed for complete blood count and standard biochemical measurements, including electrolytes (potassium, sodium), urea, creatinine, lipid profile (total cholesterol, triglycerides, LDL cholesterol, and HDL cholesterol), blood glucose, and high-sensitivity C-reactive protein (hsCRP). These standard biochemical analyses were performed at the Department of Clinical Laboratory Diagnostics, Osijek University Hospital, Osijek, Croatia.

2.5. Analysis of Fatty Acid Profile, Selenium, Vitamin E, and Lutein Levels in Serum

According to the protocol described in a previous work of our research group, serum samples were analyzed for serum fatty acid profile (37 fatty acids in total) by gas chromatography-tandem mass spectrometry (GC-MS/MS system Thermo Fisher GC Trace 1300 coupled with a TSQ 9000 Triple Quadrupole) in the Bioanalytical Laboratory of BIOCentre, BIOCentre—Incubation Center for biosciences, Zagreb, Croatia.

Serum vitamin E concentration was determined according to a standardized protocol [26], in which serum proteins were first denatured with absolute ethanol, and the supernatant was separated from the proteins with xylene. After separation of the supernatant, 2,2-bipyridyl and FeCl3 were added to the mixture, resulting in pink staining. After 2 min of incubation, absorbance was measured at 492 nm using an ELISA READER. The absorbance obtained was proportional to the vitamin E concentration in serum.

The protocol for measuring Se concentration in serum samples was optimized by the partner institutions in the project [27]. Serum samples were digested in ultrapure HNO_3 and H_2O_2 (5:1 ratio) for 60 min at 180 °C in a CEM Mars 6 closed microwave system (CEM, Matthew, NC, USA). Inductively coupled plasma mass spectrometry (ICP-MS) (ICP-MS,

Agilent 7500a Agilent Technologies Inc., Santa Clara, CA, USA) was used to determine the Se concentration in the solution of the digested serum samples. Each serum sample was analyzed by ICP, and the analytical method was controlled by the reference material NIST 1567b (wheat flour, National Institute of Standards and Technology, Gaithersburg, MD, USA).

Determination of lutein concentrations in serum samples was performed according to the existing protocol [28]. In 200 μL of serum, 1 mL of deionized water and 0.01% ascorbic acid dissolved in absolute ethanol were added and stirred in the mixture. Then, 2 mL of hexane was added, stirred, and centrifuged at 2500 RPM for 20 min. After centrifugation, the supernatant was separated and then evaporated, and the lutein concentration was determined by high-performance liquid chromatography (HPLC). HPLC analysis was performed at the Department of Chemistry, Josip Juraj Strossmayer University of Osijek.

2.6. Evaluation of Peripheral Microvascular Reactivity

Microvascular reactivity was assessed by laser Doppler flowmetry (LDF) (MoorVMS-LDF, Axminister, UK). Post-occlusive reactive hyperemia (PORH) and acetylcholine iontophoresis (ACh) were performed to assess endothelium-dependent responses, whereas sodium nitroprusside iontophoresis (SNP) was performed for endothelium-independent responses. The PORH assay involved measurement of microvascular blood flow before, during, and after the release of a 1-min vascular occlusion. During baseline flow (B), occlusion (O), and reperfusion (R), microcirculatory blood flow was determined using software that calculated the area under the curve (AUC) for 1-min intervals, and the result was expressed as the difference between the percent change in flow during reperfusion ($R\% = R/B*100$) and occlusion ($O\% = O/B*100$) relative to baseline (R-O% increase = $R\% - O\%$). In addition, after baseline microvascular blood flow using anode current, either the positively charged ACh (1%) or the negatively charged SNP (1%) was iontophorected as a vasodilator to obtain the stable plateau of maximal LDF response according to adapted established protocols [29]. Software to calculate AUC (1 min intervals) was also used to determine baseline microcirculatory blood flow and during ACh or SNP administration. The result was expressed as the increase in blood flow relative to baseline after administration of ACh or SNP (ACh or SNP blood flow increase).

2.7. Protocol for Acute Strenuous Training Sessions

Participants underwent an acute strenuous training session (ASTS) in the form of a rowing protocol to examine the effects of a specific dietary protocol on the adaptation of microvascular function to a specific challenge. The rowing protocol was performed on the Dynamic Indoor Rower Concept 2 rowing ergometer (Concept2 Inc., Morrisville, VT, USA). The ASTS included a progressively incremental rowing protocol modified to include 5×4 min submaximal stages and a single maximal stage [29]. The initial workload of the training session was 150 W, with increments of 40 W in the stages. The submaximal stages were separated by 1 min recovery periods, and a 5 min rest was given before the maximal stage, during which subjects were instructed to row at maximal power until exhaustion. For the subjective assessment of physical activity by individuals, we used the category ratio scale (CR10) introduced by Borg, with values ranging from 0 (not at all) to 10 (very strenuous activity). The difference in microvascular reactivity to administration of vascular occlusion (PORH), ACh (AChID), and SNP (SNPID) before and after ASTS indicates the range of microvascular reactivity and is expressed as ΔPORH, ΔAChID, and ΔSNPID. Delta value (Δ) was determined at each study visit before and after completion of each dietary protocol.

2.8. Measurement of Biomarkers of Oxidative Stress and Antioxidant Protection

Serum activity and concentration of the enzyme myeloperoxidase (MPO), which catalyzes the formation of various reactive oxygen species (ROS) and is produced mainly by polymorphonuclear neutrophils, were measured. A colorimetric human myeloperoxidase assay kit (ab105136, abcam9) and a commercially available ELISA kit (Human Myeloperoxidase Kit ab272101, Abcam, Cambridge, UK) were used to determine serum MPO activity and concentration, respectively, according to the manufacturer's instructions.

Serum activity of antioxidant enzymes catalase (CAT), glutathione peroxidase (GPx), and superoxide dismutase (SOD) was tested using the Lambda 25UV-Vis spectrophotometer equipped with the UV WinLab 6.0 software package (PerkinElmer for the Better, Waltham, MA, USA), according to the protocol established in the Laboratory of Biochemistry, Department of Biology, University of Osijek [25].

Serum protein concentration of 8-iso-prostaglandin F2α (8-iso-PGF2α), a product of non-enzymatic peroxidation of arachidonic acid in membrane phospholipids, was measured using a commercially available ELISA kit (Elabscience, catalog number: E-EL-0041) according to the manufacturer's instructions.

Assessment of intracellular ROS production in peripheral blood mononuclear cells (PBMCs) was performed with the FACS Canto II flow cytometer (BD Bioscience; 488 excitation laser and 530/30 BP analysis filter) and Flow Logic software V8 (Inivai Technologies, Mentone, Australia) using previously described laboratory protocols [30]. Dichlorofluorescein diacetate (DCF-DA) was used to determine H_2O_2 and peroxynitrite (ONOO-) content, and dihydroethidium (DHE) was used to determine superoxide (O2-) content in PBMCs (lymphocytes). Phorbol 12-myristate 13-acetate (PMA) was added to each sample to stimulate ROS production. Data are expressed as fold change of DCF fluorescence units with respect to control.

2.9. Statistical Analysis

All results are reported as arithmetic mean and standard deviation (SD). Preliminary results of 10 subjects were considered in the sample size calculation. The calculated sample size was 14 per group to detect differences in primary outcomes reported in this study (e.g., LDF measurement), with a significance level of 0.05 and statistical power of 80% for the paired *t*-test. The Kolmogorov–Smirnov normality test was used to assess the normality of the data distribution. To assess differences within groups (measurements before and after each dietary protocol), the paired *t*-test was used. To assess differences between groups at baseline, the Student's *t*-test was used. Differences between groups in post-intervention measurements were tested using analysis of covariance (ANCOVA) with baseline (pre-measurement) as a covariate. $p < 0.05$ was considered statistically significant. SigmaPlot, version 11.2 (Systat Software, Inc., Chicago, IL, USA) was used for statistical analysis.

3. Results

Table 1 lists the initial anthropometric, hemodynamic, and biochemical characteristics of the subjects. All subjects were lean and normotensive and had normal complete blood counts, serum electrolytes, renal function, hsCRP, fasting blood glucose, and lipid levels. There were no significant differences in any of the measured parameters (e.g., age, BMI, HR, BP, and biochemical parameters) between the athletes in the control and Nutri4 groups at the time of entry into the study protocol. The dietary protocol was completed by all participants.

Table 1. Anthropometric, hemodynamic, and biochemical parameters in control and Nutri4 group of competitive male athletes.

Parameter	Control		Nutri4	
	Before	After	Before	After
N	17		14	
Age (years)	22 ± 3		23 ± 4	
BMI (kg/m^2)	24.9 ± 4.0	24.7 ± 3.9	23.6 ± 2.4	23.5 ± 2.4
WHR	0.83 ± 0.09	0.84 ± 0.08	0.83 ± 0.04	0.82 ± 0.04

Table 1. *Cont.*

Parameter	Control		Nutri4	
	Before	After	Before	After
SBP (mmHg)	122.1 ± 12.2	119.4 ± 12.8	124.1 ± 14.6	118.4 ± 12.0
DBP (mmHg)	70.6 ± 9.2	70.8 ± 8.7	71.0 ± 11.2	70.4 ± 10.8
MAP (mmHg)	87.2 ± 9.4	87.4 ± 9.0	88.8 ± 10.7	85.0 ± 9.1 *
HR (beats per minute)	72.2 ± 9.3	73.8 ± 11.6	70.1 ± 8.9	69.4 ± 11.7
erythrocytes ($\times 10^{12}$/L)	5.0 ± 0.3	5.0 ± 0.4	5.1 ± 0.3	5.1 ± 0.3
hemoglobin (g/L)	149.1 ± 8.2	148.4 ± 10.6	147.4 ± 7.4	148.3 ± 9.0
hematocrit (%)	42.8 ± 2.4	42.6 ± 2.9	42.6 ± 2.0	43.2 ± 2.2
leukocytes ($\times 10^{9}$/L)	5.8 ± 1.0	5.7 ± 1.0	6.0 ± 1.1	6.0 ± 1.8
thrombocytes ($\times 10^{9}$/L)	214.1 ± 57.8	214.8 ± 43.4	233.8 ± 36.4	230.2 ± 61.6 †
urea (mmol/L)	6.2 ± 1.2	6.8 ± 1.3	6.4 ± 0.9	6.3 ± 0.9
creatinine (μmol/L)	91.4 ± 9.9	91.8 ± 9.7	91.1 ± 10.0	90.4 ± 9.2
sodium (mmol/L)	139.4 ± 1.2	140.6 ± 1.9 *	139.8 ± 1.4	139.8 ± 2.1
potassium (mmol/L)	4.2 ± 0.3	4.1 ± 0.2	4.3 ± 0.3	4.1 ± 0.3 *
calcium (mmol/L)	2.6 ± 0.5	2.5 ± 0.4	2.4 ± 0.1	2.4 ± 0.1
iron (μmol/L)	17.0 ± 5.6	17.6 ± 4.6	19.2 ± 9.0	16.1 ± 7.0
transferrin (g/L)	2.5 ± 0.3	2.5 ± 0.3	2.6 ± 0.4	2.6 ± 0.3
glucose (mmol/L)	5.0 ± 0.5	4.8 ± 0.9	5.2 ± 1.0	5.2 ± 1.2 †
hsCRP (mg/L)	0.9 ± 0.8	0.9 ± 1.3	0.7 ± 0.6	0.9 ± 0.8
cholesterol (mmol/L)	4.0 ± 0.7	4.2 ± 0.8	4.4 ± 1.1	4.6 ± 1.1
triglycerides (mmol/L)	1.0 ± 0.5	1.3 ± 1.1	0.9 ± 0.5	1.1 ± 0.7
HDL cholesterol (mmol/L)	1.3 ± 0.3	1.3 ± 0.3	1.5 ± 0.3	1.4 ± 0.3
LDL cholesterol (mmol/L)	2.4 ± 0.6	2.6 ± 0.6	2.6 ± 0.8	2.7 ± 0.8

Data are presented as mean ± standard deviation (SD). N: number of participants; W: women; M: men; BMI: body mass index; WHR: waist-to-hip ratio; SBP: systolic blood pressure; DBP: diastolic blood pressure; MAP: mean arterial pressure; HR: heart rate; hsCRP: high-sensitivity C-reactive protein; HDL: high-density lipoprotein; LDL: low-density lipoprotein. * $p < 0.05$ before vs. after within the group (control or Nutri4)—paired *t*-test. † $p < 0.05$ after control vs. after Nutri4 group—analysis of covariance (ANCOVA) with the baseline value as the covariate.

3.1. Anthropometric, Cardiovascular, and Biochemical Parameters

The effects of 3 weeks of consumption of normal (control group) and enriched chicken eggs (Nutri4 group) on anthropometric, hemodynamic, and biochemical parameters are shown in Table 1. Mean arterial pressure (MAP) and serum potassium concentration were significantly decreased after the dietary protocol in the Nutri4 group compared with baseline values. An increase in serum sodium concentration was noted after 3 weeks of consumption of normal chicken eggs. Platelet and fasting glucose levels were higher in the Nutri4 group after the dietary protocol than in the control group. However, all significant changes observed were within the population reference range, so they were not physiologically significant. No significant difference was observed in other anthropometric (BMI, WHR), hemodynamic (systolic BP, diastolic BP, HR), and biochemical (full blood count, urea, creatinine, hsCRP) parameters, as well as serum lipid profile (total cholesterol, triglycerides, LDL cholesterol, and HDL cholesterol) after consumption of Nutri4 or regular chicken eggs compared to initial measurements within the Nutri4 or control group, or when these values were compared between groups.

3.2. Serum Fatty Acid Profile, Vitamin E, and Selenium Level Analysis

Serum samples were analyzed for a total of 37 free fatty acids, and only the fatty acids whose concentrations were above the limit of quantitation are listed in Table 2. The serum concentration of cis-4,7,10,13,16,19-docosahexaenoic acid (DHA) increased significantly, and the concentration of palmitic acid (C16:0) decreased significantly after consumption of Nutri4 eggs compared with the first measurement in the Nutri4 group. Serum concentrations of the other free fatty acids measured were similar before and after each dietary protocol in the Nutri4 group. Overall, consumption of Nutri4 eggs significantly decreased the serum n-6/n-3 ratio by approximately 36%. Serum concentrations of measured free fatty acids and the n6/n3 ratio remained unchanged (12% decrease) after regular egg consumption compared with baseline measurements in the control group. Serum C15:0 pentadecyl and C16:0 palmitic acid concentrations and total n6/n3 ratio were significantly lower in the Nutri4 group than in the controls following completion of the corresponding dietary protocol (adjusted for baseline values) (Table 2).

Table 2. Serum fatty acid profile in control and Nutri4 group of competitive male athletes.

	Parameter	Control		Nutri4	
		Before	After	Before	After
	SFA (μmol/L)				
	C8:0 Caprylic acid	N/F	96.9	29.5	63.8 ± 34.8
	C12:0 Lauric acid	27.2 ± 0.0	38.8 ± 8.6	36.7 ± 21.8	51.2 ± 34.0
	C14:0 Myristic acid	79.8 ± 33.1	106.6 ± 30.8	21.6 ± 23.7	70.7 ± 34.7
	C15:0 Pentadecylic acid	17.8 ± 6.3	21.0 ± 2.7	17.8 ± 2.2	16.7 ± 2.9 †
	C16:0 Palmitic Acid	1701.5 ± 330.2	1872.8 ± 187.6	1811.1 ± 131.2	1634.4 ± 185.0 *†
	C17:0 Margaric acid	21.5 ± 5.0	23.0 ± 1.9	23.0 ± 2.4	20.7 ± 3.6
	C18:0 Stearic acid	808.7 ± 116.6	802.3 ± 73.1	817.2 ± 94.0	772.1 ± 90.3
	PUFA (μmol/L)				
n-5	C14:1[cis-9] Myristoleic acid	<LOQ	11.07 ± 0.0	<LOQ	11.26 ± 0.0
n-7	C16:1[cis-9] Palmitoleic acid	87.8 ± 36.4	159.3 ± 106.1	84.7 ± 27.5	70.2 ± 16.9
	C17:1[cis-10] cis-10-Heptadecenoic acid	23.0 ± 7.8	20.4 ± 6.4	22.5 ± 8.0	20.0 ± 2.9
	C18:1[trans-9] Elaidic acid	1021.7 ± 307.6	1167.2 ± 0.0	941.2 ± 51.4	697.2 ± 0.0
n-9	C18:1[cis-9] Oleic acid	610.8 ± 595.7	1115.2 ± 694.1	494.6 ± 622.9	1047.4 ± 497.0
	C20:1[cis-11] 11-Eicosenoic acid	15.7 ± 3.3	16.5 ± 3.2	16.4 ± 2.0	14.7 ± 2.6
	C24:1[cis-15] Nervonic acid	6.5 ± 0.0	8.1 ± 1.0	6.9 ± 0.0	<LOQ
	C18:2[cis-9,12] Linoleic acid	1696.2 ± 343.8	1921.4 ± 490.5	1881.5 ± 543.2	1969.4 ± 486.9
	C18:3[cis-6,9,12] gamma-Linolenic acid	31.3 ± 5.9	46.7 ± 23.6	33.2 ± 16.5	28.4 ± 8.8
n-6	C21:2[cis-11,14] Eicosadienoic acid	15.7 ± 3.3	16.5 ± 3.2	16.4 ± 2.0	14.7 ± 2.6
	C20:3[cis-8,11,14] Dihomo-gamma-linolenic acid	94.3 ± 37.1	136.9 ± 92.3	107.6 ± 30.4	90.5 ± 27.1
	C20:4[cis-5,8,11,14] Arachidonic acid	538.8 ± 67.5	600.1 ± 103.8	578.4 ± 53.9	538.3 ± 88.7
	C18:3[cis-9,12,15] alpha-Linolenic acid	21.0 ± 9.9	24.1 ± 3.6	18.1 ± 4.2	23.0 ± 5.9
n-3	C20:4[cis-5,8,11,14] Eicosa-5,8,11,14,17-pentaenoic acid	22.6 ± 3.8	28.4 ± 2.4	20.4 ± 4.3	22.1 ± 5.1
	C22:6[cis-4,7,10,13,16,19] cis-4,7,10,13,16,19-Docosahexaenoic acid	111.1 ± 89.7	207.4 ± 81.5	108.0 ± 32.7	185.4 ± 86.6 *
	n6/n3 PUFAs	25.2 ± 19.4	22.1 ± 22.3	19.5 ± 9.3	12.4 ± 3.2 *†

Results are expressed as mean ± standard deviation (SD). SFA: saturated fatty acids; PUFAs: polyunsaturated fatty acids; <LOQ: below limit of quantification; N/F: not found. * $p < 0.05$ before vs. after within the group (control or Nutri4)—paired *t*-test. † $p < 0.05$ after control vs. after Nutri4 group—analysis of covariance (ANCOVA) with the baseline value as the covariate.

Serum concentrations of Se and vitamin E increased significantly in the Nutri4 group after the dietary protocol, whereas they remained unchanged in the control groups. Serum concentrations of Se and vitamin E were also significantly higher in the Nutri4 group compared with the control group after each nutritional protocol (adjusted for baseline). Serum lutein concentrations before and after the nutritional protocol did not change significantly in any group. These results are shown in Table 3.

Table 3. Selenium, vitamin E, and lutein serum concentration in control and Nutri4 group of competitive male athletes.

Parameter	Control (N = 17)		Nutri4 (N = 14)	
	Before	After	Before	After
Selenium (μg/L)	74.5 ± 13.9	70.1 ± 8.2	76.2 ± 15.0	89.3 ± 9.3 *†
Vitamin E (μg/mL)	7.0 ± 4.2	7.1 ± 4.3	7.6 ± 7.8	10.7 ± 5.3 *†
Lutein (μmol/L)	0.089 ± 0.027	0.105 ± 0.061	0.110 ± 0.0409	0.088 ± 0.038

Data are presented as mean ± standard deviation (SD). * $p < 0.05$ before vs. after within the group (control or Nutri4)—paired *t*-test. † $p < 0.05$ after control vs. after Nutri4 group—analysis of covariance (ANCOVA) with the baseline value as the covariate.

3.3. Endothelium-Dependent and Endothelium-Independent Vasodilation in Forearm Skin Microcirculation

Both post-occlusive reactive hyperemia (PORH) (Figure 1A) and ACh-induced dilation (AChID) (Figure 1B) of forearm skin microcirculation were significantly increased after consumption of enriched chicken eggs compared with baseline measurements. Consumption of normal chicken eggs resulted in no significant change in PORH and AChID compared with baseline in the control group (Figure 1A,B). PORH was significantly increased in the Nutri4 group compared with controls after the appropriate dietary protocol (adjusted for baseline). SNP-induced dilation (SNPID) was not significantly affected by consumption of either enriched or normal chicken eggs, and it did not differ between the Nutri4 group and controls after the corresponding dietary protocol (adjusted for baseline) (Figure 1C).

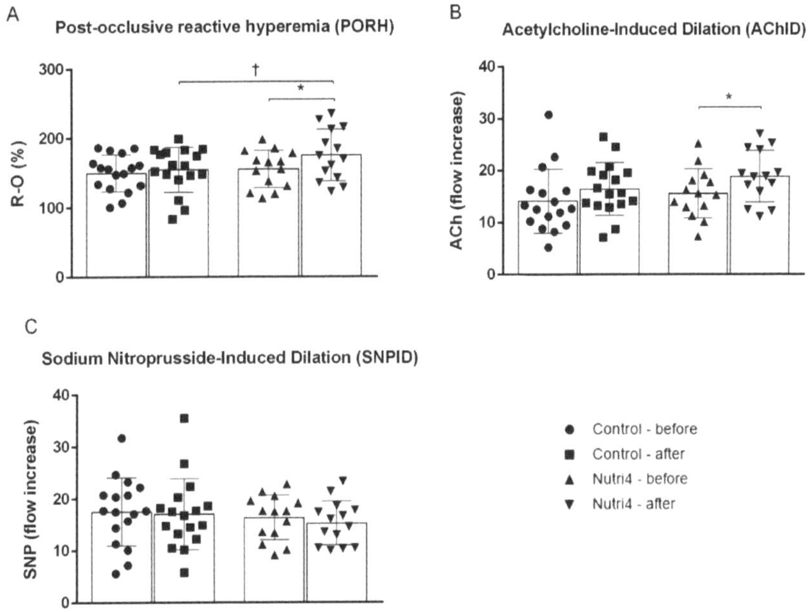

Figure 1. Microvascular endothelium dependent and endothelium-independent vasodilation of the skin in the control and Nutri4 groups of competitive male athletes. (**A**). Post-occlusive reactive hyperemia, (**B**). Acetylcholine-induced dilation (AChID), and (**C**). Sodium nitroprusside-induced dilation (SNPID). PORH measurement is expressed as the difference between the percentage of flow change during reperfusion and occlusion relative to baseline (R-O%). AChID and SNPID are expressed as the increase in flow after ACh or SNP administration relative to baseline. Control N = 17, Nutri4 N = 14. Results are expressed as arithmetic mean and standard deviation (SD). * $p < 0.05$ before vs. after in Nutri4 group (paired *t*-test); † $p < 0.05$ after control vs. after Nutri4 group—analysis of covariance (ANCOVA) with baseline as covariate.

3.4. Acute Strenuous Training Session and Microvascular Responsiveness Range

Consumption of enriched chicken eggs significantly increased the range of ΔPORH responsiveness to ASTS (Figure 2A), whereas ΔAChID responsiveness to ASTS (Figure 2B) was not significantly altered in the Nutri4 group compared with baseline measurements. No significant change in ΔPORH (Figure 2A) or ΔAChID (Figure 2B) was observed in the control group after the corresponding dietary protocol. The increase in ΔPORH, but not in ΔAChID, was significantly greater in the Nutri4 group compared with the control group after the respective dietary protocol (adjusted for baseline).

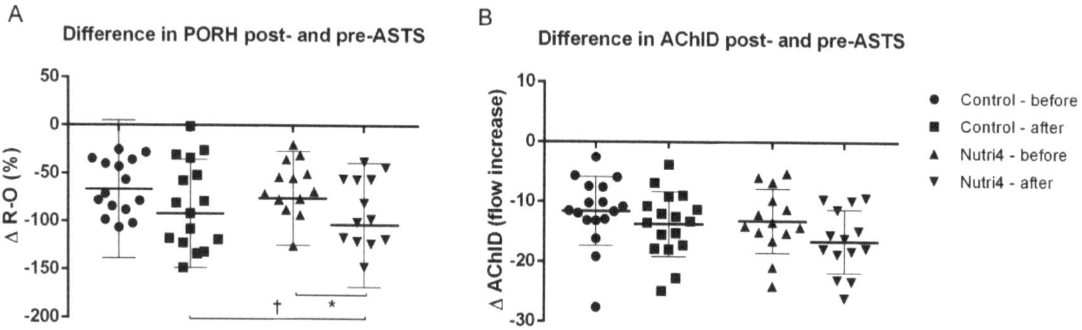

Figure 2. Range of skin microvascular responsiveness to acute strenuous training session (ASTS) in control and Nutri4 groups of competitive male athletes. (**A**). Difference in PORH responsiveness after and before ASTS, ΔPORH = PORH after ASTS—PORH before ASTS; and (**B**). Difference in AChID responsiveness after and before ASTS, ΔAChID = AChID after ASTS—AChID before ASTS. PORH measurement is expressed as the difference between the percent change in flow during reperfusion and occlusion relative to baseline (R-O%), and ΔPORH indicates the difference between the PORH value measured immediately after (post-) and before (pre-) an acute strenuous training session (ASTS). AChID is expressed as the increase in flow after ACh administration compared with baseline flow, and ΔAChID represents the difference in AChID value measured immediately after (post-) and before (pre-) an acute strenuous training session (ASTS). Control N = 17, Nutri4 N = 14. Results are expressed as arithmetic mean and standard deviation (SD). PORH: post-occlusive reactive hyperemia; AChID: acetylcholine-induced dilation; ASTS: acute strenuous training session. * $p < 0.05$ before vs. after within Nutri4 group (paired *t*-test); † $p < 0.05$ after control vs. after Nutri4 group—analysis of covariance (ANCOVA) with baseline as covariate.

3.5. *Biomarkers of Oxidative Stress and Antioxidant Defense*

Myeloperoxidase (MPO) serum protein concentration (MPO pg/mL control before 9719 ± 8623 vs. after 12,019 ± 8316, $p = 0.233$; Nutri4 before 6940 ± 5514 vs. after 11,293 ± 8973, $p = 0.199$) and serum enzyme activity (MPO pmol/mL control before 0.200 ± 0.174 vs. after 0.201 vs. 0.205, $p = 0.971$; Nutri4 before 0.205 ± 0.127 vs. after 0.131 ± 0.089, $p = 0.086$) remained unchanged before and after the respective dietary protocol in both the control and Nutri4 groups. Similarly, serum MPO protein concentration and serum enzyme activity did not differ between the Nutri4 group and controls after the respective dietary protocol (adjusted for baseline).

Serum antioxidant enzyme activity CAT (CAT U/mg protein control before 3.185 ± 0.758 vs. after 3.078 ± 0.808, $p = 0.681$; Nutri4 before 2.440 ± 0.826 vs. after 2.510 ± 0.725, $p = 0.823$), GPx (GPx U/mg protein control before 0.007 ± 0.004 vs. after 0.011 ± 0.006, $p = 0.088$; Nutri4 before 0.013 ± 0.007 vs. after 0.010 ± 0.003, $p = 0.246$) and SOD (SOD U/mg protein control before 9.811 ± 1.109 vs. after 10.061 ± 0.599, $p = 0.426$; Nutri4 before 9.562 ± 0.980 vs. after 10.279 ± 0.884, $p = 0.061$) have not been significantly changed following any of the dietary protocols. In addition, there was no significant difference in the serum activity of the respective enzymes after the dietary protocols (adjusted for baseline) between the control and Nutri4 groups.

The 8-iso-PGF2α serum protein concentration decreased significantly after the three-week protocol in the Nutri4 group, whereas it remained unchanged in the control group (Figure 3). The serum protein concentration of 8-iso-PGF2α was also significantly decreased in the Nutri4 group compared with the control group after each dietary protocol (adjusted for baseline) (Figure 3).

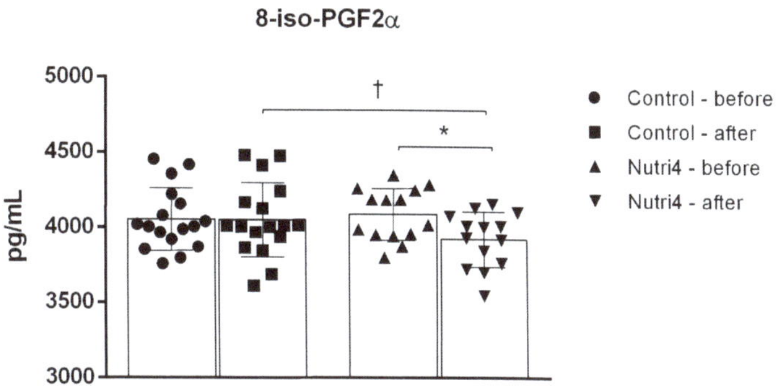

Figure 3. 8-iso prostaglandin F2α (8-iso-PGF2α) serum protein concentration in control and Nutri4 groups of competitive male athletes. Control N = 17, Nutri4 N = 14. Results are expressed as arithmetic mean and standard deviation (SD). * $p < 0.05$ before vs. after in the Nutri4 group (paired *t*-test). † $p < 0.05$ after control vs. after Nutri4 group—analysis of covariance (ANCOVA) with baseline as covariate.

Consumption of enriched chicken eggs significantly decreased the formation of hydrogen peroxide and peroxynitrite in peripheral blood mononuclear cells (PBMCs). Consumption of regular chicken eggs did not have any significant effect on the formation of hydrogen peroxide and peroxynitrite in PBMCs (Figure 4).

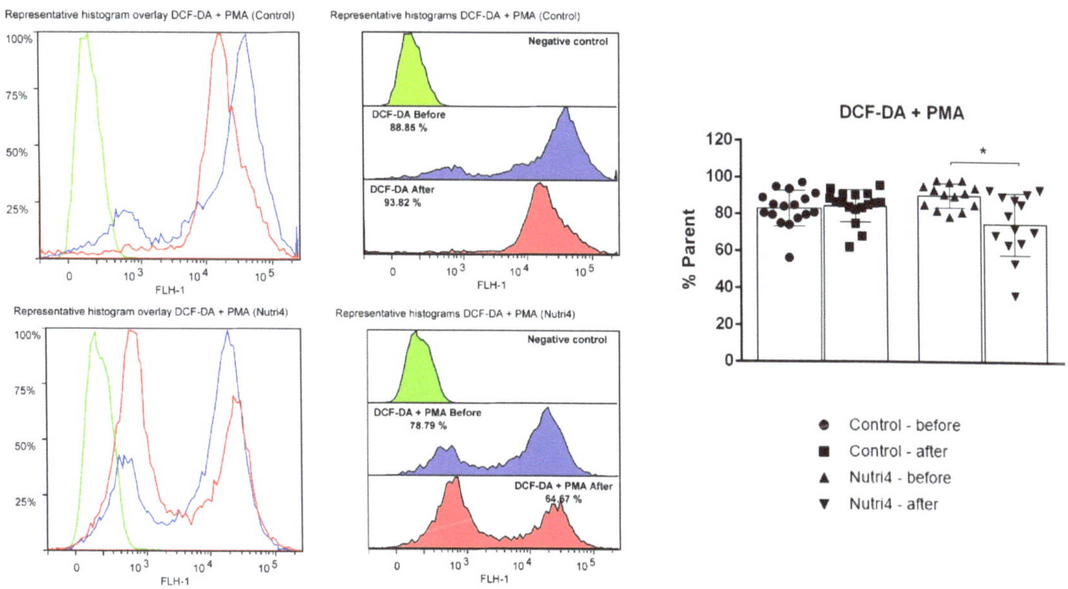

Figure 4. Formation of hydrogen peroxide and peroxynitrite (DCF-DA) in peripheral blood mononuclear cells (PBMCs) in control and Nutri4 groups of competitive male athletes. Control N = 17, Nutri4 N = 14. Results are expressed as arithmetic mean and standard deviation (SD). PMA: phorbol 12-myristate 13-acetate. * $p < 0.05$ before vs. after in the Nutri4 group (paired *t*-test).

The dietary protocol in both groups did not have any significant effect on superoxide formation in PBMCs in competitive athletes (Figure 5), nor did it differ between groups following respective dietary protocols (adjusted for baseline).

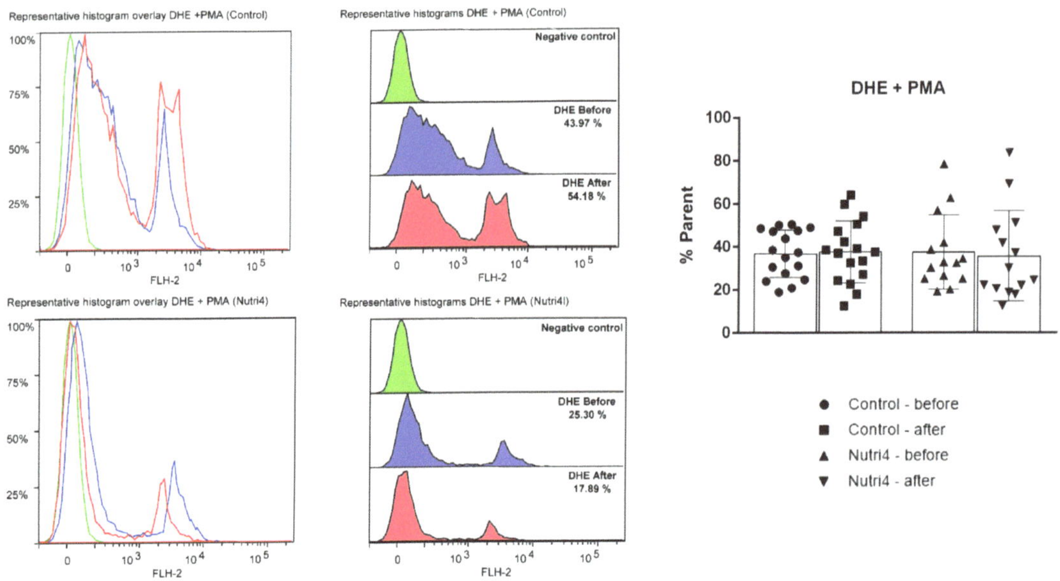

Figure 5. Formation of superoxide (DHE) in peripheral blood mononuclear cells (PBMCs) in control and Nutri4 group of competitive male athletes. Control N = 17, Nutri4 N = 14. Results are expressed as arithmetic mean and standard deviation (SD). PMA: phorbol 12-myristate 13-acetate.

4. Discussion

This randomized interventional study is the first one to investigate whether the combined supplementation of four different micronutrients (n-3 PUFAs, Se, lutein, and vitamin E) in the form of functional food (i.e., naturally enriched chicken eggs) has a beneficial effect on endothelial function and oxidative stress levels and whether it increases the capacity of the microvasculature to accommodate for acute exertional stress in a population of competitive male athletes. The most important finding of this study is the significantly improved endothelium-dependent vasodilation of forearm skin microcirculation following consumption of Nutri4 eggs. Importantly, forearm skin microvascular adaptation to the acute exercise stress challenge also increased after consumption of functionally enriched eggs. Another important finding, in addition to functional vascular changes, is a significant decrease in serum protein concentration of 8-isoPGF2α and a decrease in the formation of hydrogen peroxide and peroxynitrite in PBMCs, which suggest a reduction in oxidative stress levels following consumption of the Nutri4 eggs in competitive athletes. Finally, the present study has demonstrated that, while consumption of Nutri4 eggs resulted in increased nutrient (n-3 PUFAs, Se, and vitamin E) serum levels, serum lipid profile and BP remained within the reference range, and no noxious effects have been observed following consumption of fairly large amounts of regular chicken eggs. This is in agreement with our previous studies [5,20] and supports the conclusion that eggs could be safely consumed, particularly in young, nutritionally demanding populations such as athletes.

Approximately 85% of elite track and field athletes use supplements, mostly vitamins and antioxidants, followed by minerals, proteins, creatine, and ergogenic supplements [9]. To avoid taking large amounts of supplements and to achieve balanced nutrition necessary to minimize high-intensity training damage and to improve sports performance, the benefits of functional food in athletes have been intensively investigated [8,9]. Results of available studies suggest that functional food in athlete populations has resulted in endurance improvement, maintained immunity, reduced oxidative stress, and decreased muscle pain; e.g., a study in recreational runners reported these effects after 14 days of consuming date seed powder. Muscle size, strength, and endurance are important to professional athletes, and the role that skeletal muscle microcirculation plays in muscle health is extensive. Microcirculation in skeletal muscle serves to supply oxygen and nutrients but also to remove waste products and heat from skeletal muscle cells [31]. This role is even more pronounced during exercise, as the metabolic rate of muscle can increase up to 50-fold [31]. Therefore, endothelial dysfunction and capillary thinning contribute to exercise intolerance and muscle wasting [31]. The present study demonstrated the enhancement of microvascular endothelium-dependent reactivity after 3-week consumption of Nutri4 eggs in competitive athletes. These results are consistent with the observation that peripheral microvascular response to vascular occlusion (PORH) and ACh were increased in athletes after consumption of chicken eggs enriched with n-3 PUFAs [16]. Long-term regular exercise has a positive effect on vascular function, leading to improved macrovascular and microvascular endothelium-dependent vasodilation compared with sedentary subjects [32,33]. However, the effects of acute strenuous exercise on vascular function are less well studied, particularly with respect to microvascular function. Interestingly, some studies have shown reduced endothelium-dependent flow-mediated dilation (FMD) of the brachial artery after acute exercise [34,35], while others yielded opposite results, reporting improvement of vascular reactivity of large conductance arteries following an acute exercise session. Earlier results from our research group demonstrated that athletes exhibited enhanced baseline microvascular vasodilation compared to sedentary individuals but also that the reduction in endothelium-dependent vasodilation after ASTS was greater in athletes than in sedentary individuals [16]. Moreover, the decrease in PORH and AChID immediately after the ASTS was even more pronounced at the end of 3 weeks of consumption of chicken eggs enriched with n-3 PUFAs [16], which may be in contradiction with the expected results. However, the greater reduction in endothelium-dependent vasodilation in athletes compared with sedentary subjects may indicate increased microvascular responsiveness in the ASTS area or better utilization of vasodilatory capacity, consistent with the hormesis hypothesis [36]. According to this hypothesis, temporary reductions in the endothelial response after ASTS should lead to better long-term endothelial function in athletes, which is also demonstrated in the present study. The potential underlying mechanism for the observed enhanced microvascular reactivity and increased range of microvascular responsiveness to ASTS could be explained by changes in oxidative stress levels that have been observed in the present study due to consumption of functional food enriched in antioxidants.

In the present study, oxidative stress biomarkers (i.e., serum protein concentration of 8-iso-PGF2α and hydrogen peroxide and peroxynitrite formation in PBMCs) were significantly decreased after consumption of nutrient-enriched chicken eggs. Most of the available data on supplementation of nutrients with antioxidant properties (including n-3 PUFAs, Se, and vitamin E) indicate their beneficial effect in reducing oxidative stress levels in a population of athletes. For example, one week of n-3 PUFA supplementation ameliorated the rise in oxidative stress markers after acute resistance exercise in young athletes [37]. Also, 2-month diet supplementation with docosahexaenoic acid (DHA) increased antioxidant capabilities and reduced mitochondrial ROS production in football players [38]. Furthermore, supplementation of vitamins E and C for 7 days reduced oxidative stress after an exercise-induced oxidative stress protocol in football players [39]. And while available studies indicate that vitamin E supplementation appears to be effective at decreasing markers of exercise-induced oxidative stress, evidence of its effects on endurance performance is still lacking [40]. The combination of antioxidant microelements such as DHA and vitamin E in the form of a fortified drink has been shown to have a protective effect against oxidative damage and to increase gene expression of antioxidant enzymes in PBMCs after exercise [41]. Similarly, three weeks of treatment with 100 mcg/day of Se has shown significant changes in the peroxides and glucose-6-phosphate dehydrogenases in eighteen elite athletes, suggesting an antioxidant effect of this element [42]. A different antioxidant compound containing Se, vitamin E, glutathione, and cysteine in the form of pills has also shown an antioxidant effect in fifteen road cyclists after 3 weeks of treatment [42]. Optimal Se plasma levels have been shown to be important in minimizing chronic exercise-induced oxidative effects and modulating the exercise effect on mitochondrial changes [24]. On the other hand, enzymatic activity of CAT, GPx, and SOD remained similar after dietary protocols in both Nutri4 and controls, while MPO enzymatic activity had a tendency to decrease in the Nutri4 group after the protocol; however, that was not significant. Interestingly, it is well accepted that repeated exposure of the organism to low-grade oxidative stress from chronic exercise training leads to an upregulation in the body's antioxidant defense system, thus providing better protection from oxidative stress during subsequent training sessions [43]. Since both of our examined groups were athletes, this effect of chronic exercise training might be a plausible explanation for the observed lack of significant change in antioxidant defense but decreased oxidative stress biomarkers after an enriched chicken egg dietary regime.

5. Conclusions

The present study showed that consumption of chicken eggs functionally enriched with n-3 PUFAs, Se, lutein, and vitamin E resulted in improved endothelium-dependent microvascular reactivity in competitive athletes and an increased range of microvascular responsiveness to ASTS. This is the first study that investigated the combined effects of four nutrients in the form of functional food, and besides improvement in microvascular endothelial function, it demonstrated its effect on a decrease in oxidative stress levels in competitive athletes. Importantly, microvascular adaptation to the acute exercise stress challenge was improved after consumption of Nutri4 eggs. These beneficial pro-endothelial and antioxidant effects of enriched eggs occurred without changes in serum lipids and blood pressure levels, indicating the safety of consuming a large amount of chicken eggs in the population of young athletes. The results of this study could encourage professional athletes to enrich their diet with natural sources of micronutrients in the form of functional foods.

Author Contributions: Conceptualization, I.D. and M.S.; methodology, P.Š., M.S., A.S., Z.M., I.B., M.L. and I.D.; validation, M.S., A.S., I.B., M.L. and I.D.; formal analysis, L.K., P.Š., M.S. and A.S.; investigation, L.K., P.Š., Z.M., I.J., N.K., I.B. and D.N.; resources, I.D.; data curation, L.K., P.Š., M.S. and A.S.; writing—original draft preparation, L.K., P.Š., M.S. and A.S.; writing—review and editing, I.D. and M.S.; visualization, A.S.; supervision, I.D. All authors have read and agreed to the published version of the manuscript.

Funding: This research was funded by the European Structural and Investment Funds grant for the Croatian National Scientific Center of Excellence for Personalized Health Care, University of Josip Juraj Strossmayer Osijek, Osijek, Croatia (grant #KK.01.1.1.01.0010).

Institutional Review Board Statement: The study was conducted in accordance with the Declaration of Helsinki and approved by the Ethics Committee of the Faculty of Medicine Osijek (Cl: 602-04/21-08/07; No: 2158-61-07-21-151).

Informed Consent Statement: Informed consent was obtained from all subjects involved in the study.

Data Availability Statement: The data that support the findings of this study are available from the corresponding author upon reasonable request.

Conflicts of Interest: The authors declare no conflict of interest.

References

1. Spriet, L.L. Performance Nutrition for Athletes. *Sport. Med.* **2019**, *49*, 1–2. [CrossRef] [PubMed]
2. Baranauskas, M.; Stukas, R.; Tubelis, L.; Žagminas, K.; Šurkienė, G.; Švedas, E.; Giedraitis, V.R.; Dobrovolskij, V.; Abaravičius, J.A. Nutritional habits among high-performance endurance athletes. *Medicina (B. Aires).* **2015**, *51*, 351–362. [CrossRef] [PubMed]
3. Ritz, P.P.; Rogers, M.B.; Zabinsky, J.S.; Hedrick, V.E.; Rockwell, J.A.; Rimer, E.G.; Kostelnik, S.B.; Hulver, M.W.; Rockwell, M.S. Dietary and Biological Assessment of the Omega-3 Status of Collegiate Athletes: A Cross-Sectional Analysis. *PLoS ONE* **2020**, *15*, e0228834. [CrossRef] [PubMed]

4. Higgins, M.; Izadi, A.; Kaviani, M. Antioxidants and Exercise Performance: With a Focus on Vitamin E and C Supplementation. *Int. J. Environ. Res. Public Health* **2020**, *17*, 8452. [CrossRef] [PubMed]
5. Stupin, A.; Mihalj, M.; Kolobarić, N.; Šušnjara, P.; Kolar, L.; Mihaljević, Z.; Matić, A.; Stupin, M.; Jukić, I.; Kralik, Z.; et al. Anti-Inflammatory Potential of n-3 Polyunsaturated Fatty Acids Enriched Hen Eggs Consumption in Improving Microvascular Endothelial Function of Healthy Individuals—Clinical Trial. *Int. J. Mol. Sci.* **2020**, *21*, 4149. [CrossRef] [PubMed]
6. Devrim-Lanpir, A.; Bilgic, P.; Kocahan, T.; Deliceoğlu, G.; Rosemann, T.; Knechtle, B. Total Dietary Antioxidant Intake Including Polyphenol Content: Is It Capable to Fight against Increased Oxidants within the Body of Ultra-Endurance Athletes? *Nutrients* **2020**, *12*, 1877. [CrossRef]
7. Busquets-Cortés, C.; Capó, X.; Martorell, M.; Tur, J.A.; Sureda, A.; Pons, A. Training Enhances Immune Cells Mitochondrial Biosynthesis, Fission, Fusion, and Their Antioxidant Capabilities Synergistically with Dietary Docosahexaenoic Supplementation. *Oxid. Med. Cell. Longev.* **2016**, *2016*, 1–10. [CrossRef]
8. Rawson, E.S.; Miles, M.P.; Larson-Meyer, D.E. Dietary Supplements for Health, Adaptation, and Recovery in Athletes. *Int. J. Sport Nutr. Exerc. Metab.* **2018**, *28*, 188–199. [CrossRef]
9. Oliver, J.M. Omega-3 Fatty Acids and Student-Athletes: Is It Time for Better Education and a Policy Change? *J. Athl. Train.* **2019**, *54*, 5–6. [CrossRef]
10. Asgary, S.; Rastqar, A.; Keshvari, M. Functional Food and Cardiovascular Disease Prevention and Treatment: A Review. *J. Am. Coll. Nutr.* **2018**, *37*, 429–455. [CrossRef]
11. Kralik, Z.; Kralik, G.; Košević, M.; Galović, O.; Samardžić, M. Natural Multi-Enriched Eggs with n-3 Polyunsaturated Fatty Acids, Selenium, Vitamin E, and Lutein. *Animals* **2023**, *13*, 321. [CrossRef]
12. Oppedisano, F.; Macrì, R.; Gliozzi, M.; Musolino, V.; Carresi, C.; Maiuolo, J.; Bosco, F.; Nucera, S.; Caterina Zito, M.; Guarnieri, L.; et al. The Anti-Inflammatory and Antioxidant Properties of n-3 PUFAs: Their Role in Cardiovascular Protection. *Biomedicines* **2020**, *8*, 306. [CrossRef]
13. Sakai, C.; Ishida, M.; Ohba, H.; Yamashita, H.; Uchida, H.; Yoshizumi, M.; Ishida, T. Fish oil omega-3 polyunsaturated fatty acids attenuate oxidative stress-induced DNA damage in vascular endothelial cells. *PLoS ONE* **2017**, *12*, e0187934. [CrossRef]
14. Kyriakidou, Y.; Wood, C.; Ferrier, C.; Dolci, A.; Elliott, B. The effect of Omega-3 polyunsaturated fatty acid supplementation on exercise-induced muscle damage. *J. Int. Soc. Sports Nutr.* **2021**, *18*, 9. [CrossRef]
15. Walser, B.; Giordano, R.M.; Stebbins, C.L. Supplementation with omega-3 polyunsaturated fatty acids augments brachial artery dilation and blood flow during forearm contraction. *Eur. J. Appl. Physiol.* **2006**, *97*, 347–354. [CrossRef]
16. Kolar, L.; Stupin, M.; Stupin, A.; Šušnjara, P.; Mihaljević, Z.; Matić, A.; Jukić, I.; Kolobarić, N.; Drenjančević, I. Does the Endothelium of Competitive Athletes Benefit from Consumption of n-3 Polyunsaturated Fatty Acid-Enriched Hen Eggs? *Prev. Nutr. Food Sci.* **2021**, *26*, 388–399. [CrossRef]
17. Da Boit, M.; Hunter, A.M.; Gray, S.R. Fit with good fat? The role of n-3 polyunsaturated fatty acids on exercise performance. *Metabolism* **2017**, *66*, 45–54. [CrossRef]
18. Zehr, K.R.; Walker, M.K. Omega-3 polyunsaturated fatty acids improve endothelial function in humans at risk for atherosclerosis: A review. *Prostaglandins Other Lipid Mediat.* **2018**, *134*, 131–140. [CrossRef]
19. Ziegler, M.; Wallert, M.; Lorkowski, S.; Peter, K. Cardiovascular and Metabolic Protection by Vitamin E: A Matter of Treatment Strategy? *Antioxidants* **2020**, *9*, 935. [CrossRef]
20. Stupin, A.; Rasic, L.; Matic, A.; Stupin, M.; Kralik, Z.; Kralik, G.; Grcevic, M.; Drenjancevic, I. Omega-3 polyunsaturated fatty acids-enriched hen eggs consumption enhances microvascular reactivity in young healthy individuals. *Appl. Physiol. Nutr. Metab.* **2018**, *43*, 988–995. [CrossRef]
21. Takanami, Y.; Iwane, H.; Kawai, Y.; Shimomitsu, T. Vitamin E Supplementation and Endurance Exercise. *Sport. Med* **2000**, *29*, 73–83. [CrossRef]
22. Serov, D.A.; Khabatova, V.V.; Vodeneev, V.; Li, R.; Gudkov, S.V. A Review of the Antibacterial, Fungicidal and Antiviral Properties of Selenium Nanoparticles. *Materials* **2023**, *16*, 5363. [CrossRef]
23. Barchielli, G.; Capperucci, A.; Tanini, D. The Role of Selenium in Pathologies: An Updated Review. *Antioxidants* **2022**, *11*, 251. [CrossRef]
24. Fernández-Lázaro, D.; Fernandez-Lazaro, C.I.; Mielgo-Ayuso, J.; Navascués, L.J.; Córdova Martínez, A.; Seco-Calvo, J. The Role of Selenium Mineral Trace Element in Exercise: Antioxidant Defense System, Muscle Performance, Hormone Response, and Athletic Performance. A Systematic Review. *Nutrients* **2020**, *12*, 1790. [CrossRef]
25. Šušnjara, P.; Kolobarić, N.; Matić, A.; Mihaljević, Z.; Stupin, A.; Marczi, S.; Drenjančević, I. Consumption of Hen Eggs Enriched with n-3 Polyunsaturated Fatty Acids, Selenium, Vitamin E and Lutein Incites Anti-Inflammatory Conditions in Young, Healthy Participants—A Randomized Study. *Front. Biosci.* **2022**, *27*, 332. [CrossRef]
26. Jargar, J.G.; Hattiwale, S.H.; Das, S.; Dhundasi, S.A.; Das, K.K. A modified simple method for determination of serum α-tocopherol (vitamin E). *J. Basic Clin. Physiol. Pharmacol.* **2012**, *23*, 45–48. [CrossRef]
27. D'Ilio, S.; Violante, N.; Caimi, S.; Di Gregorio, M.; Petrucci, F.; Senofonte, O. Determination of trace elements in serum by dynamic reaction cell inductively coupled plasma mass spectrometry. *Anal. Chim. Acta* **2006**, *573–574*, 432–438. [CrossRef]
28. Tzeng, M.-S.; Yang, F.-L.; Wang-Hsu, G.-S.; Chen, B.-H. Determination of major carotenoids in human serum by liquid chromatography. *J. Food Drug Anal.* **2020**, *12*, 13. [CrossRef]

29. Stupin, M.; Stupin, A.; Rasic, L.; Cosic, A.; Kolar, L.; Seric, V.; Lenasi, H.; Izakovic, K.; Drenjancevic, I. Acute exhaustive rowing exercise reduces skin microvascular dilator function in young adult rowing athletes. *Eur. J. Appl. Physiol.* **2018**, *118*, 461–474. [CrossRef]
30. Barić, L.; Drenjančević, I.; Mihalj, M.; Matić, A.; Stupin, M.; Kolar, L.; Mihaljević, Z.; Mrakovčić-Šutić, I.; Šerić, V.; Stupin, A. Enhanced Antioxidative Defense by Vitamins C and E Consumption Prevents 7-Day High-Salt Diet-Induced Microvascular Endothelial Function Impairment in Young Healthy Individuals. *J. Clin. Med.* **2020**, *9*, 843. [CrossRef]
31. Hendrickse, P.; Degens, H. The role of the microcirculation in muscle function and plasticity. *J. Muscle Res. Cell Motil.* **2019**, *40*, 127–140. [CrossRef]
32. Green, D.J.; Smith, K.J. Effects of Exercise on Vascular Function, Structure, and Health in Humans. *Cold Spring Harb. Perspect. Med.* **2018**, *8*, a029819. [CrossRef]
33. Lavie, C.J.; Arena, R.; Swift, D.L.; Johannsen, N.M.; Sui, X.; Lee, D.; Earnest, C.P.; Church, T.S.; O'Keefe, J.H.; Milani, R.V.; et al. Exercise and the Cardiovascular System. *Circ. Res.* **2015**, *117*, 207–219. [CrossRef]
34. Jurva, J.W.; Phillips, S.A.; Syed, A.Q.; Syed, A.Y.; Pitt, S.; Weaver, A.; Gutterman, D.D. The Effect of Exertional Hypertension Evoked by Weight Lifting on Vascular Endothelial Function. *J. Am. Coll. Cardiol.* **2006**, *48*, 588–589. [CrossRef]
35. Phillips, S.A.; Das, E.; Wang, J.; Pritchard, K.; Gutterman, D.D. Resistance and aerobic exercise protects against acute endothelial impairment induced by a single exposure to hypertension during exertion. *J. Appl. Physiol.* **2011**, *110*, 1013–1020. [CrossRef]
36. Birk, G.; Dawson, E.; Batterham, A.; Atkinson, G.; Cable, T.; Thijssen, D.H.; Green, D. Effects of Exercise Intensity on Flow Mediated Dilation in Healthy Humans. *Int. J. Sports Med.* **2012**, *34*, 409–414. [CrossRef]
37. Atashak, S.; Sharafi, H.; Azarbayjani, M.A.; Stannard, S.R.; Goli, M.A.; Haghighi, M.M. Effect of omega-3 supplementation on the blood levels of oxidative stress, muscle damage and inflammation markers after acute resistance exercise in young athletes. *Kinesiology* **2013**, *45*, 22–29.
38. Capó, X.; Martorell, M.; Sureda, A.; Llompart, I.; Tur, J.A.; Pons, A. Diet supplementation with DHA-enriched food in football players during training season enhances the mitochondrial antioxidant capabilities in blood mononuclear cells. *Eur. J. Nutr.* **2015**, *54*, 35–49. [CrossRef]
39. de Oliveira, D.C.X.; Rosa, F.T.; Simões-Ambrósio, L.; Jordao, A.A.; Deminice, R. Antioxidant vitamin supplementation prevents oxidative stress but does not enhance performance in young football athletes. *Nutrition* **2019**, *63–64*, 29–35. [CrossRef]
40. Mason, S.A.; Trewin, A.J.; Parker, L.; Wadley, G.D. Antioxidant supplements and endurance exercise: Current evidence and mechanistic insights. *Redox Biol.* **2020**, *35*, 101471. [CrossRef]
41. Capó, X.; Martorell, M.; Busquets-Cortés, C.; Sureda, A.; Riera, J.; Drobnic, F.; Tur, J.A.; Pons, A. Effects of dietary almond- and olive oil-based docosahexaenoic acid- and vitamin E-enriched beverage supplementation on athletic performance and oxidative stress markers. *Food Funct.* **2016**, *7*, 4920–4934. [CrossRef] [PubMed]
42. Drâgan, I.; Ploeşteanu, E.; Cristea, E.; Mohora, M.; Dinu, V.; Troescu, V.S. Studies on selenium in top athletes. *Physiologie* **1988**, *25*, 187–190.
43. Elosua, R.; Molina, L.; Fito, M.; Arquer, A.; Sanchez-Quesada, J.; Covas, M.; Ordoñez-Llanos, J.; Marrugat, J. Response of oxidative stress biomarkers to a 16-week aerobic physical activity program, and to acute physical activity, in healthy young men and women. *Atherosclerosis* **2003**, *167*, 327–334. [CrossRef] [PubMed]

life

MDPI

Article

Cardiopulmonary Exercise Testing in Patients with Heart Failure: Impact of Gender in Predictive Value for Heart Transplantation Listing

Pedro Garcia Brás [1,*], António Valentim Gonçalves [1], João Ferreira Reis [1], Rita Ilhão Moreira [1], Tiago Pereira-da-Silva [1], Pedro Rio [1], Ana Teresa Timóteo [1,2], Sofia Silva [1], Rui M. Soares [1] and Rui Cruz Ferreira [1]

1 Cardiology Department, Santa Marta Hospital, Central Lisbon Hospital University Center, 1169-024 Lisbon, Portugal
2 NOVA Medical School, Faculdade de Ciências Médicas (NMS | FCM), 1169-056 Lisbon, Portugal
* Correspondence: pedrobras3@gmail.com

Abstract: Background: Exercise testing is key in the risk stratification of patients with heart failure (HF). There are scarce data on its prognostic power in women. Our aim was to assess the predictive value of the heart transplantation (HTx) thresholds in HF in women and in men. Methods: Prospective evaluation of HF patients who underwent cardiopulmonary exercise testing (CPET) from 2009 to 2018 for the composite endpoint of cardiovascular mortality and urgent HTx. Results: A total of 458 patients underwent CPET, with a composite endpoint frequency of 10.5% in females vs. 16.0% in males in 36-month follow-up. Peak VO_2 (pVO_2), VE/VCO_2 slope and percent of predicted pVO_2 were independent discriminators of the composite endpoint, particularly in women. The International Society for Heart Lung Transplantation recommended values of $pVO_2 \leq 12$ mL/kg/min or ≤ 14 if the patient is intolerant to β-blockers, VE/VCO_2 slope > 35, and percent of predicted $pVO_2 \leq 50\%$ showed a higher diagnostic effectiveness in women. Specific pVO_2, VE/VCO_2 slope and percent of predicted pVO_2 cut-offs in each sex group presented a higher prognostic power than the recommended thresholds. Conclusion: Individualized sex-specific thresholds may improve patient selection for HTx. More evidence is needed to address sex differences in HF risk stratification.

Keywords: gender; heart failure; heart transplantation; cardiopulmonary exercise testing; peak O_2 consumption

Citation: Garcia Brás, P.; Gonçalves, A.V.; Reis, J.F.; Moreira, R.I.; Pereira-da-Silva, T.; Rio, P.; Timóteo, A.T.; Silva, S.; Soares, R.M.; Ferreira, R.C. Cardiopulmonary Exercise Testing in Patients with Heart Failure: Impact of Gender in Predictive Value for Heart Transplantation Listing. *Life* **2023**, *13*, 1985. https://doi.org/10.3390/life13101985

Academic Editors: Ines Drenjančević and Helena Lenasi

Received: 30 August 2023
Revised: 22 September 2023
Accepted: 27 September 2023
Published: 29 September 2023

1. Introduction

Cardiopulmonary exercise testing (CPET) is a critical complementary test in the evaluation of patients with heart failure (HF) with reduced ejection fraction (HFrEF), particularly in selectin patients who may benefit from heart transplantation (HTx) [1,2]. Peak O_2 consumption (pVO_2) [3–5] and the VE/VCO_2 slope (minute ventilation–CO_2 production relationship) [3,5,6] are reliable indicators of heart failure events. A cut-off for pVO_2 of ≤ 12 mL/kg/min is recommended to guide HTx listing for patients receiving β-blocker therapy, and a cut-off of 14 mL/kg/min may be used for patients intolerant to β-blockers, according to the 2016 International Society for Heart Lung Transplantation (ISHLT) listing criteria for HTx [7,8]. In female patients, alternative parameters such as a VE/VCO_2 slope of >35 and a percent of predicted $pVO_2 \leq 50\%$ may be considered to guide HTx listing [7]. However, the data supporting these values come from studies that enrolled mostly male patients, with a sample that was between 80 and 90 percent male [2,5].

Indeed, female patients are underrepresented in HFrEF trials, although they account for around half of the adult HFrEF population [9]. Notably, in studies exploring CPET parameters in HFrEF, this gap in female representation is even larger [1,10–12]. Thus, the current evidence on female HFrEF pathophysiology and exercise testing prognostic power is scarce, and therapy changes, risk stratification, and recommendations for advanced HF therapies may be impacted by our insufficient comprehension of potential sex variations in HF [13,14]. Several trials evaluating pVO_2 have reported lower values in female patients compared to male patients, which can be explained by anatomical and physiological differences [9,15]. Female patients exhibit lower left ventricular dimensions, with lower stroke volume and lower diastolic compliance [16,17]; women show a higher prevalence of iron deficiency, have lower hemoglobin levels [18], and have inferior lean mass compared to male patients [9,19].

The HF-ACTION trial [10] assessed the prognostic power of CPET variables to predict all-cause mortality in HFrEF and reported that the prognosis associated with a given pVO_2 differed by sex. Female patients generally present a more favorable outcome, and have a lower pVO_2 and a higher percent of predicted pVO_2 [1,10]. Taking into account the sex-based variations in the pathophysiology and development of HFrEF, several authors proposed that prognostic values for pVO_2 and VE/VCO_2 slope should be tailored for different patient populations [12,14,20]. According to the ISHLT [7], different CPET variables can be used for risk stratification in women. Nevertheless, there is insufficient evidence to support these at this time, as unbiased data are not available [1].

This study's objective was to assess the predictive power of the traditional HTx CPET cut-off values in HF patients, comparing women and men.

2. Materials and Methods

2.1. Study Population

From 2009 to 2018, we performed a retrospective study of a prospective database in our center. We assessed consecutive HfrEF patients who underwent CPET, were in New York Heart Association (NYHA) classes II or III, and presented left ventricular (LV) dysfunction (LV ejection fraction $\leq$ 40%). Patients were referred to the Heart Failure team for evaluation to determine whether HTx or mechanical circulatory support (MCS) were indicated.

2.2. Study Protocol

The patient's comorbidities, HF etiology, medication, NYHA class, HFSS (Heart Failure Survival Score) [21], laboratory tests, CPET data, and electrocardiographic and echocardiographic results were evaluated.

2.3. Patients Were Excluded If One of the following Was Present

Age under 18 years; submaximal CPET (peak RER of $\leq$1.05 [7]); previous HTx or elective HTx during follow-up; coronary revascularization in the last six months; concomitant conditions limiting maximal exercise, including previous stroke, peripheral arterial disease, or musculoskeletal conditions.

2.4. Cardiorespiratory Exercise Testing

The modified Bruce protocol was employed to assess maximal exercise tolerance on a GE Marquette Series 2000 treadmill, with equipment calibration before each exercise exam. The VE, VO_2, and VCO_2 values were acquired with a Vmax 229 (SensorMedics, Yorba Linda, CA, USA) gas analyzer. Continuous ECG monitoring was used to assess the heart rate (HRt). Blood pressure (BP) was obtained with a sphygmomanometer, and O_2 saturation was tracked with pulse oximetry. An exercise test was considered maximal if the RER (respiratory exchange ratio) was above 1.05 [7].

The pVO_2 was defined as the highest achieved 30 s average in maximal exercise, which was then normalised for body mass. The standard methods (combining V-slope preferentially and ventilatory equivalents) were used to determine the gas exchange threshold (GET). The VE/VCO_2 slope was determined with the least squares linear regression. The minimum ventilatory equivalent for oxygen (minimum VE/VO_2) was employed to calculate the COP (cardiorespiratory optimal point). The partial pressure of end-tidal carbon dioxide ($PetCO_2$) was recorded both before exercise and at GET. Peak O_2 pulse, measured in millilitres per beat, was computed by dividing the derived pVO_2 by the highest HRt during exercise. The peak systolic BP was divided by the VE/VCO_2 slope to determine the ventilatory power. The circulatory power was estimated by multiplying the peak systolic BP by pVO_2. The heart rate reserve was calculated using the difference between the highest HRt attained during maximal effort and the resting heart rate. The difference between the maximal heart rate attained with exercise and the heart rate one minute in recovery was used to determine the HRt recovery.

2.5. Follow-Up and Endpoint

All patients with HF were under follow-up for a 36-month period. The composite endpoint was defined as the combination of cardiovascular mortality or urgent HTx. Data were collected from medical records from inpatient and outpatient visits.

2.6. Statistical Analysis

All analytical tests compared patients according to female or male sex. Statistical analyses were performed with the Statistical Package for the Social Sciences (SPSS) v23.0.

Regarding categorical variables, results were reported as absolute frequency (number) and relative frequency (%). Continuous variables were presented as mean and standard deviation if normal distribution, or as median and interquartile range (IQR) if non-normal distribution. Normality assumptions were tested using the Kolmogorov–Smirnov test and a visual histogram analysis.

The comparison of categorical variables was performed using the Pearson's X^2 test. The Mann–Whitney U test was used to compare variables with non-normal distribution and the Student's t-test was used to compare variables with normal distribution.

The correlation between the CPET parameters and the composite endpoint was evaluated using a Cox hazards regression analysis. Variables presenting a p-value < 0.200 in the univariate analysis were included in a multivariate analysis, adjusted for potential confounders, in order to identify independent predictors of the composite endpoint and calculate adjusted hazard ratios (HR) in each sex subgroup. The HR and the 95% confidence interval (CI) were used to report the results.

A receiver operating characteristic (ROC) curve analysis was used to examine the sensitivity and specificity of each CPET parameter in predicting the composite endpoint, in accordance with the thresholds defined by the ISHLT [7]: $pVO_2 \leq 12$ mL/Kg/min ($pVO_2 \leq 14$ in patients not tolerant to β-blockers), VE/VCO_2 slope > 35 and percent of predicted $pVO_2 \leq 50\%$.

The threshold with the highest combination of specificity and sensitivity was estimated using the Youden index (J). The DeLong test [22] was employed to evaluate the difference in area under the curve (AUC) between groups. Additionally, the Kaplan–Meier analysis was used to assess the event-free survival rate. A log-rank test was performed to compare the sex subgroups based on the different pVO_2, VE/VCO_2 slope, and percent of predicted pVO_2 thresholds indicated by the ISHLT [7] and based on the proposed cut-offs. A significance threshold of $\alpha = 5\%$ was considered whenever a statistical hypothesis was being tested.

3. Results

3.1. Patient Characteristics

Our study included 458 patients who underwent maximal exercise testing (Figure 1). Of these patients, 79% were men, 57% had ischemic etiology, 76% were in NYHA II and 24% in NYHA III, with a mean LVEF of 29.7 ± 8.0%, and 24% had atrial fibrillation (AF). In addition, 79% were taking either an ACEi (angiotensin-converting enzyme inhibitor) or an ARB (angiotensin receptor blockers), with 17% on an angiotensin receptor/neprilysin inhibitor. Mineralocorticoid receptor antagonists (MRAs) were being taken by 73% and β-blockers by 86%. Additionally, sodium-glucose cotransporter-2 inhibitors (SGLT2i) were being taken by 10% of the patients; 64% of patients had an ICD and 22% had a cardiac resynchronization device (CRT-D). Moreover, there was no difference in the mean Heart Failure Survival Score (HFSS). Compared to male patients, female patients had a similar

pVO_2 and a higher percent of predicted pVO_2. The mean respiratory exchange ratio (RER) was 1.14 ± 0.07. Table 1 lists the baseline characteristics of both groups as well as the CPET values.

Table 1. Baseline characteristics of the study population (*n* = 458).

	Overall (*n* = 458)	Female (*n* = 95)	Male (*n* = 363)	*p*-Value
Clinical and demographic data				
Age (years)	56 ± 12	54 ± 14	56 ± 12	0.328
Body mass index (kg/m^2)	27.1 ± 4.3	26.3 ± 4.6	27.4 ± 4.2	0.335
Ischemic etiology (*n*, %)	261 (57)	47 (49)	214 (59)	0.092
ACEi/ARB (*n*, %)	361 (79)	77 (81)	284 (78)	0.199
ARNI (*n*, %)	80 (17)	13 (14)	67 (18)	0.273
β-blocker (*n*, %)	392 (86)	81 (85)	311 (86)	0.726
MRA (*n*, %)	336 (73)	72 (76)	264 (73)	0.789
iSGLT2 (*n*, %)	47 (10)	8 (8)	39 (11)	0.164
Digoxin (*n*, %)	129 (28)	23 (24)	106 (29)	0.372
Diabetes	104 (23)	15 (16)	89 (25)	0.094
CKD (*n*, %)	145 (32)	25 (26)	120 (34)	0.138
AF (*n*, %)	109 (24)	14 (15)	95 (26)	0.021
ICD * (*n*, %)	293 (64)	59 (62)	234 (64)	0.617
Cardiac resynchronization therapy (*n*, %)	102 (22)	27 (28)	75 (21)	0.128
NYHA II	347 (76)	74 (78)	273 (75)	0.485
NYHA III	111 (24)	21 (22)	90 (25)	0.485
HFSS	8.6 ± 1.1	8.8 ± 0.9	8.6 ± 1.2	0.109
Laboratory data				
eGFR, ml/min/1.73 m^2	75.3 ± 29.2	77.1 ± 30.9	74.8 ± 28.7	0.517
Na^+, mEq/l	138.0 ± 3.0	138.4 ± 2.8	137.9 ± 3.1	0.108
N-terminal pro b-type natriuretic peptide, pg/mL	2196 ± 2101	2204 ± 1724	2193 ± 2099	0.979
Echocardiographic data				
LVEDD, mm/m^2	67.4 ± 10.3	63.8 ± 9.7	68.0 ± 10.3	0.064
LVEF, %	29.7 ± 8.0	31.3 ± 7.9	29.0 ± 7.5	0.213
Mitral regurgitation severity III–IV, %	67 (14)	19 (20)	48 (13)	0.097
Right ventricular dysfunction (*n*, %)	69 (15)	9 (10)	60 (16)	0.630
Exercise testing data				
Peak Respiratory Exchange Ratio	1.14 ± 0.07	1.13 ± 0.08	1.14 ± 0.07	0.566
Delta heart rate during exercise	51 (37–68)	48 (34–67)	52 (38–69)	0.819
HHR1	17 (11–27)	19 (14–29)	16 (11–26)	0.058
pVO_2, mL/kg/min	18.5 ± 5.8	18.0 ± 5.6	18.6 ± 5.9	0.363
Percent of predicted pVO_2 (%)	63.8 ± 18.7	67.4 ± 16.7	62.8 ± 19.1	0.021
VE/VCO_2 slope	33.9 ± 9.6	33.0 ± 8.9	34.2 ± 9.8	0.246
pVO_2, mL/kg/min at GET	13.6 ± 4.6	10.9 ± 2.8	14.2 ± 4.7	0.001
Peak O_2 pulse	0.14 ± 0.06	0.13 ± 0.03	0.14 ± 0.07	0.149
Circulatory power	2883 ± 1543	2715 ± 1035	2927 ± 1649	0.235
Ventilatory power	4.8 ± 1.7	4.8 ± 1.5	4.8 ± 1.7	0.739
COP	28.9 ± 7.2	29.5 ± 7.9	28.8 ± 7.0	0.630
$PetCO_2$ at rest, mmHg	33.6 ± 4.8	33.9 ± 5.1	33.5 ± 4.7	0.558
$PetCO_2$ at GET, mmHg	36.8 ± 6.0	37.5 ± 5.9	36.6 ± 6.1	0.262

* including patients with a cardiac resynchronization therapy device. CPET: Cardiopulmonary exercise test; ACEi: Angiotensin-converting enzyme inhibitors; ARNI: Angiotensin receptor neprilysin inhibitors; ARB: Angiotensin receptor blockers; MRA: Mineralocorticoid receptor antagonists; CKD: Chronic kidney disease; AF: Atrial fibrillation; ACEi: Angiotensin-converting enzyme inhibitors; ICD: Implantable cardioverter-defibrillator; HFSS: Heart Failure Survival Score; eGFR: estimated glomerular filtration rate; LVEF: Left ventricular ejection fraction; LVEDD: Left ventricular end-diastolic diameter; pVO_2: Peak O_2 consumption; VE/VCO_2 slope: Minute ventilation-carbon dioxide production relationship; GET: Gas exchange threshold; COP: Cardiorespiratory optimal point; HRR1: Heart rate recovery in the first minute after finishing CPET; $PetCO_2$: Partial pressure of end-tidal carbon dioxide.

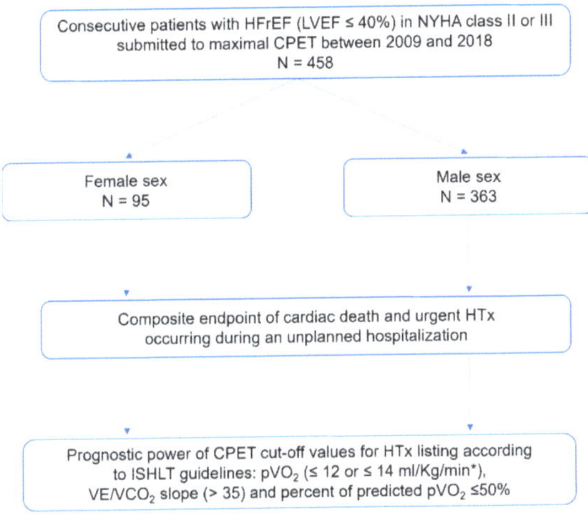

Figure 1. Flowchart of the study population. * in patients intolerant to β-blockers. HFrEF: Heart failure with reduced ejection fraction; LVEF: Left ventricular ejection fraction; NYHA: New York Heart Association; CPET: Cardiopulmonary exercise test; HTx: Heart transplantation; ISHLT: International Society for Heart and Lung Transplantation; pVO_2: Peak oxygen consumption; VE/VCO_2 slope: Minute ventilation–carbon dioxide production relationship.

3.2. Composite Endpoint

The composite endpoint occurred in 68 (14.8%) patients in 36 months of follow-up, with cardiovascular death occurring in 54 individuals and urgent HTx occurring in 14 patients (Table 2). No urgent MCS was required; 10.5% of female patients and 16.0% of male patients experienced the composite endpoint, with no significant difference between groups.

Table 2. Total adverse events during follow-up.

	Total Cohort (n = 458)	Female (n = 95)	Male (n = 363)	p-Value
Composite endpoint (n, %)	68 (14.8%)	10 (10.5%)	58 (16.0%)	0.199
Total mortality (n, %)	67 (14.6%)	13 (13.7%)	54 (14.9%)	0.597
Cardiac mortality (n, %)	54 (11.8%)	8 (8.4%)	46 (12.7%)	0.098
Sudden cardiac death (n, %)	19 (4.1%)	2 (2.1%)	17 (4.7%)	0.147
Death from worsening HF (n, %)	35 (7.6%)	6 (6.3%)	29 (7.9%)	0.638
Urgent HTx (n, %)	14 (3.1%)	2 (2.1%)	12 (3.3%)	0.744

HF: Heart failure; HTx: Heart transplantation.

3.3. Prognostic Power of CPET Parameters

The pVO_2 (HR 0.856), the VE/VCO_2 slope (HR 1.064), and the percent of predicted pVO_2 (HR 0.955) were associated with the composite endpoint in a multivariable Cox regression analysis, regardless of the sex subgroup. Table 3 displays the results of the uni- and multivariable models. The correlations in the multivariable model were independent of potential confounders such as body mass index, LVEF, age, sex, smoking, diabetes mellitus, or estimated glomerular filtration rate. In the multivariable analysis, most of the other exercise testing variables were not linked with the primary endpoint. The peak O_2 pulse was associated with the endpoint in both female and male patients. The ventilatory power, the circulatory power, and the $PetCO_2$ at GET were linked with the primary endpoint in male patients, as shown in Table 3.

In an ROC curve analysis, the pVO_2, the VE/VCO_2 slope, and the percent of predicted pVO_2 were linked to the composite endpoint, both in females and males. The predictive ability of these variables was significantly higher in women compared to males, including for pVO_2, VE/VCO_2 slope, and the percent of predicted pVO_2, as presented in Table 4. The ROC curves for these subgroups are illustrated in Figure 2 and Supplementary Figure S1. In addition, the predictive power of the peak O_2 pulse was also significantly higher in female patients compared to males (AUC 0.816 vs. AUC 0.616, $p = 0.023$).

Table 3. Univariable and multivariable analysis of the composite endpoint.

		Total Cohort				
Model	Univariable HR	95% CI	*p*-value	Multivariable HR	95% CI	*p*-value
Male sex	1.547	0.791–3.026	0.203			
Age	1.002	0.983–1.021	0.829			
BMI	0.953	0.897–1.013	0.121	0.954	0.887–1.027	0.210
LVEF	0.927	0.900–0.955	<0.001	0.935	0.905–0.966	<0.001
eGFR	0.979	0.969–0.989	<0.001	0.986	0.976–0.996	0.009
Diabetes	1.196	0.254–5.632	0.821			
Smoker	1.716	1.405–2.820	0.033	1.395	0.835–2.328	0.203
Peak VO_2	0.835	0.789–0.883	<0.001	0.856	0.804–0.912	<0.001
Percent of predicted pVO_2	0.948	0.934–0.963	<0.001	0.955	0.939–0.971	<0.001
VE/VCO_2 slope	1.058	1.041–1.075	<0.001	1.064	1.039–1.090	<0.001
Peak VO_2 at GET, mL/kg/min	0.854	0.737–0.989	0.035	0.879	0.687–1.124	0.305
O_2 pulse, mL/kg/beat	0.858	0.791–0.932	<0.001	0.865	0.780–0.961	0.007
Circulatory power, mmHg.mL/kg/min	0.999	0.999–0.999	<0.001	0.999	0.998–1.000	<0.001
Ventilatory power, mmHg	0.575	0.483–0.684	<0.001	0.632	0.521–0.768	<0.001
COP	1.118	1.054–1.186	<0.001	1.060	0.956–1.174	0.268
$PetCO_2$ at rest, mmHg	0.887	0.839–0.937	<0.001	0.948	0.889–1.011	0.102
$PetCO_2$ at GET, mmHg	0.862	0.826–0.900	<0.001	0.890	0.845–0.993	<0.001
		Female sex				
Model	Univariable HR	95% CI	*p*-value	Multivariable HR	95% CI	*p*-value
Age	1.003	0.960–1.048	0.888			
BMI	0.897	0.770–1.045	0.162	0.861	0.694–1.067	0.171
LVEF	0.893	0.820–0.973	0.010	0.941	0.864–1.016	0.160
eGFR	0.977	0.952–1.003	0.086	0.991	0.966–1.016	0.459
Diabetes	1.135	0.629–2.053	0.674			
Smoker	0.940	0.199–4.436	0.937	1.565	0.178–13.699	0.686
Peak VO_2	0.704	0.583–0.850	<0.001	0.746	0.604–0.922	0.007
Percent of predicted pVO_2	0.911	0.875–0.948	<0.001	0.913	0.858–0.972	0.004
VE/VCO_2 slope	1.093	1.052–1.135	<0.001	1.143	1.039–1.257	0.006
Peak VO_2 at GET, mL/kg/min	0.223	0.010–5.159	0.350			
O_2 pulse, mL/kg/beat	0.493	0.346–0.703	<0.001	0.458	0.261–0.802	0.006
Circulatory power, mmHg.mL/kg/min	0.998	0.997–0.999	0.002	0.999	0.998–1.000	0.069
Ventilatory power, mmHg	0.405	0.240–0.684	0.001	0.565	0.297–1.072	0.080
COP	1.775	0.100–3.450	0.903			
$PetCO_2$ at rest, mmHg	0.903	0.792–1.028	0.123	0.981	0.841–1.144	0.807
$PetCO_2$ at GET, mmHg	0.814	0.715–0.927	0.002	0.871	0.736 –1.031	0.108
		Male sex				
Model	Univariable HR	95% CI	*p*-value	Multivariable HR	95% CI	*p*-value
Age	1.001	0.979–1.022	0.963			
BMI	0.960	0.898–1.027	0.240			
LVEF	0.933	0.905–0.963	<0.001	0.938	0.905–0.971	<0.001
eGFR	0.980	0.969–0.991	<0.001	0.987	0.976–0.998	0.020
Diabetes	1.211	0.639–2.230	0.558			
Smoker	1.791	1.024–3.133	0.041	1.425	0.805–2.521	0.224
Peak VO_2	0.854	0.806–0.905	<0.001	0.869	0.813–0.928	<0.001
Percent of predicted pVO_2	0.956	0.941–0.971	<0.001	0.960	0.943–0.977	<0.001
VE/VCO_2 slope	1.051	1.032–1.070	<0.001	1.056	1.030–1.084	<0.001
Peak VO_2 at GET, mL/kg/min	0.862	0.746–0.996	0.044	0.880	0.691–1.121	0.302
O_2 pulse, mL/kg/beat	0.873	0.802–0.949	0.001	0.884	0.794–0.985	0.026
Circulatory power, mmHg.mL/kg/min	0.999	0.999–0.999	<0.001	0.999	0.999–1.000	<0.001
Ventilatory power, mmHg	0.611	0.510–0.733	<0.001	0.645	0.526–0.792	<0.001
COP	1.095	1.027–1.167	0.005	1.062	0.962–1.173	0.230
$PetCO_2$ at rest, mmHg	0.886	0.834–0.942	<0.001	0.937	0.873–1.005	0.070
$PetCO_2$ at GET, mmHg	0.870	0.831–0.911	<0.001	0.887	0.839–0.939	<0.001

BMI: Body mass index; eGFR: Estimated glomerular filtration rate; LVEF: Left ventricular ejection fraction; pVO_2: Peak oxygen consumption; VE/VCO_2 slope: Minute ventilation–carbon dioxide production relationship; GET: Gas exchange threshold; COP: Cardiorespiratory optimal point; $PetCO_2$: Partial pressure of end-tidal carbon dioxide.

Table 4. Receiver operating characteristic (ROC) curve analysis of the composite endpoint.

CPET Parameters	Female (n = 95)			Male (n = 363)			
	AUC	95% CI	p-Value	AUC	95% CI	p-Value	p-Value (Interaction)
pVO_2, mL/kg/min	0.849	0.740–0.958	<0.001	0.701	0.629–0.773	<0.001	0.031
Predicted pVO_2 (%)	0.918	0.860–0.975	<0.001	0.701	0.628–0.774	<0.001	<0.001
VE/VCO_2 slope	0.894	0.803–0.986	<0.001	0.688	0.615–0.761	<0.001	<0.001
pVO_2, mL/kg/min at GET	0.648	0.464–0.832	0.096	0.635	0.451–0.820	0.140	0.594
O_2 pulse, mL/kg/beat	0.816	0.669–0.962	0.001	0.616	0.537–0.695	0.005	0.023
Circulatory power, mmHg.ml/kg/min	0.788	0.642–0.935	0.003	0.713	0.646–0.780	<0.001	0.444
Ventilatory power, mmHg	0.782	0.597–0.967	0.004	0.711	0.641–0.780	<0.001	0.504
COP	0.626	0.482–0.770	0.095	0.704	0.560–0.848	0.019	0.372
$PetCO_2$ at rest, mmHg	0.606	0.390–0.822	0.275	0.654	0.580–0.728	<0.001	0.694
$PetCO_2$ at GET, mmHg	0.784	0.638–0.930	0.004	0.719	0.644–0.794	<0.001	0.461

CPET: Cardiopulmonary exercise testing; pVO_2: Peak oxygen consumption; VE/VCO_2 slope: Minute ventilation–carbon dioxide production relationship; GET: Gas exchange threshold; COP: Cardiorespiratory optimal point; $PetCO_2$: Partial pressure of end-tidal carbon dioxide.

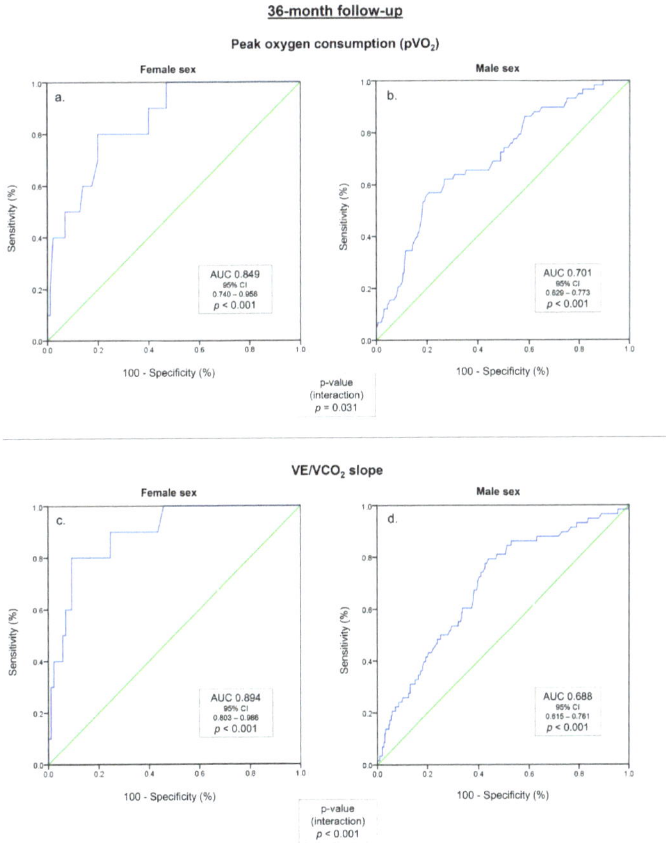

Figure 2. ROC curves for the composite endpoint in a 36-month follow up. (**a**) Peak oxygen consumption (pVO_2) in female patients. (**b**) pVO_2 in male patients. (**c**) Minute ventilation–carbon dioxide production relationship (VE/VCO_2 slope) in female patients. (**d**) VE/VCO_2 slope in male patients.

The circulatory power presented a slightly higher prognostic power than the recommended exercise testing parameters in men (AUC 0.713 vs. AUC 0.701, $p = 0.161$), albeit with no statistically significant differences in predictive power. Despite being significant predictors of the composite endpoint, additional CPET variables such as peak O_2 pulse, ventilatory power, COP, $PetCO_2$ at rest, and $PetCO_2$ at GET had an inferior predictive power than the traditional CPET parameters (Table 4).

3.4. ISHLT Recommended Thresholds for HTx Listing

A pVO_2 of $\leq$ 12 mL/kg/min ($\leq$14 if the patient is intolerant to β-blockers) was present in 49 (11%) patients. This threshold was linked with poor HF outcomes (HR 3.487, $p < 0.001$). This pVO_2 cut-off showed a sensitivity of 40% and a specificity of 94% in women, presenting a higher Youden index compared to men (*J* 0.34 vs. *J* 0.12), with a sensitivity of 21% and a specificity of 91%, as shown in Table 5. This cut-off was shown to be a strong discriminator of HF outcomes for both sex subgroups in a Kaplan–Meier analysis (Figure 3a).

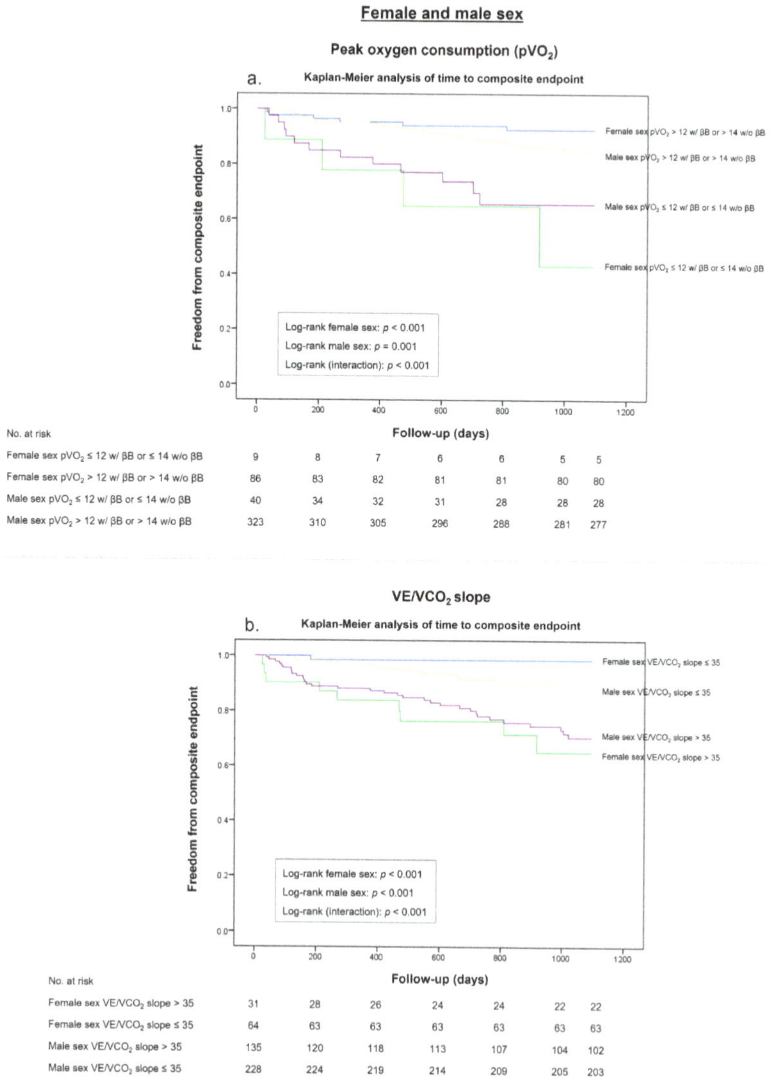

Figure 3. Survival analysis for the composite endpoint according to the International Society for Heart and Lung Transplantation (ISHLT) thresholds in female patients and male patients. (**a**) Peak oxygen consumption (pVO_2) $\leq$ 12 mL/Kg/min ($\leq$14 mL/kg/min if intolerant to β-blockers [βB]). (**b**) Minute ventilation–carbon dioxide production ratio (VE/VCO_2 slope) of >35.

Table 5. Evaluation of traditional and alternative thresholds cut-off values of the composite endpoint.

	Female (n = 95)			Male (n = 363)		
Exercise Testing Parameters	**Specificity**	**Sensitivity**	**Youden (J) Index**	**Specificity**	**Sensitivity**	**Youden (J) Index**
$pVO_2 \leq 12$ mL/kg/min *	94%	40%	0.34	91%	21%	0.12
$pVO_2 \leq 14$ mL/kg/min	**80%**	**80%**	**0.60**	82%	47%	0.29
$pVO_2 \leq 15$ mL/kg/min	67%	80%	0.47	**79%**	**57%**	**0.36**
VE/VCO_2 slope > 35	**75%**	**90%**	**0.65**	66%	57%	0.23
VE/VCO_2 slope > 32	68%	90%	0.58	**57%**	**78%**	**0.35**
Percent of predicted $pVO_2 \leq 50\%$	89%	60%	0.49	78%	48%	0.26
Percent of predicted $pVO_2 \leq 55\%$	**86%**	**90%**	**0.76**	69%	60%	0.29
Percent of predicted $pVO_2 \leq 58\%$	81%	90%	0.71	**63%**	**69%**	**0.32**

* $pVO_2 \leq 12$ mL/kg/min ($\leq$14 if the patient is intolerant to β-blockers). The highest Youden index (*J*) of each CPET variable is highlighted in bold. pVO_2: Peak O_2 consumption; VE/VCO_2 slope: Minute ventilation–CO_2 production relationship.

A total of 166 (36%) patients showed a VE/VCO_2 slope value higher than 35. The composite endpoint occurred at a higher rate in individuals over this threshold as well (HR 3.587, 95% CI 2.194–5.864, $p < 0.001$). This threshold revealed a substantially higher Youden index in women (*J* 0.65 vs. *J* 0.23), with sensitivity of 90% and a specificity of 75%, in comparison with male patients, with a sensitivity of 57% and a specificity of 66%. In the survival analysis, this VE/VCO_2 slope cut-off was a reliable indicator of the composite endpoint in both sex categories (Figure 3b).

In our cohort, a percent of predicted pVO_2 of less than 50% was present in 120 (26%) patients. This cut-off was associated with the composite endpoint (HR 4.355, 95% CI 2.694–7.039, $p < 0.001$). This cut-off showed a sensitivity of 60% and a specificity of 89% in females, while it had a sensitivity of 48% and specificity of 78% in males. As a result, the Youden index in females was higher than in male patients (*J* 0.49 vs. *J* 0.26). This threshold was a reliable discriminator in both subgroups according to the survival curve analysis (Supplementary Figure S2a).

3.5. Alternative Thresholds for pVO_2 and VE/VCO_2 Slope

In an assessment of potential alternative thresholds, a $pVO_2 \leq 14$ mL/kg/min yielded a higher Youden index in female patients compared to the $pVO_2 \leq 12$ mL/kg/min cut-off (*J* 0.60 vs. *J* 0.34) (Table 5). Similarly, a $pVO_2 \leq 15$ mL/kg/min value showed a higher overall diagnostic effectiveness in male patients compared to the traditional cut-off (*J* 0.36 vs. *J* 0.12). The predictive value of this cut-off was supported by the Kaplan–Meier analysis (Figure 4a).

In males, a VE/VCO_2 slope threshold of > 32 demonstrated sensitivity of 78% and a specificity of 57%, exhibiting a higher Youden index than the traditional VE/VCO_2 slope cut-off (*J* 0.35 vs. *J* 0.23). Regarding female patients, the traditional VE/VCO_2 slope > 35 threshold was associated with the highest overall diagnostic effectiveness (*J* 0.65). Additionally, it was demonstrated in the survival analysis that these cut-off values accurately predicted worse outcomes (Figure 4b).

A percent of predicted pVO_2 of $\leq 55\%$ yielded a significantly higher Youden index in female patients compared to the $\leq 50\%$ threshold (*J* 0.76 vs. *J* 0.49) while a percent of predicted pVO_2 of $\leq 58\%$ showed a higher Youden index in comparison to the traditional thresholds (*J* 0.32 vs. *J* 0.26) (Table 5). These cut-offs were accurate discriminators of the composite endpoint in both sex subgroups (log-rank $p < 0.001$) (Supplementary Figure S2b).

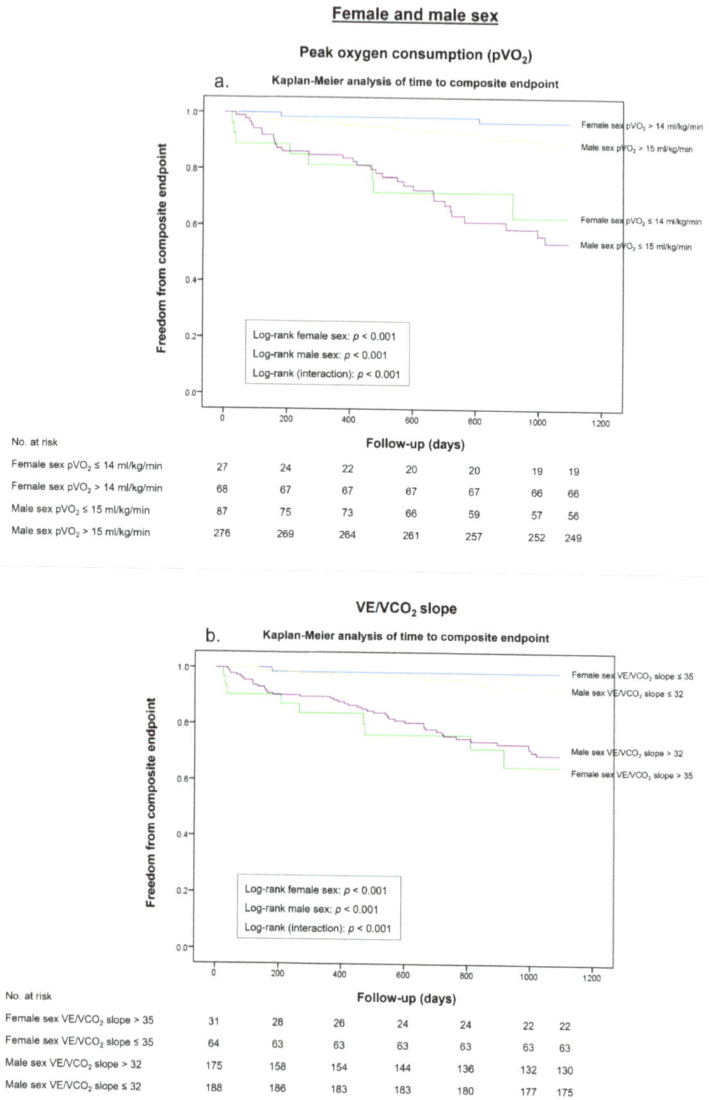

Figure 4. Survival analysis for the composite endpoint in female patients and male patients according to (**a**) Peak O_2 consumption (pVO_2) of $\leq$ 14 and $\leq$ 15 mL/Kg/min, respectively. (**b**) Minute ventilation–CO_2 production ratio (VE/VCO_2 slope) of > 35 and > 32, respectively.

4. Discussion

Our study's key conclusion was that the traditional CPET variables had a considerably higher predictive power for HF outcomes in women compared to men. Furthermore, the ISHLT recommended thresholds for pVO_2 ($\leq$12 mL/kg/min, or $\leq$ 14 mL/kg/min if intolerant to β-blockers), VE/VCO_2 slope (>35), and percent of predicted $pVO_2 \leq 55\%$ showed a significantly higher overall diagnostic effectiveness in women compared to men. Additionally, our study assessed the predictive capacity of various CPET variables and proposed sex-specific cut-offs for pVO_2, VE/VCO_2 slope, and the percent of predicted pVO_2, which may assist in a more precise risk assessment in women and men with HFrEF.

However, one of the main limitations of our study was that 79% of the enrolled patients were male; thus, further studies should include a higher proportion of female patients.

The current evidence on the predictive value of CPET in women with HFrEF was evaluated in a recent article by the Heart Failure Association's Committee on Exercise Physiology and Training [1,14,23–27]. The mean age of female patients enrolled in these studies was slightly lower than that of male patients, and one of the explanations for female underrepresentation in HfrEF trials was a larger proportion of older women who were excluded due to the policy of non-inclusion of elderly patients [28].

pVO_2 is influenced by gender, age, motivation, pulmonary status, and muscle mass [29], which raised concerns that this parameter's role as a prognostic indicator in female patients may lead to premature cardiac transplantation in women [14].

However, several observational studies showed that pVO_2 is a reliable discriminator for HF events in female patients [24,26] and a large trial [12] showed that predictive pVO_2 cut-offs for men and women with HfrEF should be independent. Although thresholds such as the GET were described to provide incremental value in the assessment of cardiorespiratory fitness in healthy controls [30,31], VO_2 measured at GET did not show a significant prognostic power compared to pVO_2 in our HfrEF cohort.

Women generally exhibited a lower corrected pVO_2 than male patients; however, female patients presented a lower rate of HF events. Notably, female patients showed a nearly 10% higher percent of predicted pVO_2 compared to men [23,27]. However, a study by Corrà et al. [11] postulated that HF outcomes in women may not actually be better than in men, as the female prognostic advantage is lost when sex-specific variations are properly taken into account with propensity score matching. Therefore, adjusting for sex-related characteristics should be undertaken. Indeed, female patients in our cohort showed a significantly higher percent of predicted pVO_2 despite having a similar absolute pVO_2 value, with a numerically inferior frequency of the composite endpoint.

The VE/VCO_2 slope is an alternative CPET parameter with proven prognostic power, and the HF event risk is constant throughout a large range of VE/VCO_2 slope values [32–34]. A study by Guazzi et al. [25] demonstrated that in both men and women with HfrEF, the predictive power of pVO_2 and the VE/VCO_2 slope are similar. Notably, the discriminative power of the VE/VCO_2 slope was greater than that of pVO_2 in female patients. Our findings are in keeping with these results, as the VE/VCO_2 slope also showed a slightly higher prognostic power compared to pVO_2 in the ROC curve analysis in the female subgroup.

The percent of predicted pVO_2, an age- and gender-adjusted parameter assessing exercise capacity, was shown to stratify the risk for HF events with a higher accuracy compared to pVO_2 in women [27]. The role of CPET in pre-surgical risk stratification in women has also been studied. In a study by Rose et al. [35,36], sex-specific CPET thresholds improved surgical risk stratification and thus may contribute to optimise clinical decision-making.

There is a paucity of randomized clinical trials evaluating the value of CPET variables in women with HFrEF. The HF-ACTION [10], a randomized trial with 2100 patients, also concluded that women presented a better clinical outcome, showing a lower pVO_2 and a higher percent of predicted pVO_2 compared to men. The parameter with the highest predictive power in women was the percent of predicted pVO_2. Our study had similar findings, as the percent of predicted pVO_2 was the CPET parameter with the highest predictive power for HF outcomes in the female subgroup. This result is in keeping with the ISHLT guidelines [7], which recommend that alternative parameters such as percent of predicted pVO_2 may be considered in conjunction with pVO_2 to guide HTx listing in female patients. In contrast, in the male subgroup, pVO_2 and percent of predicted pVO_2 had a similar prognostic power for risk stratification of HF events. Moreover, our study showed that, in a cohort with similar pVO_2 values between sexes, the predictive power of the traditional CPET parameters was notably lower in men than in women, which is in contrast with the results reported in a previous trial by Elmariah et al. [14].

The position paper by Corrà et al. [1] proposes three different threshold values of pVO_2 for male HFrEF patients: <10 mL/kg/min, 10 to 18, or > 18 mL/kg/min. However, there are still limited data to define an accurate cut-off for other subgroups of patients, women or elderly patients in particular [9]. Extrapolating these three advocated thresholds of pVO_2 in male patients to other subgroups may lead to misconceptions and inaccuracies of the objective pVO_2 [9]. Consequently, further studies are necessary to define an accurate threshold to guide patient selection for HTx listing in women.

In a trial by Green et al. [26], the proposed pVO_2 thresholds in females with HFrEF for high- ($\leq$10), medium- (10.1 to 14), and low-risk (>14 mL/kg/min) showed a one-year event-free survival of 80%, 84%, and 93%, respectively. Elmariah et al. [14] reported that, in the current era, HTx may be deferred if the pVO_2 is over 10 mL/kg/min. However, this study had several disparities between sexes in the baseline characteristics and it did not consider patients with CRT, which can affect pVO_2 values [37,38].

In our cohort, the ISHLT recommended thresholds of $pVO_2 \leq 12$ or ≤ 14 mL/kg/min, VE/VCO_2 slope > 35 and percent of predicted $pVO_2 \leq 50\%$ showed a higher overall diagnostic effectiveness in women compared to men, in keeping with the higher prognostic power these parameters showed in female patients. We proposed alternative thresholds that may improve risk discrimination among female patients. A threshold of $pVO_2 \leq 14$ mL/kg/min (including patients on β-blockers) and a percent of predicted $pVO_2 \leq 55\%$ showed a slightly lower specificity but a higher sensitivity, with an overall higher overall diagnostic effectiveness. The recommended cut-off of VE/VCO_2 slope > 35 was the strongest predictor of HF events in women. Regarding male patients, a pVO_2 threshold of $\leq$ 15 mL/kg/min, a VE/VCO_2 slope of > 32, and a percent of predicted pVO_2 of $\leq 58\%$ may also provide an improved diagnostic effectiveness compared to the traditional thresholds for HTx listing.

Circulatory power is a surrogate of left ventricular stroke work index, incorporating pVO_2 and peak systolic BP [39]. Circulatory power was a significant predictor of HF events in our cohort, especially in males. In a recent study by Martinez et al. [40] evaluating patients with advanced HF, circulatory power presented the highest discriminative power for HF outcomes and mortality, concluding that this parameter should also be considered for risk stratification in conjunction with the traditional CPET variables. However, further research is needed to determine whether circulatory power can contribute to the decision of the optimal timing for HTx in women.

In our study, peak O_2 pulse presented a significantly higher predictive power for HF outcomes in women. Peak O_2 pulse, a non-invasive measure of stroke volume and arteriovenous O_2 differential, represents the pVO_2 corrected for HRt [41]. Several CPET measures, including the pVO_2, are corrected for total weight rather than lean body mass. There is a high variability in body fat as a percentage of total body weight [27,42,43] which can also contribute to the lower pVO_2 reported in women [44]. The use of corrected pVO_2 adjusted for lean body mass may be a more accurate measurement of exercise intolerance, particularly in groups with a greater body fat percentage such as women [42,45].

A trial by Lavie [45] et al. found that pVO_2 lean and peak O_2 pulse lean outperformed pVO_2 as predictors of major HF events, including among obese patients and women. The authors noted that, when combined with conventional CPET variables, peak O_2 pulse and lean body mass-adjusted O_2 pulse were powerful predictors of HF outcomes in patients with HFrEF, particularly in populations with a higher percent of body fat.

Prognostic risk scores such as a high to medium risk HFSS [21] or a Seattle Heart Failure Model (SHFM) [46,47] <80% are also recommended as alternative parameters to consider HTx listing [7]. Although the SHFM was also an accurate predictor of HF outcomes in female patients, the Meta-Analysis Global Group in Chronic Heart Failure (MAGGIC) [48,49] score outperformed the SHFM owing to improved risk classification, presenting a similar discriminatory ability in both sexes, despite an overestimation of death in female patients at the 3-year follow-up [50].

Limitations

Firstly, since this is a retrospective study of a prospective database in our center, our findings require confirmation in larger, randomized studies. Additionally, the majority of patients enrolled were men (79%), which is a high proportion, particularly in a study evaluating disparities between female and male patients.

Secondly, our study enrolled unmatched patient subgroups. However, consecutive patient enrollment attenuated the lack of randomization. Furthermore, most baseline characteristics were comparable between sex groups. Men had a numerically higher proportion of ischemic HF, although there was no statistical difference among groups.

Most patients in each subgroup were receiving optimal disease-modifying pharmacological therapies for HF. However, only 10% of patients in our cohort were taking SGLT2i, as they were included in our study between 2009 and 2018, when this drug class was not yet considered as an optimised standard of care medication for patients without diabetes [51]. Future studies should include more patients taking SGLT2i, as they have shown to significantly reduce HF events [52,53]. Less than 25% of patients were taking angiotensin receptor/neprilysin inhibitors, as this therapy was not available for patients enrolled before 2016. Future trials should include more patients receiving sacubitril/valsartan. Moreover, new therapies such as selective cardiac myosin activators or guanylate cyclase stimulators were not available at the time of patient enrolment.

Thirdly, our research lacked the statistical power to infer a new pVO_2 threshold for patients who were intolerant to β-blockers, as the majority of patients (86%) were taking β-blockers. As a result, the proposed pVO_2 thresholds might not be reproducible in this subgroup of HFrEF patients.

Our study only included patients who had a maximal CPET. There is no current agreement on the best peak RER cut-off to determine maximal effort, especially in patients with HFrEF. A number of cut-offs ranging from 1.0 to 1.10 were suggested [1,54–56]. As our aim was to assess the recommendations for HTx, a peak RER of 1.05 was considered to determine a maximal CPET, as recommended by the ISHLT [7]. Consequently, our proposed cut-offs might not be applicable to an HF population with submaximal exercise testing, particularly considering the lower prognostic power of pVO_2 in submaximal exercise capacity [57]. In patients with submaximal exercise capacity, VE/VCO_2 slope and percent of predicted pVO_2 may assist in the clinical stratification [7,34,55]. Indeed, the reliability of RER based assessment of maximal exercise is suboptimal as there are methodological issues thwarting the accurate assessment of VO_2max in submaximal exercise. Pool and Jones [58] caution against the acceptance of pVO_2 measured during ramp incremental exercise as a maximum value in patients with submaximal exercise and proposed the inclusion of a second short constant work rate CPET, completed at a higher work rate than that previously achieved during the ramp test, in order to accurately verify the VO_2max. Therefore, serial CPET may be more informative than a single cardiopulmonary exercise test and thus provide a more accurate assessment of the VO_2max.

Our study evaluated the GET, as described by Beaver et al. [59]. However, it is now recognized that insufficient O_2 is not the primary basis for lactatemia. Critical power likely represents the threshold above which there is a sustained glycolytic contribution with lactate accumulation. Although lactate is a key energy source, there is no evidence that the muscle becomes dysoxic or anoxic [60]. Thus, instead of the GET, critical power may potentially be a more accurate predictor of exercise capacity [60].

Lastly, our study had a lower rate of HF outcomes, especially urgent HTx, compared to other studies [32]. As all the recruited patients were reviewed by the specialized Heart Failure team for a possible indication for HTx, our results may not be applicable to the overall HFrEF population encompassing older patients or patients with significant comorbidities.

5. Conclusions

In an HFrEF cohort undergoing CPET, pVO_2, VE/VCO_2 slope, and the percent of predicted pVO_2 were the variables with the highest discriminative power for HF events, with a higher predictive power in female patients compared to male patients. The ISHLT guideline thresholds for pVO_2 and VE/VCO_2 slope showed a higher diagnostic effectiveness in women. Sex-specific pVO_2, VE/VCO_2 slope, and percent of predicted pVO_2 cut-offs presented a higher prognostic power than the recommended thresholds. Our results indicate that sex-specific cut-offs may assist in patient selection for HTx. However, more data are necessary to help close the gap in evidence between sexes.

Supplementary Materials: The following supporting information can be downloaded at: https://www.mdpi.com/article/10.3390/life13101985/s1, Figure S1: ROC curves for the composite endpoint in a 36-month follow up. (a) Percent of predicted peak oxygen consumption (pVO_2) in female patients. (b) Percent of predicted pVO_2 in male patients; Figure S2: Survival analysis for the composite endpoint in female patients and male patients according to (a) the International Society for Heart and Lung Transplantation (ISHLT) thresholds of percent of predicted peak O_2 consumption (pVO_2) $\leq 50\%$ and (b) thresholds of percent of predicted $pVO_2 \leq 55\%$ in females and percent of predicted $pVO_2 \leq 58\%$ in males.

Author Contributions: Conceptualization, P.G.B. and A.V.G.; methodology, J.F.R. and P.G.B.; software, S.S. and A.V.G.; validation, P.R.; formal analysis, P.G.B.; investigation, R.I.M. and T.P.-d.-S.; resources, A.T.T. and J.F.R.; data curation, P.G.B.; writing—original draft preparation, P.G.B., J.F.R.; writing—review and editing, A.T.T., T.P.-d.-S., A.V.G., R.M.S. and P.G.B.; visualization, R.I.M.; supervision, R.M.S.; project administration, R.M.S. and R.C.F.; funding acquisition, none. All authors have read and agreed to the published version of the manuscript.

Funding: This study did not receive any funding.

Institutional Review Board Statement: The National Committee for Patient Information Protection and the Institutional Ethics Committee (authorization number 1232/2022 by the Comissão Nacional de Protecção de Dados) approved the study, which was conducted in conformity with the Declaration of Helsinki. The study did not involve any experiments on humans or animals.

Informed Consent Statement: All participants in the study provided written informed consent for publishing of this study.

Data Availability Statement: The data presented in this study are available upon request from the corresponding author. The data are not publicly available due to patient consent regarding availability of individual patient data, applicable only to the local investigation team.

Acknowledgments: The authors would like to acknowledge the invaluable contribution of the Heart Failure team and the Cardiorespiratory Exercise Testing Laboratory team of Santa Marta Hospital in this study.

Conflicts of Interest: The authors declare no conflict of interest. The authors declare that no funds, grants, or other support were received during the preparation of this manuscript.

References

1. Corrà, U.; Agostoni, P.G.; Anker, S.D.; Coats, A.J.; Leiro, M.G.C.; de Boer, R.A.; Harjola, V.-P.; Hill, L.; Lainscak, M.; Lund, L.H.; et al. Role of cardiopulmonary exercise testing in clinical stratification in heart failure. A position paper from the Committee on Exercise Physiology and Training of the Heart Failure Association of the European Society of Cardiology. *Eur. J. Heart Fail.* **2018**, *20*, 3–15. [CrossRef]
2. Chua, T.P.; Ponikowski, P.; Harrington, D.; Anker, S.D.; Webb-Peploe, K.; Clark, A.L.; A Poole-Wilson, P.; Coats, A.J. Clinical Correlates and Prognostic Significance of the Ventilatory Response to Exercise in Chronic Heart Failure. *J. Am. Coll. Cardiol.* **1997**, *29*, 1585–1590. [CrossRef]
3. Francis, D.; Shamim, W.; Davies, L.; Piepoli, M.; Ponikowski, P.; Anker, S.; Coats, A. Cardiopulmonary exercise testing for prognosis in chronic heart failure: Continuous and independent prognostic value from VE/VCO2slope and peak VO2. *Eur. Heart J.* **2000**, *21*, 154–161. [CrossRef]
4. Arena, R.; Humphrey, R. Comparison of ventilatory expired gas parameters used to predict hospitalization in patients with heart failure. *Am. Heart J.* **2002**, *143*, 427–432. [CrossRef]
5. Arena, R.; Myers, J.; Aslam, S.S.; Varughese, E.B.; Peberdy, M.A. Peak VO2 and VE/VCO2 slope in patients with heart failure: A prognostic comparison. *Am. Heart J.* **2004**, *147*, 354–360. [CrossRef]
6. Kleber, F.X.; Vietzke, G.; Wernecke, K.D.; Bauer, U.; Opitz, C.; Wensel, R.; Sperfeld, A.; Glaser, S. Impairment of Ventilatory Efficiency in Heart Failure. *Circulation* **2000**, *101*, 2803–2809. [CrossRef]

7. Mehra, M.R.; Canter, C.E.; Hannan, M.M.; Semigran, M.J.; Uber, P.A.; Baran, D.A.; Danziger-Isakov, L.; Kirklin, J.K.; Kirk, R.; Kushwaha, S.S.; et al. The 2016 International Society for Heart Lung Transplantation listing criteria for heart transplantation: A 10-year update. *J. Heart Lung Transpl.* **2016**, *35*, 1–23. [CrossRef]
8. Peterson, L.R.; Schechtman, K.B.; Ewald, G.A.; Geltman, E.M.; Fuentes, L.d.L.; Meyer, T.; Krekeler, P.; Moore, M.L.; Rogers, J.G. Timing of cardiac transplantation in patients with heart failure receiving β-adrenergic blockers. *J. Heart Lung Transpl.* **2003**, *22*, 1141–1148. [CrossRef]
9. Palau, P.; Domínguez, E.; Núñez, J. Sex differences on peak oxygen uptake in heart failure. *ESC Heart Fail.* **2019**, *6*, 921–926. [CrossRef]
10. Keteyian, S.J.; Patel, M.; Kraus, W.E.; Brawner, C.A.; McConnell, T.R.; Piña, I.L.; Leifer, E.S.; Fleg, J.L.; Blackburn, G.; Fonarow, G.C.; et al. Variables Measured During Cardiopulmonary Exercise Testing as Predictors of Mortality in Chronic Systolic Heart Failure. *J. Am. Coll. Cardiol.* **2016**, *67*, 780–789. [CrossRef]
11. Corrà, U.; Agostoni, P.; Giordano, A.; Cattadori, G.; Battaia, E.; La Gioia, R.; Scardovi, A.B.; Emdin, M.; Metra, M.; Sinagra, G.; et al. Sex Profile and Risk Assessment with Cardiopulmonary Exercise Testing in Heart Failure: Propensity Score Matching for Sex Selection Bias. *Can. J. Cardiol.* **2015**, *32*, 754–759. [CrossRef]
12. Ehrman, J.K.; Brawner, C.A.; Shafiq, A.; Lanfear, D.E.; Saval, M.; Keteyian, S.J. Cardiopulmonary Exercise Measures of Men and Women with HFrEF Differ in Their Relationship to Prognosis: The Henry Ford Hospital Cardiopulmonary Exercise Testing (FIT-CPX) Project. *J. Card. Fail.* **2018**, *24*, 227–233. [CrossRef]
13. Piña, I.L.; Kokkinos, P.; Kao, A.; Bittner, V.; Saval, M.; Clare, B.; Goldberg, L.; Johnson, M.; Swank, A.; Ventura, H.; et al. Baseline differences in the HF-ACTION trial by sex. *Am. Heart J.* **2009**, *158*, S16–S23. [CrossRef]
14. Elmariah, S.; Goldberg, L.R.; Allen, M.T.; Kao, A. Effects of Gender on Peak Oxygen Consumption and the Timing of Cardiac Transplantation. *J. Am. Coll. Cardiol.* **2006**, *47*, 2237–2242. [CrossRef]
15. Bruce, R.A.; Kusumi, F.; Hosmer, D. Maximal oxygen intake and nomographic assessment of functional aerobic impairment in cardiovascular disease. *Am. Heart J.* **1973**, *85*, 546–562. [CrossRef]
16. Beale, A.L.; Meyer, P.; Marwick, T.H.; Lam, C.S.; Kaye, D.M. Sex Differences in Cardiovascular Pathophysiology. *Circulation* **2018**, *138*, 198–205. [CrossRef]
17. Redfield, M.M.; Jacobsen, S.J.; Borlaug, B.A.; Rodeheffer, R.J.; Kass, D.A. Age- and Gender-Related Ventricular-Vascular Stiffening. *Circulation* **2005**, *112*, 2254–2262. [CrossRef]
18. Martens, P.; Nijst, P.; Verbrugge, F.H.; Smeets, K.; Dupont, M.; Mullens, W. Impact of iron deficiency on exercise capacity and outcome in heart failure with reduced, mid-range and preserved ejection fraction. *Acta Cardiol.* **2017**, *73*, 115–123. [CrossRef]
19. Karastergiou, K.; Smith, S.R.; Greenberg, A.S.; Fried, S.K. Sex differences in human adipose tissues—The biology of pear shape. *Biol. Sex Differ.* **2012**, *3*, 13. [CrossRef]
20. Jessup, M.; Piña, I.L. Is it important to examine gender differences in the epidemiology and outcome of severe heart failure? *J. Thorac. Cardiovasc. Surg.* **2004**, *127*, 1247–1252. [CrossRef]
21. Aaronson, K.D.; Schwartz, J.S.; Chen, T.-M.; Wong, K.-L.; Goin, J.E.; Mancini, D.M. Development and Prospective Validation of a Clinical Index to Predict Survival in Ambulatory Patients Referred for Cardiac Transplant Evaluation. *Circulation* **1997**, *95*, 2660–2667. [CrossRef]
22. DeLong, E.R.; DeLong, D.M.; Clarke-Pearson, D.L. Comparing the Areas under Two or More Correlated Receiver Operating Characteristic Curves: A Nonparametric Approach. *Biometrics* **1988**, *44*, 837–845. [CrossRef]
23. Corrà, U.; Mezzani, A.; Giordano, A.; Pistono, M.; Gnemmi, M.; Caruso, R.; Giannuzzi, P. Peak oxygen consumption and prognosis in heart failure. *Int. J. Cardiol.* **2013**, *167*, 157–161. [CrossRef]
24. Hsich, E.; Chadalavada, S.; Krishnaswamy, G.; Starling, R.C.; Pothier, C.E.; Blackstone, E.H.; Lauer, M.S. Long-Term Prognostic Value of Peak Oxygen Consumption in Women Versus Men with Heart Failure and Severely Impaired Left Ventricular Systolic Function. *Am. J. Cardiol.* **2007**, *100*, 291–295. [CrossRef]
25. Guazzi, M.; Arena, R.; Myers, J. Comparison of the prognostic value of cardiopulmonary exercise testing between male and female patients with heart failure. *Int. J. Cardiol.* **2006**, *113*, 395–400. [CrossRef]
26. Green, P.; Lund, L.H.; Mancini, D. Comparison of Peak Exercise Oxygen Consumption and the Heart Failure Survival Score for Predicting Prognosis in Women Versus Men. *Am. J. Cardiol.* **2007**, *99*, 399–403. [CrossRef]
27. Richards, D.R.; Mehra, M.R.; O Ventura, H.; Lavie, C.J.; Smart, F.W.; Stapleton, D.D.; Milani, R.V. Usefulness of Peak Oxygen Consumption in Predicting Outcome of Heart Failure in Women Versus Men. *Am. J. Cardiol.* **1997**, *80*, 1236–1238. [CrossRef]
28. Sciomer, S.; Moscucci, F.; Salvioni, E.; Marchese, G.; Bussotti, M.; Corrà, U.; Piepoli, M.F. Role of gender, age and BMI in prognosis of heart failure. *Eur. J. Prev. Cardiol.* **2020**, *27*, 46–51. [CrossRef]
29. Stelken, A.M.; Younis, L.T.; Jennison, S.H.; Miller, D.D.; Miller, L.W.; Shaw, L.J.; Kargl, D.; Chaitman, B.R. Prognostic value of cardiopulmonary exercise testing using percent achieved of predicted peak oxygen uptake for patients with ischemic and dilated cardiomyopathy. *J. Am. Coll. Cardiol.* **1996**, *27*, 345–352. [CrossRef]
30. Meyler, S.; Bottoms, L.; Wellsted, D.; Muniz-Pumares, D. Variability in exercise tolerance and physiological responses to exercise prescribed relative to physiological thresholds and to maximum oxygen uptake. *Exp. Physiol.* **2023**, *108*, 581–594. [CrossRef]
31. Collins, J.; Leach, O.; Dorff, A.; Linde, J.; Kofoed, J.; Sherman, M.; Proffit, M.; Gifford, J.R. Critical power and work-prime account for variability in endurance training adaptations not captured by Vo_{2max}. *J. Appl. Physiol.* **2022**, *133*, 986–1000. [CrossRef] [PubMed]

32. Arena, R.; Myers, J.; Abella, J.; Peberdy, M.A.; Bensimhon, D.; Chase, P.; Guazzi, M. Development of a Ventilatory Classification System in Patients with Heart Failure. *Circulation* **2007**, *115*, 2410–2417. [CrossRef] [PubMed]
33. Tabet, J.-Y.; Beauvais, F.; Thabut, G.; Tartière, J.-M.; Logeart, D.; Cohen-Solal, A. A critical appraisal of the prognostic value of the VE/VCO2 slope in chronic heart failure. *Eur. J. Cardiovasc. Prev. Rehabil.* **2003**, *10*, 267–272. [CrossRef] [PubMed]
34. Ingle, L.; Goode, K.; Carroll, S.; Sloan, R.; Boyes, C.; Cleland, J.G.; Clark, A.L. Prognostic value of the VE/VCO2 slope calculated from different time intervals in patients with suspected heart failure. *Int. J. Cardiol.* **2007**, *118*, 350–355. [CrossRef] [PubMed]
35. Rose, G.A.; Davies, R.G.; Torkington, J.; Berg, R.M.G.; Appadurai, I.R.; Poole, D.C.; Bailey, D.M. Assessing cardiorespiratory fitness relative to sex improves surgical risk stratification. *Eur. J. Clin. Investig.* **2023**, *53*, e13981. [CrossRef] [PubMed]
36. Rose, G.A.; Davies, R.G.; Appadurai, I.R.; Williams, I.M.; Bashir, M.; Berg, R.M.G.; Poole, D.C.; Bailey, D.M. 'Fit for surgery': The relationship between cardiorespiratory fitness and postoperative outcomes. *Exp. Physiol.* **2022**, *107*, 787–799. [CrossRef]
37. Anand, I.S.; Carson, P.; Galle, E.; Song, R.; Boehmer, J.; Ghali, J.K.; Jaski, B.; Lindenfeld, J.; O'Connor, C.; Steinberg, J.S.; et al. Cardiac Resynchronization Therapy Reduces the Risk of Hospitalizations in Patients with Advanced Heart Failure. *Circulation* **2009**, *119*, 969–977. [CrossRef]
38. Abraham, W.T.; Fisher, W.G.; Smith, A.L.; Delurgio, D.B.; Leon, A.R.; Loh, E.; Kocovic, D.Z.; Packer, M.; Clavell, A.L.; Hayes, D.L.; et al. Cardiac Resynchronization in Chronic Heart Failure. *N. Engl. J. Med.* **2002**, *346*, 1845–1853. [CrossRef]
39. Cohen-Solal, A.; Tabet, J.; Logeart, D.; Bourgoin, P.; Tokmakova, M.; Dahan, M. A non-invasively determined surrogate of cardiac power ('circulatory power') at peak exercise is a powerful prognostic factor in chronic heart failure. *Eur. Heart J.* **2002**, *23*, 806–814. [CrossRef]
40. Martinez, L.; Cacciottolo, P.; Barnes, J.; Sylvester, K.; Oates, K.; Kydd, A.; Lewis, C.; Parameshwar, J.; Pettit, S.; Bhagra, S. Circulatory Power is Superior to Peak Oxygen Consumption in Predicting Adverse Outcomes in Ambulatory Patients Assessed for Heart Transplantation. *J. Heart Lung Transplant.* **2022**, *41*, S214. [CrossRef]
41. Cohen-Solal, A.; Barnier, P.; Pessione, F.; Seknadji, P.; Logeart, D.; Laperche, T.; Gourgon, R. Comparison of the long term prognostic value of peak exercise oxygen pulse and peak oxygen uptake in patients with chronic heart failure. *Heart* **1997**, *78*, 572–576. [CrossRef]
42. Osman, A.F.; Mehra, M.R.; Lavie, C.J.; Nunez, E.; Milani, R.V. The incremental prognostic importance of body fat adjusted peak oxygen consumption in chronic heart failure. *J. Am. Coll. Cardiol.* **2000**, *36*, 2126–2131. [CrossRef] [PubMed]
43. Milani, R.V.; Lavie, C.J. The Effects of Body Composition Changes to Observed Improvements in Cardiopulmonary Parameters After Exercise Training with Cardiac Rehabilitation. *Chest* **1998**, *113*, 599–601. [CrossRef] [PubMed]
44. Toth, M.J.; Gardner, A.W.; Ades, P.A.; Poehlman, E.T. Contribution of body composition and physical activity to age-related decline in peak VO2 in men and women. *J. Appl. Physiol.* **1994**, *77*, 647–652. [CrossRef] [PubMed]
45. Lavie, C.J.; Milani, R.V.; Mehra, M.R. Peak exercise oxygen pulse and prognosis in chronic heart failure. *Am. J. Cardiol.* **2004**, *93*, 588–593. [CrossRef]
46. Gorodeski, E.Z.; Chu, E.C.; Chow, C.H.; Levy, W.C.; Hsich, E.; Starling, R.C. Application of the Seattle Heart Failure Model in Ambulatory Patients Presented to an Advanced Heart Failure Therapeutics Committee. *Circ. Heart Fail.* **2010**, *3*, 706–714. [CrossRef]
47. Kalogeropoulos, A.P.; Georgiopoulou, V.V.; Giamouzis, G.; Smith, A.L.; Agha, S.A.; Waheed, S.; Laskar, S.; Puskas, J.; Dunbar, S.; Vega, D.; et al. Utility of the Seattle Heart Failure Model in Patients with Advanced Heart Failure. *J. Am. Coll. Cardiol.* **2009**, *53*, 334–342. [CrossRef]
48. Pocock, S.J.; Ariti, C.A.; McMurray, J.J.; Maggioni, A.; Køber, L.; Squire, I.B.; Swedberg, K.; Dobson, J.; Poppe, K.K.; Whalley, G.A.; et al. Predicting survival in heart failure: A risk score based on 39 372 patients from 30 studies. *Eur. Heart J.* **2012**, *34*, 1404–1413. [CrossRef]
49. Sartipy, U.; Dahlström, U.; Edner, M.; Lund, L.H. Predicting survival in heart failure: Validation of the MAGGIC heart failure risk score in 51 043 patients from the Swedish Heart Failure Registry. *Eur. J. Heart Fail.* **2013**, *16*, 173–179. [CrossRef]
50. Vishram-Nielsen, J.K.; Foroutan, F.; Ross, H.J.; Gustafsson, F.; Alba, A.C. Performance of Prognostic Risk Scores in Heart Failure Patients: Do Sex Differences Exist? *Can. J. Cardiol.* **2019**, *36*, 45–53. [CrossRef]
51. Ponikowski, P.; Voors, A.A.; Anker, S.D.; Bueno, H.; Cleland, J.G.F.; Coats, A.J.S.; Falk, V.; González-Juanatey, J.R.; Harjola, V.-P.; Jankowska, E.A.; et al. 2016 ESC Guidelines for the diagnosis and treatment of acute and chronic heart failure. *Eur. J. Heart Fail.* **2016**, *18*, 891–975. [CrossRef] [PubMed]
52. McMurray, J.J.V.; Solomon, S.D.; Inzucchi, S.E.; Køber, L.; Kosiborod, M.N.; Martinez, F.A.; Ponikowski, P.; Sabatine, M.S.; Anand, I.S.; Bělohlávek, J.; et al. Dapagliflozin in Patients with Heart Failure and Reduced Ejection Fraction. *N. Engl. J. Med.* **2019**, *381*, 1995–2008. [CrossRef] [PubMed]
53. Packer, M.; Anker, S.D.; Butler, J.; Filippatos, G.; Pocock, S.J.; Carson, P.; Januzzi, J.; Verma, S.; Tsutsui, H.; Brueckmann, M.; et al. Cardiovascular and Renal Outcomes with Empagliflozin in Heart Failure. *N. Engl. J. Med.* **2020**, *383*, 1413–1424. [CrossRef] [PubMed]
54. Guazzi, M.; Arena, R.; Halle, M.; Piepoli, M.F.; Myers, J.; Lavie, C.J. 2016 focused update: Clinical recommendations for cardiopulmonary exercise testing data assessment in specific patient populations. *Eur. Heart J.* **2016**, *39*, 1144–1161. [CrossRef]
55. Malhotra, R.; Bakken, K.; D'elia, E.; Lewis, G.D. Cardiopulmonary Exercise Testing in Heart Failure. *JACC Heart Fail.* **2016**, *4*, 607–616. [CrossRef]

56. Guazzi, M.; Bandera, F.; Ozemek, C.; Systrom, D.; Arena, R. Cardiopulmonary Exercise Testing: What is its value? *J. Am. Coll. Cardiol.* **2017**, *70*, 1618–1636. [CrossRef]
57. Balady, G.J.; Arena, R.; Sietsema, K.; Myers, J.; Coke, L.; Fletcher, G.F.; Forman, D.; Franklin, B.; Guazzi, M.; Gulati, M.; et al. Clinician's Guide to Cardiopulmonary Exercise Testing in Adults. *Circulation* **2010**, *122*, 191–225. [CrossRef]
58. Poole, D.C.; Jones, A.M. Measurement of the maximum oxygen uptake Vo_{2max}: Vo_{2peak} is no longer acceptable. *J. Appl. Physiol.* **2017**, *122*, 997–1002. [CrossRef]
59. Beaver, W.L.; Inglis, E.C.; Iannetta, D.; Murias, J.M.; Salvadego, D.; Sartorio, A.; Agosti, F.; Tringali, G.; Patrizi, A.; Isola, M.; et al. A new method for detecting anaerobic threshold by gas exchange. *J. Appl. Physiol.* **1986**, *60*, 2020–2027. [CrossRef]
60. Poole, D.C.; Rossiter, H.B.; Brooks, G.A.; Gladden, L.B. The anaerobic threshold: 50+ years of controversy. *J. Physiol.* **2020**, *599*, 737–767. [CrossRef]

life

MDPI

Article

The Effect of Caffeine on Heart Rate Variability in Newborns: A Pilot Study

Helena Lenasi [1], Eva Rihar [2], Jerneja Filipič [2], Matjaž Klemenc [3] and Petja Fister [4,*]

[1] Institute of Physiology, Medical Faculty, University of Ljubljana, Zaloška cesta 4, 1000 Ljubljana, Slovenia; helena.lenasi.ml@mf.uni-lj.si

[2] Children's Hospital, University Medical Centre Ljubljana, Zaloška cesta 2, 1000 Ljubljana, Slovenia; eva.rihar@kclj.si (E.R.); jerneja.filipic@kclj.si (J.F.)

[3] Department of Cardiology, General Hospital Dr. Franc Derganc, Ulica Padlih Borcev 13A, 5290 Šempeter pri Gorici, Slovenia; matjaz.klemenc@bolnisnica-go.si

[4] Children's Hospital, Pediatric Intensive Care Unit, University Medical Centre Ljubljana, Bohoričeva ulica 20, 1000 Ljubljana, Slovenia

* Correspondence: petja.fister@kclj.si

Abstract: Neonatal apnoea can be treated with caffeine, which affects the central nervous and cardiovascular systems. Heart rate variability (HRV) reflects the activity of the autonomic nervous system (ANS) and might be used as a measure of ANS maturation in newborns. We aimed to establish the effect of caffeine on HRV in newborns and investigated the potential correlation between HRV and postmenstrual age (PMA). In 25 haemodynamically stable newborns hospitalized due to apnoea and treated with caffeine (2.5 mg/kg), we assessed breathing frequency, arterial oxygen saturation, body temperature, and the heart rate while they were sleeping. We assessed HRV by spectral analysis using fast Fourier transformation. The same protocol was reapplied 100 h after caffeine withdrawal to assess the control parameters. Caffeine increased breathing frequency ($p = 0.023$) but did not affect any other parameter assessed including HRV. We established a positive correlation between postmenstrual age and HRV during treatment with caffeine as well as after caffeine had been withdrawn (total power: $p = 0.044$; low-frequency band: $p = 0.039$). Apparently, the maintenance dose of caffeine is too low to affect the heart rate and HRV. A positive correlation between PMA and HRV might reflect maturation of the ANS, irrespective of caffeine treatment.

Keywords: newborn; heart rate variability; caffeine; apnoea; apnea; autonomic nervous system

Citation: Lenasi, H.; Rihar, E.; Filipič, J.; Klemenc, M.; Fister, P. The Effect of Caffeine on Heart Rate Variability in Newborns: A Pilot Study. *Life* **2023**, *13*, 1459. https://doi.org/10.3390/life13071459

Academic Editor: Emanuela Locati

Received: 24 May 2023
Revised: 17 June 2023
Accepted: 23 June 2023
Published: 27 June 2023

1. Introduction

Neonatal apnoea is a life-threatening complication in newborns that might successfully be treated with methylxanthines (aminophylline, theophylline), decreasing its incidence and the need for mechanical ventilation [1,2]. Caffeine has become the drug of choice to treat neonatal apnoea due to its efficacy, tolerability, large therapeutic window, and safety margin [3]. Caffeine is a xanthine exerting many complex and pleiotropic effects on various organ systems mediated by a variety of mechanisms, including antagonism of the adenosine and GABA receptors, inhibition of phosphodiesterase enzymes, and sensitizing of various ryanodine-sensitive calcium release channels [4,5]. Its interference with the sympathetic nervous system has long been appreciated [4], although the results remain controversial [6]. As for improvement of apnoea, its effects include stimulation of the respiratory drive, enhancement of minute ventilation, increased response to hypercapnia, increased skeletal muscle tone, decreased diaphragmatic fatigue, and relaxation of airway smooth muscle [2,7]. Its effects on the cardiovascular system (CVS) are complex, including direct and indirect effects on the vessels and the heart. In the heart, it increases the heart rate (HR), cardiac contractility and stroke volume, consequently increasing cardiac output and mean arterial blood pressure [6,8,9]. As for the vascular system, caffeine exerts vasodilative effects in the pulmonary and most vessels of the systemic circulation but induces vasoconstriction in the cerebral circulation [8]. Caffeine also increases the metabolic rate, neuromuscular transmission and catecholamine release [10].

HR is modulated by an interplay between the sympathetic and the parasympathetic branches of the autonomic nervous system (ANS), and is subjected to beat-to-beat variability which can be assessed by measuring the time interval between consecutive heartbeats [11,12]. While the effect of the parasympathetic nervous system (PNS) on the HR is expressed quickly, the effect of the sympathetic nervous system (SNS) takes longer for its full expression. These different timeframes of the action of the PNS and SNS on the sinoatrial node may partly be explained by the corresponding neurotransmitter kinetics [13].

A relevant clinical tool for a non-invasive assessment of the ANS maturation in newborns, as well as the effects of the ANS on a newborn's heart, is spectral analysis of heart rate variability (HRV) [13]. Accordingly, HRV can be separated into typical frequency-spectra, most usually a high-frequency (HF) spectrum and low-frequency (LF) spectrum, using various methods. While there is no uniform consensus on the relevant parameters of HRV that might adequately reflect the balance between PNS and SNS, one of the usually applied parameters is the ratio between the LF and HF spectra (LF/HF ratio). Although not unequivocal, higher LF/HF implies higher SNS activity, while a larger HF spectrum is suggested to reflect PNS activity [11,14,15].

In general, higher HRV is suggested to be linked to decreased cardiovascular risk and consequently decreased mortality in adults [16–19], and with well-being and decreased mortality in newborns [20–22]. Studies performed in newborns found an increased HRV in term compared with preterm newborns [23–26], which might point to a more mature ANS, especially the PNS branch, in terms. It has been reported that PNS develops optimally after 37 weeks of postmenstrual age (PMA; the time elapsed between the first day of the last menstrual period and birth (gestational age) plus the time elapsed after birth (chronological age) [27–29]. Moreover, it has been suggested that HRV in newborns is affected by sleeping position. We have previously shown an increased HRV in supine compared with prone position suggesting that supine is more favourable regarding well-being of the newborns [30].

The effect of caffeine on HRV remains controversial. Studies conducted in healthy adults mostly showed an increase in HRV after caffeine administration [31–34]. Only a few studies have assessed HRV in newborns; even less is known about the potential influence of caffeine on the ANS as assessed by spectral analysis of HRV. To this end, the aim of our study was to evaluate the impact of caffeine treatment on HRV in newborns. In addition, we investigated the potential effect of caffeine treatment on breathing frequency (BF), the HR, arterial oxygen saturation (SaO_2), and body temperature (T). As caffeine might differently affect the ANS in newborns of various ages and as HRV is one of the few available tools for in vivo assessment of the maturity of the ANS [26], we further correlated the parameters of HRV with PMA, during as well as after caffeine treatment. We hypothesized that due to its known effects on the CVS, caffeine might affect the HRV in newborns.

2. Materials and Methods

2.1. Patients

A prospective clinical intervention study was performed in 25 haemodynamically stable newborns admitted to the Neonatal Department of the University Medical Centre Ljubljana, Division of Paediatrics due to neonatal apnoea that had been treated with caffeine citrate. The diagnostic criteria of apnoea, defined as absence of breathing for 20 s or longer or shorter accompanied by bradycardia or hypoxemia [35], applied to all our patients.

Newborns included in our study had the following (often coexisting) diagnoses: neonatal respiratory distress syndrome, mild dysmorphic features, mild non-optimal neurological signs, congenital heart defect (haemodynamically insignificant patent ductus arteriousu), newborn jaundice, congenital anomalies of the kidney and urinary tract, bronchopulmonary dysplasia, anaemia of prematurity, pneumonia, grade 1 intraventricular haemorrhage, macrosomy, cryptorchidism, hypoglycaemia, intra-amniotic bleeding, or bilateral pneumothorax. None of the listed diagnoses had any apparent impact on the haemodynamic parameters.

Exclusion criteria were severe perinatal hypoxia, infection, liver or renal insufficiency, neurological disorders, and congenital anomalies. Newborns whose data could not be used for spectral analysis due to artefacts of the recordings were also excluded from the study.

The caffeine treatment regimen consisted of a loading dose of 20 mg/kg body mass of caffeine citrate (i.e., 10 mg/kg caffeine), followed by a daily maintenance dose of 5 mg/kg of caffeine citrate (i.e., 2.5 mg/kg caffeine) after 24 h. Recordings on caffeine treatment were collected 48 h after the loading dose of caffeine. The newborns were treated for ten days on average, either orally or intravenously, according to their clinical state; after the cessation of caffeine, the newborns had no remining respiratory distress. It has previously been shown that the route of administration does not affect the pharmacokinetics of caffeine as there is almost complete bioavailability after oral or intravenous administration [36].

The study was approved by the National Medical Ethics Committee of the Republic of Slovenia (0120-458/2016-3 KME 67/09/16) and complies with the principles of the Declaration of Helsinki, the European Convention on Human Rights and Biomedicine, and the Slovenian Code of Medical Deontology. Written parental consent was obtained for all participants. Our trial was retrospectively registered on 27 April 2021, with reference number NCT04869176.

2.2. Study Setting

Each newborn underwent two measurement settings: the first set of experiments was performed while receiving the caffeine citrate. The second set of experiments was repeated 100 ± 26 h after the treatment with caffeine had been withdrawn. These newborns served as controls. In eight newborns, we could not perform the control measurements due to technical difficulties.

Measurements were performed while the newborns were sleeping in a supine position. Their state of alertness was scored one or two as determined according to Prechtl [37]. We simultaneously assessed each newborn's BF, SaO_2, T, and ECG, in a 20 min interval (Figure 1).

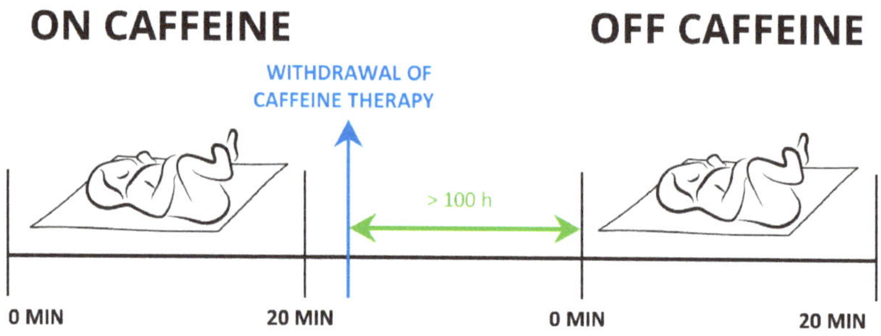

Figure 1. Timeline of the study protocol.

BF, SaO_2 and T were measured three times during a suitable alertness state of the newborn and an average over three measurements was reported. BF was determined manually by observing the chest movement. SaO_2 was measured by a pulse oximeter (IntelliVue MP 50, Philips, Eindhoven, the Netherlands) attached to the right hand. T was measured by a frontal non-contact infrared thermometer (Veratemp, Weston, FL, USA).

As for the ECG tracing, five precordial ECG electrodes (ECG Holter, Vision 5L, Burdick, Milwaukee, WI, USA) were attached to the newborn's chest prior to feeding. After feeding, the newborn was placed supine in a bed and a 20 min tracing was obtained during the suitable alertness state. During the recordings, the heating was turned off to avoid potential interference with the ECG signal.

2.3. Data Analysis

Data were extracted from the ECG recordings using the programmes Vision Premier ver. 3.4 (Cardiac Science Corp., Waukesha, WI, USA) and Nevrokard (Nevrokard, Izola, Slovenia). For the analysis of each recording, a 5 min segment was used. Before and after an ectopic beat, RR intervals (intervals between two subsequent R-waves of the QRS complex) were measured and replaced by two interpolated RR intervals, which were calculated from a proceeding and a succeeding sinus interval. If the programme (Nevrokard) did not correctly determine the R-wave due to the artefact, the exact position of the R-wave was determined by the investigator who performed the spectral analysis. Data containing artefacts in more than 1% of the corresponding segment were removed from subsequent analyses.

Interpolated RR intervals were analysed using fast Fourier transformation, a frequency domain linear method of assessing HRV. Fast Fourier transformation enclosed 1024 points, and a Hamming window was used for the calculation of spectral density. In addition to the total power spectrum (TP), two frequency bands were assessed: one for the LF (in the range of 0.04–0.15 Hz) and one for the HF (in the range of 0.15–1.0 Hz), LF and HF being represented also as a ratio (LF/HF) and in normalized units (LFnu, HFnu) [11]. We selected a segment which corresponded to the suitable alertness state of each newborn. A mean HR value was obtained from the corresponding analysed segment.

2.4. Statistical Analysis

Statistical analysis was performed by Microsoft Excel 2010 and IBM SPSS Statistics 24. The data distribution was tested by the Shapiro–Wilk normality test. The numeric variables are shown either as an arithmetic mean and standard deviation (SD) for a normal (HR, BF, T), or as a median and interquartile range (IQR) for an abnormal distribution (SaO_2, HRV parameters), respectively.

We compared variables according to the presence of caffeine ('on caffeine' or 'off caffeine').

We assessed potential correlation between PMA and HRV parameters regarding the presence of caffeine. Using univariate linear regression, we correlated the data from the first measurement (PMA 'on caffeine' and HRV parameters 'on caffeine') and, separately, for the second measurement (PMA 'off caffeine' and HRV parameters 'off caffeine').

Student's *t*-test was used for comparisons of normally distributed variables, and the Wilcoxon signed-rank test for non-normally distributed data. The correlation between HRV parameters and PMA was tested with the Pearson correlation coefficient. The significance level was set at $p \leq 0.05$.

3. Results

The demographic data and baseline characteristics of the newborns are shown in Table 1. The newborns had been treated with caffeine after being diagnosed with neonatal apnoea. The treatment was discontinued at 37 ± 2 weeks of PMA. In the subsequent analysis on the potential effects of caffeine, we included only 17 newborns in whom the measurements could be repeated after withdrawal of caffeine.

Table 1. Demographic data of the newborns.

	At Birth	Measurement at the Time of Loading Dose of Caffeine	Measurement While on Maintenance Dose of Caffeine	Measurement While off Caffeine
Postmenstrual age (weeks)	34 ± 5	37 ± 4	37 ± 3	37 ± 2
Body mass (g)	2353 ± 914	2659 ± 676	2786 ± 560	2745 ± 512
Head circumference (cm)	31 ± 4	33 ± 3	34 ± 2	34 ± 3

Table 1. *Cont.*

	At Birth	Measurement at the Time of Loading Dose of Caffeine	Measurement While on Maintenance Dose of Caffeine	Measurement While off Caffeine
Caffeine dose (mg/kg BM/day)		9.84 (5.65–9.93)	2.55 (2.31–2.67)	
Apgar score 1 min	8.0 (7.5–9.0)			
Apgar score 5 min	9.0 (7.0–9.0)			
Body length (cm)	46 ± 7			

Data are presented as an arithmetic mean and standard deviation (postmenstrual age (PMA), body mass (BM), head circumference, body length), or as a median and interquartile range (caffeine dose, Apgar score), where appropriate. PMA is the time elapsed between the first day of the last menstrual period and birth (gestational age) plus the time elapsed after birth (chronological age) [29].

3.1. *The Effect of Caffeine on the Heart Rate, Breathing Frequency, Arterial Oxygen Saturation and Body Temperature*

The BF was significantly higher during caffeine treatment. No significant differences in the HR, SaO_2 or T were found between the treatment ('on caffeine') and post-treatment ('off caffeine') (Table 2).

Table 2. Physiological variables during ('on caffeine') and after ('off caffeine') caffeine treatment.

	HR (Beats/min)		BF (Breaths/min)		SaO_2 (%)		T (°C)	
		p		*p*		*p*		*p*
On caffeine	138.6 ± 12.0	1	56.2 ± 12.5	0.023 *	99 (97–100)	0.477	36.7 ± 0.4	0.332
Off caffeine	138.6 ± 13.1		50.7 ± 13.2		99 (97–100)		36.8 ± 0.3	

Data are presented as an arithmetic mean and standard deviation (HR, heart rate; BF, breathing frequency; T, body temperature), or as a median and interquartile range (SaO_2, arterial oxygen saturation). * $p < 0.05$; Student *t*-test (HR, BF, T) or Wilcoxon signed-rank test (SaO_2) were used.

3.2. *The Effect of Caffeine on HRV*

No associations between caffeine treatment and any of the HRV parameters were found (Table 3).

Table 3. The HRV parameters during ('on caffeine') and after ('off caffeine') caffeine treatment.

	TP (ms^2)		LF (ms^2)		LFnu		HF (ms^2)		HFnu		LF/HF	
		p		*p*		*p*		*p*		*p*		*p*
On caffeine	522 (286–1399)	0.653	219 (99–357)	0.435	63.9 (54.5–72.2)	0.868	107 (66–272)	0.523	36.1 (27.1–43.1)	0.619	1.8 (1.3–2.7)	0.877
Off caffeine	732 (228–1270)		232 (85–598)		69.1 (52.8–73)		145 (57–268)		30.9 (27–47.2)		2.2 (1.1–2.7)	

Data are presented as a median and interquartile range. TP, total power; LF, low-frequency spectrum; HF, high-frequency spectrum; LF/HF, the ratio between LF and HF spectra; N, number of newborns; *p*, *p* value; Student *t*-test (TP, LF, HF, LF/HF) or Wilcoxon signed-rank test (LFnu, HFnu) were used.

3.3. *The Correlation between Postmenstrual Age and the Parameters of HRV*

We found a positive correlation between some of the HRV parameters and PMA. During caffeine treatment, we found a moderate positive correlation between PMA and LF (Pearson correlation coefficient = 0.42; $p = 0.039$), and between PMA and TP (Pearson correlation coefficient = 0.41; $p = 0.044$). After cessation of the caffeine treatment, a positive and moderately strong significant correlation was found between PMA and LF (Pearson correlation coefficient = 0.57; $p = 0.017$), and between PMA and TP (Pearson correlation coefficient = 0.5; $p = 0.041$).

4. Discussion

The main finding of our study is that treatment with caffeine does not affect any of the spectral indices of HRV in newborns. Moreover, caffeine did not induce any significant changes in HR, SaO_2 or T but did increase BF. Apparently, the maintenance dose of 2.5 mg/kg body mass to treat apnoea is not sufficient to exert any measurable effects on the above parameters, except for an expected increase in BF. To the best of our knowledge, our study is the first to have assessed the effect of caffeine on HRV in newborns at 37 weeks PMA. The preterm newborns included in the only two available similar studies were significantly younger [1,9].

The only parameter that was affected by caffeine was BF. The increase in BF after the caffeine treatment was expected, as caffeine is a known stimulant of the respiratory centre.

In our study, the values of SaO_2 were comparable during and after the cessation of caffeine treatment. As the values of SaO_2 were mostly at their physiological limits, caffeine could have hardly induced an additional increase.

Caffeine reportedly increases the rate of metabolism and could thus potentially induce an increase in T. Nevertheless, the maintenance dose used in our study apparently was not sufficient to induce any measurable effects of caffeine on T since we found no differences in T during versus after the caffeine treatment. In fact, this is a favourable outcome regarding the interpretation of the HRV data; namely, an elevated T increases HR [38], and if this had been the case in our study, it would have interfered with the interpretation of the effects of caffeine on HRV. Moreover, increased HR implies additional burden to the heart as it significantly increases the heart work [39].

In our study, the HR and HRV parameters obtained during caffeine treatment were comparable with the parameters after discontinuation of treatment. Since cardiogenic effects described in the literature in terms of tachycardia only occur when applying toxic doses of caffeine (plasma level exceeding 20 mg/L) [40], we may conclude that our maintenance dose was not sufficient to impact the HR in newborns. Our results regarding the effect of caffeine on HR and HRV are in accordance with the study of Ulanovsky et al. who also did not observe any impact of caffeine (applied in a loading dose of 15–20 mg/kg/day, followed by a maintenance dose of 5–10 mg/kg/day) either on the HR or HRV in premature newborns [1]. Yet, their sample may not be comparable with ours, as the newborns in their study [1], as well as in that of Huvanandanas et al. [9], were younger than the newborns in our study (gestational age 30.3 ± 2.5 weeks and 27.0 (23.6–33.3) weeks, compared with 34 ± 5 weeks in our study) and also had significantly lower birth weight (1397 ± 458 g and 934 (552–2100) g compared with 2353 ± 914 g in our study) [1,9]. On the other hand, Huvanandana et al., who compared the results of the linear and non-linear analyses of HRV in preterm newborns prior to and two hours after a loading dose of caffeine, reported an increased HRV after caffeine administration when using non-linear, but not when using linear modelling, as it was analysed in our study. They suspected that linear metrics might not adequately capture potentially altered dynamics in the HR control [9]. Indeed, a lack of consistency in the analysis methods applied has already been exposed in the review by [41]. Moreover, the influences of caffeine on the heart and the activity of ANS seem controversial as caffeine has been shown to increase the HF component of HRV in adults, apparently increasing the PNS activity [31,42]. On the other hand, caffeine has been suggested to modulate the response to stress by increasing the levels of circulating catecholamines and cortisol, both of which strongly impact the heart and the sinoatrial node [4].

Although all parameters in our study were measured after the newborns were fed, and during the first 20 min of sleep, the HRV measurements could also be dependent on the sleep phase, which we did not assess. Namely, sleep onset in newborns corresponds to a REM phase. One sleep cycle lasts for about 50 min, and consists of equal subsequent proportions of REM and non-REM sleep [43,44]. Yiallourou et al., who assessed HRV in preterm and term newborns, found significantly increased values of LF/HF, and both LF and TP spectra during REM compared with non-REM sleep [24]. Similarly, Takatani et al. showed higher

LF, HF and LF/HF during REM than during non-REM in newborns [25]. To this end, assessing the phase of sleep in our newborns might be valuable for further interpretation.

We found a positive correlation between HRV (LF and TP but not HF or LF/HF) and PMA, irrespective of caffeine treatment. Nevertheless, HRV is known to be dependent on mean HR—the lower the mean HR the higher the HRV [27]. In our study, the HR did not significantly change during the two measurements. A positive correlation between PMA and HRV supports the idea of ANS maturation with increasing age; however, due to the narrow time frame (100 $\pm$ 26 h) between the consecutive measurements, the results should be interpreted with caution. Our results are accordant with our previous study [30] and with the study of Sahni et al., who showed a significant increase in HRV with increasing PMA in growing low birth weight infants, and implied an important role of ANS maturation in the control of cardiac activity [45]. Based on our observations, it seems that the maturation of the ANS proceeds independently of caffeine. Contrary to our hypothesis and our observation from our previous study [30], where the study population was on average four weeks older, the HF did not significantly increase with increasing PMA in the present study. The discrepant findings could be due to limitations of the spectral analysis. In the present study, the mean HR was about 139 beats per minute. The Nyquist frequency of our sample was, therefore, 69.5 per minute, which is approximately 1.16 Hz. The upper limit of the HF spectrum used for our analysis was 1 Hz. Anything above this value was, therefore, a part of HF score that fell out of the analysis.

It is also important to note that in humans, if the BF is outside the HF band, the power of HF cannot be considered as a marker of vagal modulation. Although in the developing newborn we may expect that the PNS is still maturating, we probably should use the same rationale for the definition of the HF band. The highest BF measured in our study was 56 beats per minute (with an SD of 12), which is borderline for the upper limit of the HF band. This indicates a limitation in the use of the HF band in our sample.

In addition, the influence of respiratory sinus arrhythmia on HRV should be taken into consideration. As the PNS has a much shorter delay than the SNS on the heart to covariate with respiration, the PNS represents the main factor contributing to respiratory sinus arrhythmia. Many reports have confirmed that a considerable portion of the HRV affected by PNS actually arises from respiratory sinus arrhythmia [27,46], yet this phenomenon has not been adequately studied in newborns to date. We can speculate that the PNS, which is not fully developed in newborns, thus contributes less to HRV. It is also possible that HF did not significantly increase with PMA due to higher BF in newborns as compared with BF in adults. Moreover, our newborns presented with neonatal apnoea, which is related to a respiratory disorder or abnormal breathing patterns; therefore, their breathing pattern might have differently affected the HF than the pattern in healthy newborns; nevertheless, at least after cessation of therapy when the newborns exhibited no respiratory distress, this explanation seems less probable.

A limitation of our study is the small sample size with rather heterogenous underlying diagnoses, yet it was homogenous regarding PMA, and the haemodynamic and breathing parameters assessed. Another limitation is the rather long half-life of caffeine in preterm newborns (87 $\pm$ 25 h at 35 weeks PMA [47]). Due to its prolonged half-life, caffeine could persist up to 38 weeks PMA due to immature liver function [48]. Accordingly, caffeine concentration might still have been elevated in some newborns at the time of the control measurements. Unfortunately, we could not exceed this time frame due to the limited duration of hospitalizations; however, by ensuring that 100 h on average passed between the last dose of caffeine and the second measurement, this scenario seems rather unlikely. It would be more meaningful to assess the relationship between caffeine and HRV based on caffeine concentrations in blood rather than just on caffeine dose, yet our newborns did not have central venous access, and it would, therefore, be unethical to collect the blood samples for the purposes of our study only.

As stated above, HF values above 1 Hz could not be included in the analyses due to the upper limit of the frequency spectrum, therefore narrowing the available data.

Additional insight into potential effect of caffeine on haemodynamic parameters would be obtained by continuous measurement of arterial blood pressure, which was not performed in our study. Our study could be improved by continuous measurement of BF, as well as blood pressure monitoring, and by also measuring EEG to determine the phase of sleep, which might have impacted the outcome.

Finally, a note should be given on the study design: a more reliable estimation of the potential impact of caffeine on HRV would be obtained causally, i.e., by first assessing the parameters before caffeine was applied, and subsequently during the application of caffeine. As our newborns needed prompt treatment for apnoea, such a design was not feasible. Another approach would be to check the effect of caffeine in healthy newborns which would enable a successive regimen of measurements, yet the application of caffeine in healthy newborns is not acceptable from an ethical point of view.

5. Conclusions

This is the first study addressing the impact of caffeine on HRV in newborns at 37 weeks PMA. The results of our study did not show any effects of caffeine on any of the parameters of HRV, but increased the BF, as expected. We might speculate that the maintenance dose of caffeine in newborns is too low to affect the HR and HRV. Furthermore, we showed a positive correlation between HRV and PMA, regardless of caffeine intake, pointing to ANS maturation with age that apparently is not subjected to any alterations induced by a standard treatment dose of caffeine. Considering the data on the physiological effects of caffeine shown in this study, as well as in some other studies, we may conclude that caffeine at the prescribed dosage is a suitable treatment for apnoea in newborns.

Author Contributions: Conceptualization, H.L., E.R., J.F., M.K. and P.F.; Data curation, E.R. and J.F.; Formal analysis, E.R., J.F. and M.K.; Investigation, E.R. and J.F; Methodology, P.F.; Project administration, H.L., E.R., J.F. and P.F.; Resources, M.K. and P.F.; Software, M.K.; Supervision, H.L., M.K. and P.F.; Validation, H.L., E.R., J.F., M.K. and P.F.; Visualization, H.L., E.R., J.F., M.K. and P.F.; Writing—original draft, E.R., J.F. and P.F.; Writing—review and editing, H.L., E.R., J.F., M.K. and P.F. All authors have read and agreed to the published version of the manuscript.

Funding: This research received no external funding.

Institutional Review Board Statement: The study was conducted in accordance with the Declaration of Helsinki, and was approved by the National Medical Ethics Committee of the Republic of Slovenia (0120-458/2016-3 KME 67/09/16).

Informed Consent Statement: Written informed parental consent was obtained for all subjects involved in the study.

Data Availability Statement: The data presented in this study are available on request from the corresponding author.

Acknowledgments: We thank the statistician Vanja Erčulj for providing thorough statistical analysis.

Conflicts of Interest: The authors declare no conflict of interest.

References

1. Ulanovsky, I.; Haleluya, N.S.; Blazer, S.; Weissman, A. The Effects of Caffeine on Heart Rate Variability in Newborns with Apnea of Prematurity. *J. Perinatol.* **2014**, *34*, 620–623. [CrossRef]
2. Synnes, A.; Grunau, R.E. Neurodevelopmental Outcomes after Neonatal Caffeine Therapy. *Semin. Fetal Neonatal Med.* **2020**, *25*, 101160. [CrossRef] [PubMed]
3. Long, J.-Y.; Guo, H.-L.; He, X.; Hu, Y.-H.; Xia, Y.; Cheng, R.; Ding, X.-S.; Chen, F.; Xu, J. Caffeine for the Pharmacological Treatment of Apnea of Prematurity in the NICU: Dose Selection Conundrum, Therapeutic Drug Monitoring and Genetic Factors. *Front. Pharmacol.* **2021**, *12*, 681842. [CrossRef]
4. Echeverri, D.; Montes, F.R.; Cabrera, M.; Galán, A.; Prieto, A. Caffeine's Vascular Mechanisms of Action. *Int. J. Vasc. Med.* **2010**, *2010*, 834060. [CrossRef] [PubMed]
5. Faudone, G.; Arifi, S.; Merk, D. The Medicinal Chemistry of Caffeine. *J. Med. Chem.* **2021**, *64*, 7156–7178. [CrossRef] [PubMed]
6. Mahmud, A.; Feely, J. Acute Effect of Caffeine on Arterial Stiffness and Aortic Pressure Waveform. *Hypertension* **2001**, *38*, 227–231. [CrossRef] [PubMed]

7. Early Caffeine in Preterm Neonates. Available online: https://clinicaltrials.gov/ct2/show/NCT03086473 (accessed on 28 March 2023).
8. Shrestha, B.; Jawa, G. Caffeine Citrate—Is It a Silver Bullet in Neonatology? *Pediatr. Neonatol.* **2017**, *58*, 391–397. [CrossRef]
9. Huvanandana, J.; Thamrin, C.; McEwan, A.L.; Hinder, M.; Tracy, M.B. Cardiovascular Impact of Intravenous Caffeine in Preterm Infants. *Acta Paediatr.* **2019**, *108*, 423–429. [CrossRef]
10. Natarajan, G.; Lopes, J.M.; Aranda, J.V. Pharmacologic Treatment of Neonatal Apnea. In *Neonatal and Pediatric Pharmacology: Therapeutic Principles in Practice*, 4th ed.; Yaffe, S.J., Aranda, J.V., Eds.; Lippincott Williams & Wilkins: Philadelphia, PA, USA, 2011; Chapter 20; pp. 241–251.
11. Task Force of the European Society of Cardiology and the North American Society of Pacing and Electrophysiology. Heart Rate Variability: Standards of Measurement, Physiological Interpretation and Clinical Use. *Circulation* **1996**, *93*, 1043–1065. [CrossRef]
12. ChuDuc, H.; NguyenPhan, K.; NguyenViet, D. A Review of Heart Rate Variability and Its Applications. *APCBEE Procedia* **2013**, *7*, 80–85. [CrossRef]
13. Berntson, G.G.; Bigger, J.T.J.; Eckberg, D.L.; Grossman, P.; Kaufmann, P.G.; Malik, M.; Nagaraja, H.N.; Porges, S.W.; Saul, J.P.; Stone, P.H.; et al. Heart Rate Variability: Origins, Methods, and Interpretive Caveats. *Psychophysiology* **1997**, *34*, 623–648. [CrossRef] [PubMed]
14. Billman, G.E. The LF/HF Ratio Does Not Accurately Measure Cardiac Sympatho-Vagal Balance. *Front. Physiol.* **2013**, *4*, 26. [CrossRef]
15. Reyes del Paso, G.A.; Langewitz, W.; Mulder, L.J.M.; van Roon, A.; Duschek, S. The Utility of Low Frequency Heart Rate Variability as an Index of Sympathetic Cardiac Tone: A Review with Emphasis on a Reanalysis of Previous Studies. *Psychophysiology* **2013**, *50*, 477–487. [CrossRef]
16. Billman, G.E. Heart Rate Variability—A Historical Perspective. *Front. Physiol.* **2011**, *2*, 86. [CrossRef]
17. Soares-Miranda, L.; Sattelmair, J.; Chaves, P.; Duncan, G.E.; Siscovick, D.S.; Stein, P.K.; Mozaffarian, D. Physical Activity and Heart Rate Variability in Older Adults: The Cardiovascular Health Study. *Circulation* **2014**, *129*, 2100–2110. [CrossRef] [PubMed]
18. Lai, S.; Mangiulli, M.; Perrotta, A.M.; Di Lazzaro Giraldi, G.; Testorio, M.; Rosato, E.; Cianci, R.; Gigante, A. Reduction in Heart Rate Variability in Autosomal Dominant Polycystic Kidney Disease. *Kidney Blood Press. Res.* **2019**, *44*, 1142–1148. [CrossRef]
19. Draghici, A.E.; Taylor, J.A. The Physiological Basis and Measurement of Heart Rate Variability in Humans. *J. Physiol. Anthropol.* **2016**, *35*, 22. [CrossRef]
20. Tuzcu, V.; Nas, S.; Ulusar, U.; Ugur, A.; Kaiser, J.R. Altered Heart Rhythm Dynamics in Very Low Birth Weight Infants with Impending Intraventricular Hemorrhage. *Pediatrics* **2009**, *123*, 810–815. [CrossRef]
21. Israeli-Mendlovic, H.; Mendlovic, J.; Zuk, L.; Katz-Leurer, M. Reproducibility of 24-h Heart Rate Variability Measures in Preterm Infants Born at 28–32 weeks of Gestation. *Early Hum. Dev.* **2020**, *148*, 105117. [CrossRef] [PubMed]
22. Javorka, K.; Lehotska, Z.; Kozar, M.; Uhrikova, Z.; Kolarovszki, B.; Javorka, M.; Zibolen, M. Heart Rate Variability in Newborns. *Physiol. Res.* **2017**, *66*, S203–S214. [CrossRef]
23. Selig, F.A.; Tonolli, E.R.; da Silva, É.V.C.M.; de Godoy, M.F. Heart Rate Variability in Preterm and Term Neonates. *Arq. Bras. Cardiol.* **2011**, *96*, 443–449. [CrossRef] [PubMed]
24. Yiallourou, S.R.; Witcombe, N.B.; Sands, S.A.; Walker, A.M.; Horne, R.S.C. The Development of Autonomic Cardiovascular Control Is Altered by Preterm Birth. *Early Hum. Dev.* **2013**, *89*, 145–152. [CrossRef]
25. Takatani, T.; Takahashi, Y.; Yoshida, R.; Imai, R.; Uchiike, T.; Yamazaki, M.; Shima, M.; Nishikubo, T.; Ikada, Y.; Fujimoto, S. Relationship between Frequency Spectrum of Heart Rate Variability and Autonomic Nervous Activities during Sleep in Newborns. *Brain Dev.* **2018**, *40*, 165–171. [CrossRef] [PubMed]
26. Iyer, K.K.; Leitner, U.; Giordano, V.; Roberts, J.A.; Vanhatalo, S.; Klebermass-Schrehof, K.; Stevenson, N.J. Bedside Tracking of Functional Autonomic Age in Preterm Infants. *Pediatr. Res.* **2022**. [CrossRef]
27. Cardoso, S.; Silva, M.J.; Guimarães, H. Autonomic Nervous System in Newborns: A Review Based on Heart Rate Variability. *Child's Nerv. Syst.* **2017**, *33*, 1053–1063. [CrossRef] [PubMed]
28. Gagnon, R.; Campbell, K.; Hunse, C.; Patrick, J. Patterns of Human Fetal Heart Rate Accelerations from 26 Weeks to Term. *Am. J. Obstet. Gynecol.* **1987**, *157*, 743–748. [CrossRef]
29. Engle, W.A. Age Terminology during the Perinatal Period. *Pediatrics* **2004**, *114*, 1362–1364. [CrossRef]
30. Fister, P.; Nolimal, M.; Lenasi, H.; Klemenc, M. The Effect of Sleeping Position on Heart Rate Variability in Newborns. *BMC Pediatr.* **2020**, *20*, 156. [CrossRef]
31. Monda, M.; Viggiano, A.; Vicidomini, C.; Viggiano, A.; Iannaccone, T.; Tafuri, D.; De Luca, B. Espresso Coffee Increases Parasympathetic Activity in Young, Healthy People. *Nutr. Neurosci.* **2009**, *12*, 43–48. [CrossRef]
32. Notarius, C.F.; Floras, J.S. Caffeine Enhances Heart Rate Variability in Middle-Aged Healthy, but Not Heart Failure Subjects. *J. Caffeine Res.* **2012**, *2*, 77–82. [CrossRef]
33. Dömötör, Z.; Szemerszky, R.; Köteles, F. Subjective and Objective Effects of Coffee Consumption—Caffeine or Expectations? *Acta Physiol. Hung.* **2015**, *102*, 77–85. [CrossRef] [PubMed]
34. Yeragani, V.K.; Krishnan, S.; Engels, H.J.; Gretebeck, R. Effects of Caffeine on Linear and Nonlinear Measures of Heart Rate Variability before and after Exercise. *Depress. Anxiety* **2005**, *21*, 130–134. [CrossRef] [PubMed]
35. Eichenwald, E.C. Apnea of Prematurity. *Pediatrics* **2016**, *137*, e20153757. [CrossRef] [PubMed]

36. Abdel-Hady, H.; Nasef, N.; Shabaan, A.E.; Nour, I. Caffeine Therapy in Preterm Infants. *World J. Clin. Pediatr.* **2015**, *4*, 81–93. [CrossRef]
37. Prechtl, H.F. The Behavioural States of the Newborn Infant (a Review). *Brain Res.* **1974**, *76*, 185–212. [CrossRef]
38. Davies, P.; Maconochie, I. The Relationship between Body Temperature, Heart Rate and Respiratory Rate in Children. *Emerg. Med. J.* **2009**, *26*, 641–643. [CrossRef]
39. Koskela, J.K.; Tahvanainen, A.; Haring, A.; Tikkakoski, A.J.; Ilveskoski, E.; Viitala, J.; Leskinen, M.H.; Lehtimäki, T.; Kähönen, M.A.; Kööbi, T.; et al. Association of Resting Heart Rate with Cardiovascular Function: A Cross-Sectional Study in 522 Finnish Subjects. *BMC Cardiovasc. Disord.* **2013**, *13*, 102. [CrossRef]
40. Ainsworth, S.B. Drugs and the Body. In *Neonatal Formulary 7: Drug Use in Pregnancy and the First Year of Life*; John Wiley & Sons Inc.: Chichester, UK, 2015; pp. 12–13.
41. Latremouille, S.; Lam, J.; Shalish, W.; Sant'Anna, G. Neonatal Heart Rate Variability: A Contemporary Scoping Review of Analysis Methods and Clinical Applications. *BMJ Open* **2021**, *11*, e055209. [CrossRef]
42. Koenig, J.; Jarczok, M.N.; Kuhn, W.; Morsch, K.; Schäfer, A.; Hillecke, T.K.; Thayer, J.F. Impact of Caffeine on Heart Rate Variability: A Systematic Review. *J. Caffeine Res.* **2013**, *3*, 22–37. [CrossRef]
43. Jenni, O.J.; Carskadon, M.A. Infants to Adolescents. In *SRS Basics of Sleep Guide*; Opp, M.R., Ed.; Sleep Research Society: Westchester, IL, USA, 2005; pp. 11–20.
44. Anders, T.F.; Sadeh, A.; Appareddy, V. Normal Sleep in Neonates and Children. In *Principles and Practice of Sleep Medicine in the Child*; Ferber, R., Kryger, M., Eds.; W. B. Saunders: Philadelphia, PA, USA, 1995; pp. 7–18.
45. Sahni, R.; Schulze, K.F.; Kashyap, S.; Ohira-Kist, K.; Myers, M.M.; Fifer, W.P. Body Position, Sleep States, and Cardiorespiratory Activity in Developing Low Birth Weight Infants. *Early Hum. Dev.* **1999**, *54*, 197–206. [CrossRef]
46. Karemaker, J.M. An Introduction into Autonomic Nervous Function. *Physiol. Meas.* **2017**, *38*, R89–R118. [CrossRef] [PubMed]
47. Doyle, J.; Davidson, D.; Katz, S.; Varela, M.; Demeglio, D.; DeCristofaro, J. Apnea of Prematurity and Caffeine Pharmacokinetics: Potential Impact on Hospital Discharge. *J. Perinatol.* **2016**, *36*, 141–144. [CrossRef] [PubMed]
48. Aranda, J.V.; Beharry, K.D. Pharmacokinetics, Pharmacodynamics and Metabolism of Caffeine in Newborns. *Semin. Fetal Neonatal Med.* **2020**, *25*, 101183. [CrossRef] [PubMed]

life

MDPI

Article

Effects of Home-Based Exercise Training on Cardiac Autonomic Neuropathy and Metabolic Profile in Diabetic Hemodialysis Patients

Vassiliki Michou [1], Vassilios Liakopoulos [2], Stefanos Roumeliotis [2], Athanasios Roumeliotis [2], Maria Anifanti [1], Georgios Tsamos [3], Aikaterini Papagianni [4], Pantelis Zempekakis [1], Asterios Deligiannis [1] and Evangelia Kouidi [1,*]

[1] Sports Medicine Laboratory, School of Physical Education & Sport Science, Aristotle University, 57001 Thessaloniki, Greece
[2] Division of Nephrology and Hypertension, 1st Department of Internal Medicine, Medical School, AHEPA Hospital, Aristotle University, 57001 Thessaloniki, Greece
[3] Laboratory of Hygiene, Department of Internal Medicine, Agios Dimitrios Hospital, Aristotle University, 57001 Thessaloniki, Greece
[4] Department of Nephrology, Hippokration Hospital, Aristotle University, 57001 Thessaloniki, Greece
* Correspondence: kouidi@phed.auth.gr; Tel.: +30-69-3704-0265

Abstract: Background: This study aimed to investigate the effects of a home-based exercise training program on Cardiac Autonomic Neuropathy (CAN) and metabolic profile in Diabetic Kidney Disease (DKD) patients undergoing maintenance hemodialysis (HD). Method: Twenty-eight DKD patients undergoing hemodialysis were randomly assigned into two groups. The exercise (EX) group followed a 6-month combined exercise training program at home, while the control (CO) group remained untrained. All participants at baseline and the end of the study underwent cardiopulmonary exercise testing (CPET), biochemical tests for glucose and lipid profile, and 24-h electrocardiographic monitoring for heart rate variability (HRV) analysis and heart rate turbulence (HRT). Results: At the end of the study, compared to the CO, the EX group showed a significant increase in serum high-density lipoprotein (HDL) by 27.7% ($p = 0.01$), peak oxygen uptake (VO_2peak) by 9.3% ($p < 0.05$), the standard deviation of R-R intervals (SDNN) by 34.3% ($p = 0.03$), percentage of successive RR intervals higher than 50ms (pNN50) by 51.1% ($p = 0.02$), turbulence slope (TS) index by 18.4% ($p = 0.01$), and decrease in (glycated hemoglobin) HbA1c by 12.5% ($p = 0.04$) and low-frequency power LF (ms^2) by 29.7% ($p = 0.01$). Linear regression analysis after training showed that VO_2peak was correlated with SDNN ($r = 0.55$, $p = 0.03$) and HF ($r = 0.72$, $p = 0.02$). Multiple regression analysis indicated that the improvement of sympathovagal balance and aerobic capacity depended on patients' participation in exercise training. Conclusion: In conclusion, a 6-month home-based mixed-type exercise program can improve cardiac autonomic function and metabolic profile in DKD patients on HD.

Keywords: diabetic kidney disease; hemodialysis; cardiac autonomic neuropathy; exercise; physical activity; metabolic profile

Citation: Michou, V.; Liakopoulos, V.; Roumeliotis, S.; Roumeliotis, A.; Anifanti, M.; Tsamos, G.; Papagianni, A.; Zempekakis, P.; Deligiannis, A.; Kouidi, E. Effects of Home-Based Exercise Training on Cardiac Autonomic Neuropathy and Metabolic Profile in Diabetic Hemodialysis Patients. *Life* **2023**, *13*, 232. https://doi.org/10.3390/life13010232

Academic Editor: Emilio Nardi

Received: 2 December 2022
Revised: 26 December 2022
Accepted: 12 January 2023
Published: 13 January 2023

1. Introduction

Cardiac Autonomic Neuropathy (CAN) is the most frequent complication of Diabetes Mellitus (DM). CAN is a strong predictor of cardiovascular events, reduces the quality of life, [1] and increases the risk for mortality in these patients up to 5 times more [2] The prevalence of CAN varies and ranges on average from 2% to 91% in type 1 DM (DM-1) and from 25% to 75% in type 2 DM (DM-2) [3,4] and has been suggested to promote the progression of kidney disease. Diabetic Kidney Disease (DKD) is the most common cause of Chronic Kidney Disease (CKD), which increases the risk of arrhythmias and sudden cardiac death [5,6]. However, mechanisms linking CAN, DKD, and cardiovascular mortality are not yet fully elucidated [6].

Since CAN is closely related to reduced work tolerance and inappropriate heart rate, blood pressure and cardiac output in response to exercise, patients with increased CAN are having difficulties in performing their daily activities. On the other hand, increased physical activity is highly recommended for these populations [7–10], to ameliorate the CV burden by increasing heart rate variability (HRV) and improving left ventricular (LV) function (especially for heart patients) [11,12]. Pathophysiological mechanisms of exercise on cardiac autonomic function, functional capacity, lipid, and glucose profile in DKD patients are shown in Figure 1.

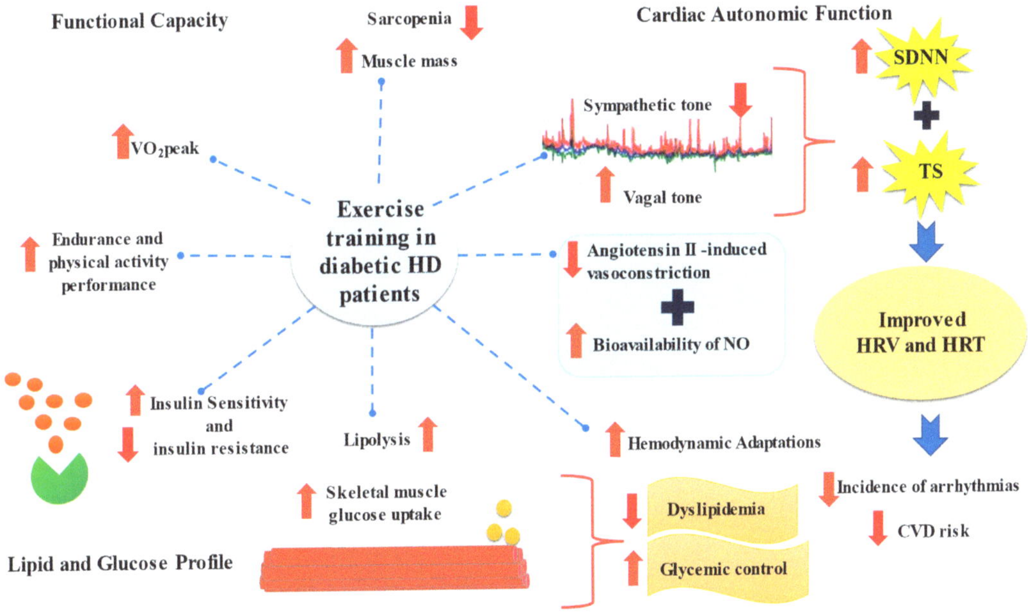

Figure 1. Pathophysiological mechanisms of exercise on cardiac autonomic function, functional capacity, lipid, and glucose profile on DKD patients. The arrows represent the effects of exercise.

Even though the effects of exercise have been thoroughly investigated in CKD and DM patients, there are no clinical trials or cohort studies regarding the impact of training on cardiac autonomic dysfunction in DKD patients. Therefore, the present study aimed to investigate the effects of a long-term, home-based, combined exercise program on CAN and metabolic profile in hemodialysis (HD) patients with DKD.

2. Materials and Methods

2.1. Patients

Initially, 60 DKD patients undergoing maintenance HD with diagnosed CAN were screened for eligibility. Forty-one patients who met the inclusion criteria and volunteered to participate in our study were randomly assigned to either EX or CO group. All patients were recruited from public and private dialysis units of the Prefecture of Thessaloniki in Greece. Inclusion criteria included: age < 80-year-old patients undergoing hemodialysis treatment and, over six months vintage of DM-2. Exclusion criteria included age > 80 years old, history of clinical coronary heart disease (CHD) within the previous six months, severe musculoskeletal problems that may limit the patient's participation in this study, no compliance with diabetes medication, receiving medication that affects the ANS and previous involvement in an exercise training program.

2.2. Study Design

Initially, patients with DKD who met the inclusion criteria underwent a review of the medical history, clinical examination, a 12-lead [electrocardiogram (ECG)], 24-h electrocardiographic monitoring for heart rate variability (HRV) and heart rate turbulence (HRT) analysis, blood sampling and cardiopulmonary exercise testing (CPET) for their functional capacity estimation, at the Sports Medicine Laboratory of the Department of Physical Education and Sports Science of the Aristotle University of Thessaloniki, Greece.

After baseline measurements, patients were randomly assigned to either an exercise group (EX group) or a control group (CO group). For the randomization process, the www.randomizer.org (accessed on 30 January 2020) website was used. Patients in the CO group were asked to continue their daily routine and avoid participating in any structured exercise program. All tests were conducted by the same researcher, blinded to group allocation. Furthermore, all patients' medications were asked to remain unchanged during the study period to exclude medication effects (Table 1). At the end of the 6-month study, all baseline assessments were repeated. This randomized controlled trial protocol was approved by the Ethics Committee of the Aristotle University of Thessaloniki (Protocol number: 117461/2019). All participants received all the necessary study information before enrollment and provided written informed consent. The clinical trial started in February 2020 and ended in June 2020.

Table 1. Patients' medication during the study.

	EX Group	CO Group	*p*-Value
Medication at home			
Calcium channel inhibitors	6 (40.00%)	5 (38.46%)	$p = 0.75$
Antidiabetic drugs (per os)	4 (26.66%)	4 (30.76%)	$p = 0.84$
Beta-adrenergic blockers	4 (26.66%)	4 (30.76%)	$p = 0.82$
Slow and/or intermediate-acting insulin	13 (86.66%)	12 (92.30%)	$p = 0.76$
Corticosteroids	2 (13.33%)	1 (7.69%)	$p = 0.73$
Lipid-lowering drugs	8 (53.33%)	9 (69.23%)	$p = 0.77$
Diuretics	4 (26.66%)	3 (23.07%)	$p = 0.80$
Antithrombotic agents	4 (26.66%)	4 (30.76%)	$p = 0.92$
Antihypertensive agents acting on the renin-angiotensin system	10 (66.66%)	9 (69.23%)	$p = 0.81$
Other	2 (13.33%)	1 (7.69%)	$p = 0.79$
Medication during hemodialysis sessions			
Erythropoietin	14 (93.33%)	12 (92.30%)	$p = 0.56$
Levocarnitine	14 (93.33%)	13 (100.00%)	$p = 0.85$
Analogues of vitamin D, Paricalcitol	12 (80.00%)	10 (76.92%)	$p = 0.61$
Vitamin complexes	7 (46.66%)	5 (38.46%)	$p = 0.59$
Other	2 (13.33%)	3 (23.07%)	$p = 0.82$

Independent *t*-test for continuous variables. Significant at the 0.05 level ($p < 0.05$).

2.3. Charlson Comorbidity Index

The original version of the Charlson Comorbidity Index (CCI), which is a worldwide indicator that predicts long-term mortality for patients with multiple comorbidities, was used in both groups. The CCI is a 19-item scale corresponding to different medical comorbid conditions combined with age. According to Charlson et al. [13] and Fraccaro et al. [14], the higher the CCI, the higher the mortality over ten years [15]. Huang et al. [16] showed that in DKD patients with DM-2 during a 5-year follow-up period, the more the mortality rate increased, the higher the CCI was.

2.4. Blood Analysis Assessment

Blood samples were taken after a 12-h fast, at least five days before the exercise training sessions, began and 24–48 h after the end of the 6-month exercise protocol. Blood analysis through biochemical auto-analyzer devices included determination of hematocrit by photometric method, hemoglobin by computational method, serum electrolytes (potassium, sodium, calcium, phosphorus, magnesium) by ion-selective electrode method, fasting plasma glucose (FBG), serum triglycerides (TG) and glycated hemoglobin (HbA1c) by enzymatic colorimetric method and serum high-density lipoprotein (HDL), low-density lipoprotein (LDL) and total cholesterol (TC) by enzymatic method.

2.5. Cardiopulmonary Exercise Testing

A symptom-limited cardiopulmonary exercise testing (CPET) on a treadmill using the Bruce protocol [17] was used to assess DKD patients' functional capacity. Patients underwent CPET during morning hours (between 9:00 and 11:00 am. The electrocardiogram (GE Medical Systems, Milwaukee, WI, USA) was continuously monitored, while blood pressure was measured at the end of each stage. In addition, the Med Graphics Breeze Suite CPX Ultima (Medical Graphics Corp, Milwaukee, WI, USA) measured breath-by-breath gas exchange. Gas indicators that were analyzed were the peak oxygen uptake (VO_2peak), pulmonary ventilation (VE), ventilatory equivalents for oxygen (VE/VO_2), carbon dioxide (VE/VCO_2), and the ratio between VO_2 and maximum HR (VO_2/HRmax). The endpoint was set as the respiratory exchange ratio $\geq$ 1.10 or the oxygen plateau during maximal exercise.

2.6. 24-H Electrocardiographic Monitoring

DKD patients underwent a 24-recording of the ECG to evaluate the cardiac autonomic function using a small portable 3-channel (with seven electrodes) ECG Holter device (SEER 1000, GE Healthcare, Chalfont St Giles, UK). In addition, patients were asked not to consume alcohol for at least 12 h and to refrain from vigorous physical activity for at least 24 h before the scheduled measurement. Data were stored and analyzed using the CardioDay software (GE Healthcare, Chalfont St Giles, UK) to estimate Heart Rate Variability (HRV) indices in the Time and Frequency Domain and Heart Rate Turbulence (HRT). Concerning HRV time domain analysis, four variables were evaluated [1]: the standard deviation of R-R (the time intervals between two successive heart beats) intervals (SDNN) [2], the standard deviation of R-R intervals calculated every 5 min (SDANN) [3], the square root of the mean sum of the squares of the differences between consecutive intervals R-R (rMSSD) and [4] the percentage of successive RR intervals higher than 50 ms (pNN50). Similarly, from the HRV frequency domain analysis, the following five indices were estimated [1]: the total frequency power (TP) [2], the very low-frequency power (VLF) (<0.003–0.04 Hz) [3], the low-frequency power (LF) (0.04–0.15 Hz) [4], the high-frequency power (HF) (0.15–0.4 Hz) and [5] the frequency ratio (LF/HF). In addition, the HRT indicators analyzed were the turbulence onset (TO) index and the turbulence slope (TS) index.

2.7. Exercise Training Program

The EX group followed a 6-month home-based exercise training program on the non-dialysis days. The exercise program consisted of 3 combined (aerobic and strengthening) exercise sessions per week. Each exercise session was 60–90 min long, of moderate intensity, i.e., at 50–70% of predicted VO_2peak achieved during CPET, and started with a 10-min warm-up and ended with 10-min recovery exercises (upper and lower extremity stretches). In addition, during the first week, all patients received an information exercise booklet and had three familiarization sessions with accredited physical education teachers with expertise in exercise rehabilitation for patients with chronic diseases.

Walking or cycling on a stationary bicycle was recommended as aerobic exercise. Patients were informed to start exercise initially for 15 min with a consequent gradual increase of time by 5 min every two weeks, reaching 40 min in the last two weeks before the end of the program. The strengthening exercise included six dynamic muscle strengthening exercises, performed in 2 sets of 8–10 repetitions (with a 1-min passive break between the sets) in a progressive sequence from sitting to standing. Initially, patients started with strengthening exercises for the upper and lower limbs using only their body weight. Afterward, they performed the same limb exercises using rubber bands, balls, and dumbbells (1 kg). These exercises included 3 phases: (a) in sitting position, two sets (8–10 repetitions) of upper limb strengthening exercise with balls and two sets with the 1 kg dumbbells, (b) in sitting position strengthening exercises (2 sets, 8–10 repetitions) by lacing the rubber bands on feet and tie them to the bottom of the bed or chair and (c) in standing position, with hands in the middle of the body and by placing the dumbbells on feet, moving the legs back and forth, right and left of the torso.

The 6-month home-based exercise training program was individualized to ensure each patient's autonomy. The researcher monitored the progress and adherence to the program via telephone at the end of each week and by a monthly home visit (wherever possible) to record improvements and provide modifications to the exercise program. Moreover, DKD patients were asked to fill in individual diaries, describing the type, frequency, and duration of each exercise session or missing sessions for any reason. The EX group included patients in the analysis, and they participated in at least 85% of the scheduled sessions.

DKD patients, after proper training, were also asked to measure blood glucose, blood pressure, and HR levels at least 20 min before starting the exercise session and note their measurements in their diaries. In case of high (over 130 mg/dL) or low (below 70 mg/dL) blood glucose levels, patients were advised not to start exercising. Patients were also advised to stop exercise in case of illness (i.e., dizziness, visual disturbance, or severe shortness of breath).

2.8. Statistical Analysis

The IBM Statistical Package for Social Sciences (IBM Corp. Released 2020. IBM SPSS Statistics for Windows, Version 27.0. Armonk, NY, USA: IBM Corp) was used for statistical analysis. The Kolmogorov-Smirnov test was used to assess the normal distribution of variables. Mean differences within time and between the two groups were analyzed using two-way ANOVA with repeated measures. The differences between the EX and CO groups regarding changes in the examined parameters were analyzed with the t-test for independent samples. Intra- and inter-observer variability were determined by intraclass correlation coefficient (CCI) and 95% confidence intervals (Cl). Linear regression was used to study the association between variables that revealed statistically significant changes over time, while multiple linear regression analysis was performed to evaluate the impact of confounding factors on results. Data were expressed as mean $\pm$ standard deviation for normally distributed variables. The significance level for accepting or not having a statistically significant difference for all statistical tests was set at $p < 0.05$.

3. Results

3.1. Patients' Characteristics

Six patients from the EX group and seven from the CO group withdrew during the follow-up period; therefore, 28 patients completed the study (Figure 2). According to per-protocol analysis, patients of the EX group performed 95 $\pm$ 2% of the scheduled sessions according to their diaries. During the six months, none of the patients showed any exercise-induced cardiovascular or musculoskeletal complications. The demographic and clinical characteristics of DKD patients are shown in Table 2.

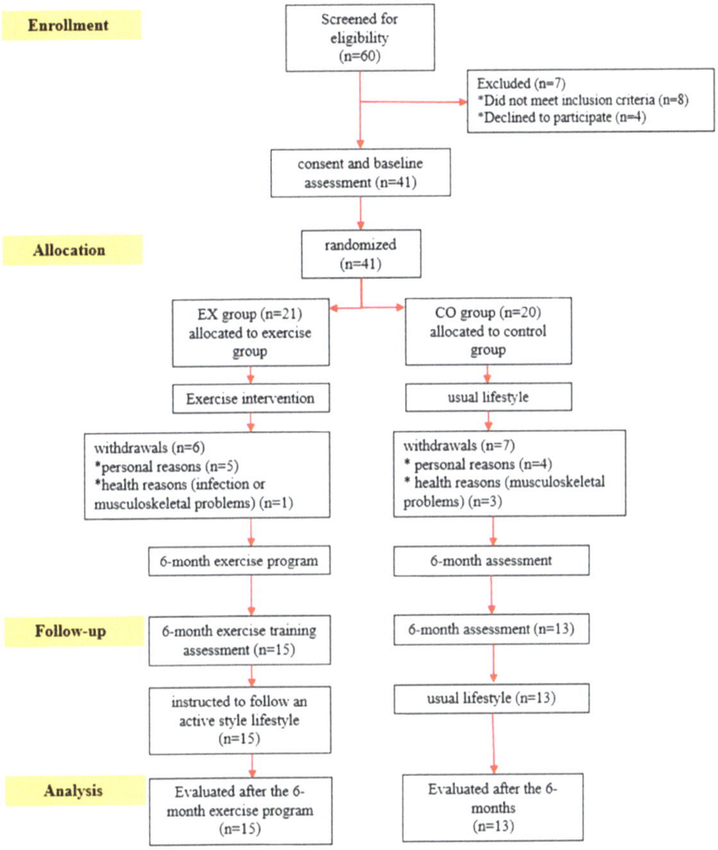

Figure 2. CONSORT diagram of the study design. * represent separately the reasons of withdrawals.

Table 2. Demographic and clinical characteristics of patients.

	EX Group (n = 15)	CO Group (n = 13)	p-Value
Sex (male/female)	10/5	7/6	$p = 0.51$
Age (years)	62.06 ± 6.34	63.30 ± 8.50	$p = 0.66$
Employment status			
Employed	3 (20.00%)	2 (15.39%)	$p = 0.76$
Unemployed	2 (13.33%)	2 (15.39%)	$p = 0.93$
Retired	10 (66.66%)	9 (69.23%)	$p = 0.81$
HD vintage (years)	6.53 ± 5.70	4.70 ± 3.17	$p = 0.31$
Dry weight (kg)	78.80 ± 17.34	79.38 ± 15.01	$p = 0.92$
BMI (kg/m^2)	28.28 ± 6.22	29.05 ± 5.71	$p = 0.73$
Dialysis access			
Arteriovenous fistula or graft	9 (60.00%)	5 (38.46%)	$p = 0.27$
Central venous catheter	6 (40.00%)	8 (61.54%)	$p = 0.45$
Previous transplantation			
Yes	3 (20.00%)	2 (15.39%)	$p = 0.76$
No	12 (80.00%)	11 (84.61%)	$p = 0.76$
Primary causes of ESKD			
Diabetes mellitus	7 (46.66%)	6 (46.15%)	$p = 0.77$
Hypertension	5 (33.33%)	3 (23.07%)	$p = 0.81$
Glomerulonephritis	2 (13.33%)	2 (15.38%)	$p = 0.79$

Table 2. *Cont.*

	EX Group (n = 15)	CO Group (n = 13)	p-Value
Other	1 (6.66%)	2 (15.38%)	$p = 0.73$
CCI	7.20 ± 1.78	7.15 ± 1.67	$p = 0.94$
Comorbidities			
Hypertension	7 (46.66%)	6 (46.15%)	$p = 0.64$
Dyslipidemia	2 (13.33%)	2 (15.38%)	$p = 0.91$
Hypothyroidism	0 (0.00%)	1 (7.69%)	$p = 0.84$
Diabetic retinopathy	2 (13.33%)	1 (7.69%)	$p = 0.71$
Peripheral neuropathy	1 (6.66%)	0 (0.00%)	$p = 0.83$
Multiple myeloma	0 (0.00%	1 (7.69%)	$p = 0.82$
Secondary hyperthyroidism	1 (6.66%)	1 (7.69%)	$p = 0.95$
Venous insufficiency	2 (13.33%)	2 (15.38%)	$p = 0.92$
Mitral valve insufficiency	1 (6.66%)	0 (0.00%)	$p = 0.82$
Varicose veins	1 (6.66%)	0 (0.00%)	$p = 0.82$
Osteoporosis	0 (0.00%)	1 (7.69%)	$p = 0.84$
COPD	1 (6.66%)	0 (0.00%)	$p = 0.82$
Other	1 (6.66%)	0 (0.00%)	$p = 0.85$

ESKD: End-stage kidney disease; HD: Hemodialysis; BMI: Body mass index; CCI: Charlson Comorbidity Index; COPD: Chronic Obstructive Pulmonary Disease. Independent t-test for continuous variables. Significant at the 0.05 level ($p < 0.05$).

3.2. Blood Analysis

At baseline, there was no statistically significant difference in any blood test indicator between the EX and CO group. After the 6-month exercise program, the EX group showed a significant decrease in FPG by 15.9% ($p < 0.05$), TC by 6.3% ($p = 0.04$), TG by 10.1% ($p = 0.01$) and HbA1c by 10.2% ($p < 0.05$), while an increase in HDL by 22.3% ($p < 0.05$) was also noticed. Regarding the inter-group changes between the EX and CO group at the end of the study, results for the EX group showed a favorable increase in HDL by 27.7% ($p = 0.01$) and a decrease in HbA1c by 12.5% ($p = 0.04$), compared to the CO group (Table 3).

3.3. Cardiorespiratory Fitness

We found that after six months, the EX group showed a significant increase in exercise time by 7.4% ($p = 0.03$), METs by 4.8% ($p < 0.05$), VO_2peak by 9.8% ($p < 0.05$) and exercise HR by 4.7% ($p < 0.05$), while a significant decrease was noticed in VE/VO_2max by 5.6% ($p < 0.05$) and VE/VCO_2max by 4.1% ($p = 0.01$) (Table 3). There were no statistically significant differences between groups at baseline. After 6 months, the EX group showed a statistically significant increase in METs by 5.4% ($p = 0.04$) and VO_2peak by 9.3% ($p < 0.05$), while lower values were observed in resting HR by 3.0% ($p < 0.05$), resting SBP by 2.4% ($p = 0.03$), exercise SBP by 4.7% ($p < 0.05$) and DBP by 3.6% ($p = 0.01$), compared to the CO group. Patients allocated to the CO group failed to demonstrate significant improvements in the above indices (Table 4).

3.4. 24-H Holter Monitoring

EX group results from the HRV analysis showed significant improvements in TP (increase by 12.5%, $p = 0.04$), SDNN (increase by 24.5%, $p < 0.05$), SDANN (increase by 19.0%, $p = 0.02$), rMSSD (increase by 21.9%, $p < 0.05$), pNN50 (increase by 29.0%, $p = 0.04$), LF [ms^2 (decrease by 30.8%, $p = 0.03$)], HF [ms^2 (increase by 29.1%, $p = 0.01$)], LF [n.u. (decrease by 22.7%, $p = 0.02$)] and TS (increase by 18.4%, $p = 0.01$) at the end of the study (Table 5). In contrast, there was no statistically significant difference in any HRV index in the CO group after six months. Moreover, inter-group changes at the end of the study showed that EX group statistically increased SDNN by 34.3% ($p = 0.03$), rMSSD by 21.5% ($p = 0.02$) and pNN50 by 51.7% ($p = 0.02$), and decreased LF (ms^2) by 29.7% ($p = 0.01$), compared to the CO group.

Table 3. Blood analysis at baseline and the end of the study.

	EX Group				CO Group				EX vs. CO Group			
	Baseline	After 6-Months	*p*-Value	Intra-Observer Variability ICC (95% CI)	Baseline	After 6-Months	*p*-Value	Intra-Observer Variability ICC (95% CI)	Pre	Inter-Observer Variability ICC (95% CI)	Post	Inter-Observer Variability ICC (95% CI)
Hematocrit (%)	35.86 ± 3.28	36.22 ± 3.54	$p = 0.19$	0.97 (0.93/0.99)	36.13 ± 2.39	36.35 ± 2.23	$p = 0.31$	0.78 (0.55/0.99)	$p = 0.80$	0.41 (−1.06/0.76)	$p = 0.78$	0.21 (−1.57/0.76)
Hemoglobin (g/dL)	11.61 ± 0.85	11.80 ± 1.35	$p = 0.38$	0.85 (0.56/0.95)	11.68 ± 0.68	11.73 ± 0.76	$p = 0.55$	0.90 (0.69/0.97)	$p = 0.81$	−0.75 (−4.75/0.46)	$p = 0.73$	−1.00 (−5.56/0.38)
Na+ (mg/dL)	138.46 ± 2.47	138.53 ± 3.06	$p = 0.68$	0.93 (0.77/0.97)	137.84 ± 4.77	137.85 ± 4.48	$p = 0.94$	0.93 (0.77/0.97)	$p = 0.66$	−1.11 (−5.92/0.35)	$p = 0.61$	−0.41 (−3.64/0.56)
K+ (mg/dL)	5.18 ± 0.71	5.16 ± 0.54	$p = 0.71$	0.87 (0.61/0.95)	5.04 ± 0.57	5.07 ± 0.55	$p = 0.69$	0.78 (0.28/0.93)	$p = 0.57$	0.04 (−2.12/0.70)	$p = 0.52$	0.41 (−0.91/0.82)
Ca+ (mg/dL)	8.81 ± 0.71	8.94 ± 0.74	$p = 0.66$	0.81 (0.44/0.93)	8.67 ± 0.83	8.52 ± 0.91	$p = 0.27$	0.95 (0.84/0.98)	$p = 0.64$	−0.60 (−4.24/0.51)	$p = 0.65$	0.009 (−2.24/0.69)
p (mg/dL)	4.88 ± 1.07	4.82 ± 1.03	$p = 0.72$	0.99 (0.97/0.99)	5.00 ± 0.59	4.90 ± 0.51	$p = 0.50$	0.66 (0.45/0.84)	$p = 0.72$	0.12 (−1.87/0.73)	$p = 0.77$	0.22 (−1.52/0.76)
Mg+ (mg/dL)	2.01 ± 0.14	2.04 ± 0.11	$p = 0.54$	0.94 (0.82/0.98)	2.02 ± 0.22	2.03 ± 0.18	$p = 0.84$	0.85 (0.53/0.95)	$p = 0.89$	0.35 (−1.11/0.80)	$p = 0.95$	0.58 (−0.34/0.87)
Fe+ (mg/dL)	72.73 ± 10.40	72.53 ± 10.82	$p = 0.97$	0.74 (0.63/0.98)	72.84 ± 11.48	72.92 ± 11.60	$p = 0.76$	0.82 (0.43/0.94)	$p = 0.79$	−0.14 (−2.74/0.65)	$p = 0.44$	0.53 (−0.53/0.85)
Urea (mg/dL)	136.20 ± 24.83	136.40 ± 15.20	$p = 0.81$	0.41 (−0.73/0.80)	136.84 ± 23.04	136.84 ± 20.49	$p = 0.98$	0.38 (0.16/0.79)	$p = 0.94$	0.06 (−2.05/0.71)	$p = 0.33$	−0.96 (−5.44/0.40)
Creatinine (mg/dL)	7.32 ± 1.70	7.37 ± 1.84	$p = 0.68$	0.85 (0.57/0.95)	7.54 ± 1.65	7.58 ± 1.60	$p = 0.92$	0.89 (0.64/0.96)	$p = 0.72$	0.65 (−0.13/0.89)	$p = 0.16$	0.55 (−0.46/0.86)
ALP (mg/dL)	76.13 ± 11.10	75.93 ± 17.13	$p = 0.33$	0.58 (−0.24/0.85)	75.92 ± 9.66	75.69 ± 11.10	$p = 0.68$	0.97 (0.92/0.99)	$p = 0.95$	0.59 (−0.32/0.87)	$p = 0.68$	0.47 (−0.71/0.84)
Uric acid (mg/dL)	6.21 ± 0.57	6.11 ± 0.62	$p = 0.46$	0.64 (−0.04/0.88)	6.26 ± 0.88	6.26 ± 0.92	$p = 0.99$	0.67 (−0.02/0.79)	$p = 0.84$	0.33 (−1.17/0.79)	$p = 0.61$	−0.37 (−3.51/0.58)
Serum albumin (g/dL)	5.02 ± 1.08	5.00 ± 1.03	$p = 0.75$	−0.15 (−2.44/0.61)	4.99 ± 0.74	5.00 ± 0.64	$p = 0.91$	0.79 (0.33/0.93)	$p = 0.93$	0.53 (−0.52/0.85)	$p = 0.98$	−0.02 (−2.37/0.68)
SGOT (IU/L)	15.46 ± 3.41	15.13 ± 3.66	$p = 0.82$	0.94 (0.82/0.98)	15.76 ± 6.32	15.69 ± 6.47	$p = 0.77$	0.97 (0.92/0.99)	$p = 0.87$	0.16 (−1.73/0.74)	$p = 0.38$	−1.07 (−5.79/0.36)
SGPT (IU/L)	16.40 ± 3.88	16.46 ± 3.24	$p = 0.76$	0.91 (0.73/0.97)	15.84 ± 6.44	16.15 ± 5.47	$p = 0.23$	0.96 (0.89/0.99)	$p = 0.78$	0.17 (−1.73/0.75)	$p = 0.35$	−0.32 (−3.35/0.59)
FPG (mg/dL)	144.73 ± 41.13	121.66 ± 36.28	$p < 0.05$	0.95 (0.87/0.98)	152.69 ± 60.55	151.76 ± 61.29	$p = 0.42$	0.55 (0.10/0.79)	$p = 0.65$	−1.07 (−5.79/0.36)	$p = 0.20$	−1.22 (−6.30/0.32)
TC (mg/dL)	239.40 ± 80.46	224.53 ± 84.05	$p = 0.04$	0.82 (0.65/0.94)	237.15 ± 95.69	237.15 ± 97.02	$p = 0.98$	0.87 (0.76/0.98)	$p = 0.94$	−0.05 (−2.44/0.68)	$p = 0.79$	0.08 (−2.00/0.72)
TG (mg/dL)	191.06 ± 62.57	171.66 ± 40.12	$p = 0.01$	0.92 (0.77/0.97)	179.69 ± 53.11	179.84 ± 54.30	$p = 0.97$	0.97 (0.93/0.99)	$p = 0.61$	0.57 (−0.41/0.86)	$p = 0.53$	0.30 (−1.27/0.78)
HDL (mg/dL)	43.53 ± 20.02	56.00 ± 22.60	$p < 0.05$	0.86 (0.58/0.96)	44.07 ± 19.92	43.84 ± 19.87	$p = 0.78$	0.89 (0.73/0.95)	$p = 0.94$	0.59 (0.33/0.96)	$p = 0.01$	0.86 (0.56/0.45)
LDL (mg/dL)	102.93 ± 16.78	97.06 ± 15.07	$p = 0.06$	0.58 (0.45/0.75)	94.53 ± 22.45	94.92 ± 22.82	$p = 0.48$	0.99 (0.99/0.99)	$p = 0.26$	0.89 (0.65/0.94)	$p = 0.66$	−0.10 (−2.63/0.66)
HbA1c (%)	6.85 ± 0.69	6.16 ± 0.70	$p < 0.05$	0.86 (0.60/0.95)	7.01 ± 1.20	7.04 ± 1.15	$p = 0.53$	0.93 (0.84/0.98)	$p = 0.67$	0.04 (−2.12/0.70)	$p = 0.04$	0.007 (−2.04/0.71)

EX exercise group; CO: control group; ICC: intraclass correlation coefficient; 95% CI: CI: 95% confidence interval (lower bound/upper bound); Na: Sodium; p: Potassium; Ca: Calcium; Mg: Magnesium; P: Phosphorus; Fe: Iron; ALP: Alkaline phosphatase; FPG: Fasting plasma glucose; HDL: High-density lipoprotein; LDL: Low-density lipoprotein; TC: Total cholesterol; TG: Triglycerides; HbA1c: glycated hemoglobin. Data are expressed as mean ± SD. $p < 0.05$: baseline vs. 6 months follow-up; $p < 0.05$: group EX vs. CO.

Table 4. Cardiorespiratory efficiency at baseline and the end of the study.

	EX Group				CO Group				EX vs. CO Group			
	Baseline	After 6-Months	p-Value	Intra-Observer Variability ICC (95% CI)	Baseline	After 6-Months	p-Value	Intra-Observer Variability ICC (95% CI)	Pre	Inter-Observer Variability ICC (95% CI)	Post	Inter-Observer Variability ICC (95% CI)
Time (min)	6.45 ± 2.04	6.96 ± 1.73	$p = 0.03$	0.94 (0.84/0.98)	6.40 ± 1.24	6.43 ± 1.32	$p = 0.66$	0.91(0.97/0.99)	$p = 0.36$	0.64 (−0.16/0.89)	$p = 0.88$	0.30 (−1.27/0.78)
METs (%pred)	67.20 ± 4.79	70.53 ± 4.18	$p < 0.05$	0.83 (0.50/0.94)	67.00 ± 4.37	66.92 ± 5.04	$p = 0.89$	0.94 (0.82/0.98)	$p = 0.95$	0.72 (0.009/0.91)	$p = 0.04$	0.66 (−0.10/0.89)
VO_2peak (mL/kg/min)	19.94 ± 2.13	21.90 ± 1.75	$p < 0.05$	0.90 (0.72/0.96)	19.80 ± 1.90	19.87 ± 1.87	$p = 0.40$	0.98 (0.77/0.99)	$p = 0.68$	0.77 (0.25/0.93)	$p < 0.05$	0.76 (0.23/0.92)
RERmax	1.12 ± 0.10	1.14 ± 0.09	$p = 0.54$	0.79 (0.38/0.93)	1.09 ± 0.07	1.09 ± 0.08	$p = 0.98$	0.73 (0.44/0.98)	$p = 0.41$	0.37 (−1.04/0.81)	$p = 0.51$	0.44 (−1.01/0.78)
VO_2/HRmax	11.81 ± 1.88	11.90 ± 1.97	$p = 0.74$	0.67 (0.21/0.88)	11.59 ± 1.32	11.64 ± 1.26	$p = 0.50$	0.96 (0.65/0.98)	$p = 0.73$	0.77 (0.26/0.93)	$p = 0.36$	0.72 (0.08/0.91)
VE/VO_2max	31.41 ± 4.46	29.67 ± 4.42	$p < 0.05$	0.82 (0.45/0.95)	31.63 ± 5.60	31.33 ± 5.27	$p = 0.22$	0.77 (0.32/0.95)	$p = 0.51$	0.70 (0.04/0.91)	$p = 0.24$	0.75 (0.20/0.92)
VE/VCO_2max	35.77 ± 4.96	34.32 ± 5.07	$p = 0.01$	0.76 (0.37/0.89)	36.66 ± 7.75	36.32 ± 7.33	$p = 0.23$	0.98 (0.94/0.99)	$p = 0.14$	0.77 (0.26/0.91)	$p = 0.11$	0.37 (−1.03/0.81)
HRrest (bpm)	71.80 ± 8.24	69.60 ± 9.18	$p < 0.05$	0.98 (0.96/0.99)	73.69 ± 6.43	73.69 ± 5.99	$p = 0.99$	0.97 (0.92/0.99)	$p = 0.75$	0.47 (−0.70/0.84)	$p = 0.54$	0.50 (0.01/0.89)
SBPrest (mmHg)	124.53 ± 7.54	121.46 ± 8.09	$p = 0.03$	0.87 (0.63/0.95)	125.00 ± 5.00	125.61 ± 4.99	$p = 0.45$	0.91 (0.70/0.97)	$p = 0.31$	0.27 (−1.37/0.77)	$p = 0.04$	0.75 (0.06/0.93)
DBPrest (mmHg)	73.66 ± 8.12	71.66 ± 6.45	$p = 0.06$	0.93 (0.80/0.97)	74.23 ± 4.93	74.23 ± 4.93	$p = 0.98$	0.95 (0.85/0.98)	$p = 0.25$	−0.21 (−2.98/0.62)	$p = 0.64$	0.08 (−2.00/0.72)
HRmax (bpm)	132.86 ± 15.58	139.46 ± 17.95	$p < 0.05$	0.97 (0.91/0.99)	131.46 ± 10.19	131.15 ± 8.90	$p = 0.64$	0.88 (0.75/0.97)	$p = 0.84$	0.77 (0.27/0.93)	$p = 0.04$	0.65 (−0.14/0.89)
SBPmax (mmHg)	163.33 ± 8.79	155.53 ± 12.18	$p < 0.05$	0.78 (0.34/0.92)	164.38 ± 8.89	164.23 ± 8.61	$p = 0.93$	0.96 (0.89/0.99)	$p = 0.42$	0.68 (−2.05/0.71)	$p = 0.03$	0.11 (−1.89/0.73)
DBPmax (mmHg)	73.66 ± 8.12	71.00 ± 8.28	$p = 0.01$	0.94 (0.83/0.98)	74.23 ± 6.72	74.23 ± 5.71	$p = 0.98$	0.95 (0.87/0.97)	$p = 0.93$	−0.09 (−2.59/0.66)	$p = 0.98$	−1.07 (−5.78/0.36)

EX exercise group; CO: control group; ICC: intraclass correlation coefficient; 95% CI: CI: 95% confidence interval (lower bound/upper bound); METs: Metabolic equivalents for physical activity; VO_2peak: Maximum oxygen consumption; RER: Respiratory exchange ratio; VO_2/HRmax: Ratio between VO_2 and maximum heart rate; VE/VO_2max: Ventilatory equivalents for oxygen; VE/VCO2max: Ventilatory equivalents for carbon dioxide; HR: Heart rate; SBP: Systolic blood pressure; DBP: Diastolic blood pressure. Data are expressed as mean ± SD. $p < 0.05$: baseline vs. 6 months follow-up; $p < 0.05$: group EX vs. CO.

Table 5. Results were derived from the 24-h electrocardiographic monitoring for HRV and HRT analysis at baseline and the end of the study.

	EX Group				CO Group				EX vs. CO Group			
	Baseline	After 6-Months	*p*-Value	Intra-Observer Variability ICC (95% CI)	Baseline	After 6-Months	*p*-Value	Intra-Observer Variability ICC (95% CI)	Pre	Inter-Observer Variability ICC (95% CI)	Post	Inter-Observer Variability ICC (95% CI)
HRV												
HR (bpm)	74.33 ± 12.31	73.13 ± 10.80	$p = 0.62$	0.81 (0.43/0.93)	73.30 ± 6.79	73.46 ± 5.99	$p = 0.67$	0.99 (0.96/0.99)	$p = 0.79$	−0.56 (−4.12/0.52)	$p = 0.92$	−0.17 (−0.64/0.39)
TP (ms^2)	978.33 ± 388.51	1118.84 ± 446.64	$p = 0.04$	0.90 (0.70/0.96)	934.10 ± 403.79	935.48 ± 404.48	$p = 0.70$	0.97 (0.89/0.99)	$p = 0.77$	−0.22 (−0.67/0.35)	$p = 0.27$	−0.71 (−4.62/0.47)
Mean 24-RR intervals (ms)	842.74 ± 128.65	870.76 ± 118.27	$p = 0.19$	0.91 (0.73/0.97)	846.07 ± 83.94	847.30 ± 92.49	$p = 0.71$	0.96 (0.92/0.99)	$p = 0.93$	0.04 (−0.50/0.56)	$p = 0.56$	−0.42 (−3.65/0.56)
Time domain variables												
SDNN (ms)	95.46 ± 15.02	126.40 ± 21.95	$p < 0.05$	0.51 (−0.44/0.83)	93.92 ± 37.99	94.11 ± 33.99	$p = 0.91$	0.98 (0.95/0.99)	$p = 0.88$	0.08 (−2.00/0.72)	$p = 0.03$	0.78 (0.55/0.98)
SDANN (ms)	73.60 ± 22.76	90.86 ± 27.05	$p = 0.02$	0.78 (0.36/0.92)	75.30 ± 20.54	74.61 ± 20.03	$p = 0.70$	0.97 (0.91/0.99)	$p = 0.83$	0.11 (−1.91/0.72)	$p = 0.17$	0.26 (−1.39/0.77)
rMSSD (ms)	55.06 ± 25.92	70.53 ± 23.73	$p < 0.05$	0.88 (0.66/0.96)	55.76 ± 13.26	55.30 ± 10.16	$p = 0.72$	0.97 (0.90/0.98)	$p = 0.93$	0.05 (−0.48/0.57)	$p = 0.02$	0.61 (−1.58/0.75)
pNN50 (%)	8.66 ± 7.20	12.20 ± 11.16	$p = 0.04$	0.91 (0.75/0.97)	8.07 ± 7.52	8.07 ± 7.87	$p = 0.99$	0.99 (0.97/0.99)	$p = 0.86$	0.45 (−0.80/0.83)	$p = 0.02$	0.65 (0.11/0.78)
Frequency domain variables												
VLF (ms^2)	1460.84 ± 950.48	1550.40 ± 1331.76	$p = 0.87$	−0.02 (−1.97/0.78)	1501.31 ± 1152.81	1526.73 ± 1148.43	$p = 0.67$	0.92 (0.76/0.97)	$p = 0.86$	0.29 (−0.28/0.71)	$p = 0.39$	0.11 (−0.44/0.61)
LF (ms^2)	155.53 ± 53.91	108.28 ± 48.04	$p = 0.03$	0.91 (0.74/0.97)	154.65 ± 69.24	154.08 ± 67.20	$p = 0.75$	0.99 (0.89/0.99)	$p = 0.98$	−0.03 (−2.39/0.68)	$p = 0.01$	0.58 (−1.78/0.69)
HF (ms^2)	305.04 ± 169.63	430.33 ± 259.49	$p = 0.01$	0.92 (0.71/0.98)	304.90 ± 108.38	304.45 ± 109.54	$p = 0.49$	0.88 (0.33/0.98)	$p = 0.69$	0.24 (−1.47/0.76)	$p = 0.18$	0.42 (0.20/0.78)
LF (n.u.)	16.34 ± 15.70	12.63 ± 8.28	$p = 0.02$	0.65 (−0.04/0.88)	11.21 ± 5.77	11.30 ± 5.90	$p = 0.37$	0.98 (0.94/0.99)	$p = 0.61$	0.13 (−0.42/0.62)	$p = 0.18$	0.56 (0.11/0.89)
HF (n.u.)	62.89 ± 17.49	81.07 ± 42.81	$p = 0.12$	0.42 (−0.70/0.80)	62.47 ± 21.76	62.66 ± 21.69	$p = 0.09$	0.83 (0.56/0.98)	$p = 0.95$	0.39 (−0.99/0.81)	$p = 0.72$	0.08 (−2.01/0.72)
LF/HF	2.08 ± 1.42	1.91 ± 1.28	$p = 0.26$	0.26 (−1.18/0.72)	2.12 ± 2.03	2.10 ± 1.90	$p = 0.71$	0.95 (0.90/0.97)	$p = 0.96$	−0.11 (−2.66/0.65)	$p = 0.08$	−0.006 (−2.48/0.67)
HRT												
TO (%)	0.02 ± 0.01	0.00 ± 0.00	$p = 0.17$	0.001 (−1.98/0.66)	0.00 ± 0.00	0.00 ± 0.00	$p = 0.33$	0.99 (0.99/1.00)	$p = 0.22$	−0.003 (−2.28/0.69)	$p = 0.12$	0.11 (−0.44/0.60)
TS (ms/RR)	6.54 ± 4.17	8.02 ± 4.65	$p = 0.01$	0.94 (0.82/0.98)	6.69 ± 5.11	6.50 ± 4.77	$p = 0.35$	0.95 (0.84/0.98)	$p = 0.88$	−0.001 (−0.53/0.53)	$p = 0.40$	−0.001 (−2.27/0.69)
Mean 24-RR intervals (ms)	1641.49 ± 154.43	2343.42 ± 151.25	$p = 0.11$	0.47 (0.11/0.96)	1639.72 ± 104.26	1641.11 ± 105.13	$p = 0.64$	0.98 (0.96/0.99)	$p = 0.50$	0.03 (−2.16/0.70)	$p = 0.52$	0.19 (−1.62/0.75)

EX exercise group; CO: control group; ICC: intraclass correlation coefficient; 95% CI: 95% confidence interval (lower bound/upper bound);HRV: Heart rate variability; HRT: Heart rate turbulence; TP: Total; RR intervals: Time intervals between two successive heartbeats; SDNN: standard deviation of RR intervals; SDANN: Standard Deviation of the 5 min Average NN intervals; rMSSD: root mean square of successive differences between normal heartbeats; pNN50: The number of pairs of successive NN (R-R) intervals that differ by more than 50 ms; VLF: very low frequency; LF: Low frequency; HF: High frequency; TO: turbulence onset; TS: turbulence slope. Data are expressed as mean ± SD. $p < 0.05$: baseline vs. 6 months follow-up; $p < 0.05$: group EX vs. CO.

3.5. Linear Regression Analysis

At the end of the study, a positive linear relationship was found in the EX group between SDNN and VO_2peak (r= 0.55, $p = 0.03$) (Figure 3), HF (ms^2) and VO_2peak (r = 0.72, $p = 0.02$) (Figure 3), time of exercise and SDNN (r = 0.62, $p = 0.04$) (Figure 4), while a negative linear relationship was noticed only between rMSSD and HbA1c (r= −0.70, $p < 0.05$) (Figure 5).

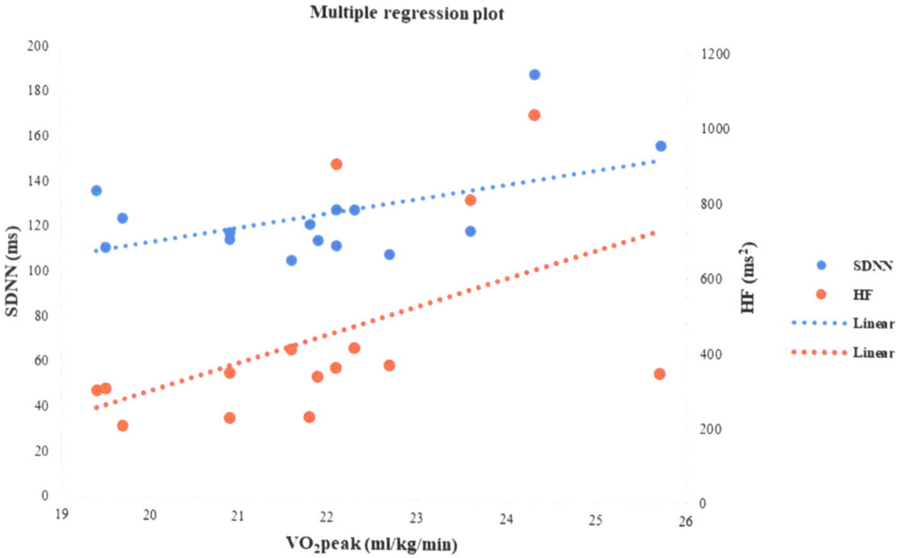

Figure 3. Linear regression analysis between the VO_2peak (ml/kg/min) and SDNN (ms) (r = 0.55, $p = 0.03$) and between VO_2peak (mL/kg/min) and HF (ms^2) (r = 0.72, $p < 0.05$), after 6 months in the EX group.

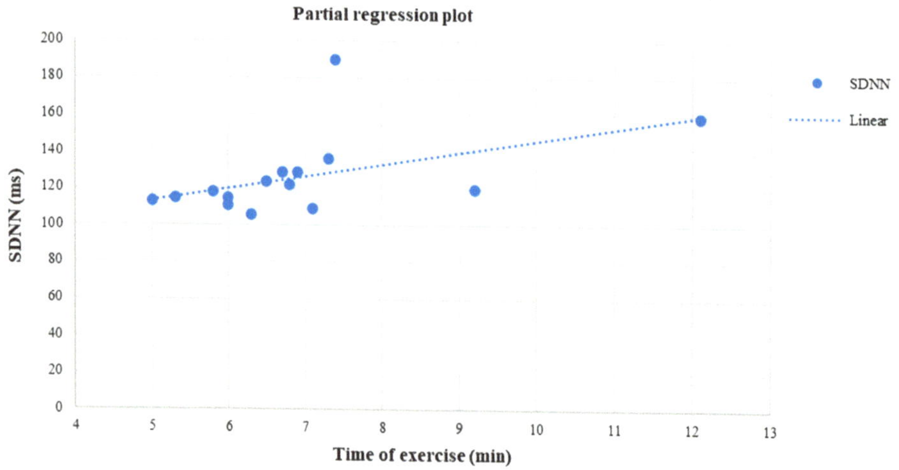

Figure 4. Linear regression analysis between the SDNN (ms) and time of exercise (min) (r = 0.62, $p = 0.04$) after six months in the EX group.

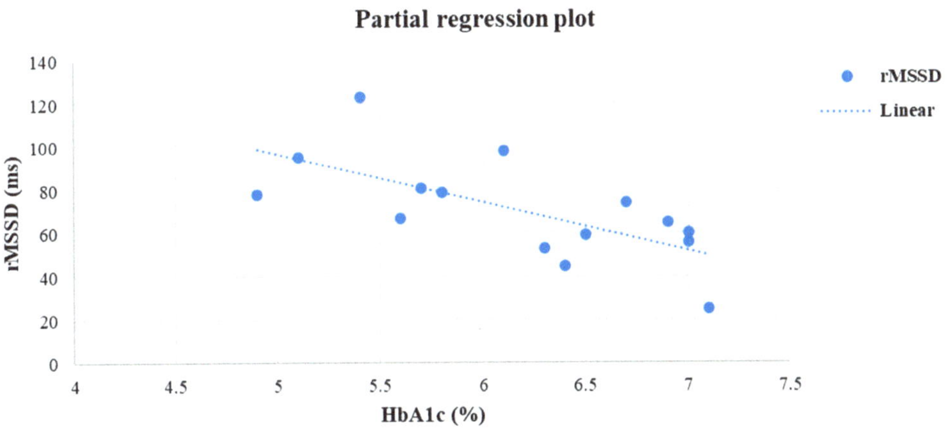

Figure 5. Linear regression analysis between HbA1c (%) and rMSSD (ms) after six months in the EX group (r =−0.70, $p < 0.05$).

3.6. Multiple Linear Regression Analysis

Finally, at the end of the study, multiple linear regression analysis was performed to examine the relationship between sympathovagal balance, as measured with SDNN, and aerobic capacity, as measured with VO_2peak, with a variety of independent variables. By using SDNN as the dependent variable, results showed that lower values of HbA1c ($p = 0.02$) and increased VO_2peak ($p < 0.05$) had a statistically significant contribution to the model (Table 6). More precisely, results revealed that 72.4% of the variability observed in SDNN was explained by the regression model ($R^2 = 0.724$, F = 3.505, $p = 0.04$). In addition, by using VO_2peak, as a subordinate variable, the analysis showed that higher values of SDNN ($p = 0.03$), HF (ms^2) ($p < 0.05$), and rMSSD ($p = 0.03$) had a significant contribution to the model (Table 7), which explained 86.3% of the total variance ($R^2 = 0.863$, F = 4.706, $p = 0.03$).

Table 6. Multiple regression analysis with SDNN as an independent variable, at the end of the study.

Model	B	β	*t*-Test	*p*
Participation to exercise	−6.142		−0.076	$p = 0.94$
Age	0.641	0.185	0.968	$p = 0.36$
HD vintage	−0.265	−0.069	−0.305	$p = 0.76$
Hb	0.201	0.012	0.065	$p = 0.94$
FPG	0.250	0.414	2.098	$p = 0.06$
HbA1c	−19.621	−0.652	−2.855	$p = 0.02$
VO_2peak	8.346	0.666	3.433	$p < 0.05$
$R^2 = 0.724$				
F = 3.505				

HD: Hemodialysis; Hb: Hemoglobin; FPG: Fasting plasma glucose; HbA1c: glycated hemoglobin; VO_2peak: Maximum oxygen consumption $p < 0.05$.

Table 7. Multiple regression analysis with VO_2peak as an independent variable, at the end of the study.

Model	B	β	*t*-Test	*p*
Participation to exercise	20.670		4373	$p < 0.05$
Age	0.017	0.063	0.261	$p = 0.80$
HD vintage	0.158	0.514	2.051	$p = 0.08$
SDNN	0.044	0.549	2.366	$p = 0.03$

Table 7. *Cont.*

Model	B	β	*t*-Test	*p*
pNN50	0.017	0.126	0.665	$p = 0.53$
SDANN	−0.008	−0.124	−0.401	$p = 0.70$
rMSSD	−0.043	−0.582	−2.833	$p = 0.03$
HF (ms^2)	0.005	0.729	3.842	$p < 0.05$
LF (ms^2)	−0.013	−0.344	−0.962	$p = 0.37$
$R^2 = 0.863F = 4.706$				

HD: Hemodialysis; SDNN: standard deviation of RR intervals; pNN50: The number of pairs of successive NN (R-R) intervals that differ by more than 50 ms; rMSSD: root mean square of successive differences between normal heartbeats; SDANN: Standard Deviation of the 5 min Average NN intervals; LF: Low frequency; HF: High frequency. $p < 0.05$

4. Discussion

Our findings showed that a 6-month home-based exercise program has favorable effects on cardiac autonomic function and functional capacity in DKD patients undergoing HD with CAN. This 6-month exercise training program significantly improved the sympathetic and vagal nerve activity and the sympathovagal balance in our cohort.

The exact mechanisms underlying the beneficial effect of exercise on cardiac autonomic nervous system (ANS) function are not yet completely fully elucidated [18]. However, it has been shown that regular exercise training improves cardiac ANS activity through specific molecular mediators, such as nitric oxide (NO) [19] and angiotensin II, a well-known inhibitor of cardiac vagal tone [20]. Physical exercise might increase NO bioavailability and suppress the activity of angiotensin II, thus leading to increased cardiac vagal tones. Therefore, subjects undergoing regular physical exercise have significantly lower plasma renin levels than those living a sedentary lifestyle [21].

An alternative, highly reproducible measure of an individual's ability to recruit cardiac vagal tone is heart rate recovery (HRR) after exercise [22–24]. In heart failure [25,26] and DM-2 [20] patients, exercise training accelerates HRR. Thus, HRR may be considered a strong indicator of improved cardiac vagal function. These references suggest that regular exercise can indeed "train" vagal tone. Still, the potential neurophysiological mechanisms (e.g., recruitment of more vagal neurons or the more efficient transmission of nerve impulses at the ganglion level) involved in the apparent plasticity of the cardiac ANS remain unknown.

Reducing 24-h-SDNN recording is considered the "gold standard" for increased cardiovascular risk, as its values can predict morbidity and mortality [27,28]. Kleiger et al. [29], showed that patients with SDNN values less than 50ms are at increased cardiovascular risk, whereas those with SDNN values higher than 100ms have an approximately 5.3 times lower mortality risk. Pagkalos et al. [30] showed that a 6-month aerobic and resistance exercise training program, at 70–85% of the HRmax, can significantly improve most HRV indices and increase the sympathovagal balance of DM-2 patients with CAN. Regardless of the exercise type, a 3-month exercise program, 3–5 times per week, was found to improve HRV indices by enhancing the vagal nerve activity and reducing the sympathetic activity in patients with Type 2 Diabetic Neuropathy [31].

Results of the present study revealed significant improvements in the indices reflecting both vagal and sympathetic activity, leading to enhanced sympathovagal balance after exercise training. Deligiannis et al. [32] investigated the effect of a 6-month exercise program on HRV indices in HD patients and found favorable improvements in the HRV triangular index. In addition, the VO_2peak increased significantly after the training, while a significant correlation between HRV index and VO_2peak was also found at the end of the study. Likewise, Kouidi et al. [33,34] highlighted substantial improvements in cardiac ANS function after a long-term supervised exercise program during the HD sessions. In disagreement with these results, in a study with a similar study design, Morais et al. [35] showed that a 3-month aerobic intradialytic exercise did not improve cardiac ANS activity. Similarly, Huppertz et al. [11], examining the effect of lifestyle change and participation of 113 patients in a gym and home exercise program, did not notice significant changes in cardiac ANS activity.

Moreover, the present study revealed a statistically significant increase in the TS index of HRT (by 18.4%) for the EX group after training. The present study is the first to evaluate the change in HRT, combined with HRV indices, after a long-term exercise program in DKD patients. Few studies have evaluated the diagnostic value of HRT on CAN in diabetic patients, although it has a significant diagnostic value so far [36]. HRT is considered an indicator of vagal activity and an independent factor of total mortality [37]. TO and TS values correlate highly with HRV parameters, such as SDNN and rMSSD [38]. Lin et al. [39] evaluated and compared the Ewing test, HRV, and HRT indices to diagnose CAN in 90 diabetic patients. They found that using both HRV and HRT analysis, the CAN rate was 56.6% and 52.2%, respectively. Combining the TS index with SDNN, the diagnostic sensitivity for CAN can increase up to 98.0%. Accordingly, Disertori et al. [40] reported that HRV strongly predicts sudden cardiac death and arrhythmic events, particularly in patients with an ejection fraction >30% after acute myocardial infarction.

Our study also revealed a statistically significant increase in VO_2peak, METs, and exercise time after a 6-month home-based combined exercise program. Several studies have shown significant improvements in cardiorespiratory efficiency and functional capacity in HD or DM patients after a home-based exercise training program. For example, Myers et al. [41] observed that a supervised 12-week home exercise program in HD patients increased the 6-min walking test performance (6MWT), VO_2peak, and exercise time. Similarly, Baggetta et al. [42] showed favorable effects of a 6-month, low-intensity exercise program at home in a total of 115 hemodialysis patients (37% with DM) in the EXCITE study. In addition, results from our study revealed positive correlations between VO_2peak and SDNN and HF after exercise training, indicating that the cardiorespiratory efficiency levels affect cardiac ANS function. Although many studies focus primarily on the beneficial effects of intradialytic exercise, there is no significant difference in the outcomes achieved by home-based exercise training programs. In fact, practice at home has advantages since it can increase patients' autonomy and subsequent compliance with exercise training [43].

Furthermore, our study revealed favorable intra-group improvements for glucose and lipid indices after six months, while a strong negative correlation between HbA1c and rMSSD was also noticed. Risk factors that increase cardiovascular mortality in DM-2 patients with CAN include dyslipidemia, duration of DM, abnormal glucose tolerance, hypertension, and obesity. Exercise programs could offset some of these risks. Mild to moderate physical activity increases lipolysis of triacylglycerols, induced by the increased response to catecholamines, contributing positively to the reduction of insulin resistance [44]. However, in HD patients, the results of previous studies are controversial. In the study of Sanavi et al. [45], after an 8-week intradialytic mixed-type exercise program, improvements were observed for CRP and creatinine levels but not for lipids. In contrast, Song and Sohng [46], after a 12-week progressive resistance training program, observed an inter-group reduction in TC and TG. Likewise, Torres et al. [47] found a significant decrease in LDL and TG levels after a 3-month exercise training program. According to Meher and Panda [48], CAN could strongly correlate with high HbA1c levels, as uncontrollable hyperglycemia could be a possible explanation for CAN deterioration. Pagkalos et al. [30] have also noticed a similar correlation between the improvements in HbA1c levels and vagal activity, as indicated with pNN50 and rMSSD, in DM-2 patients. Additionally, our results are in agreement with previous studies [49,50]. However, even though the strong correlation between CAN and poor glycemic control is well-established, more studies are needed to examine these correlations in the DKD population.

The present study has strengths and limitations that need to be acknowledged. It is the first randomized controlled trial that evaluated the cardiac ANS effects of a long-term, home-based combined exercise program in DKD patients undergoing maintenance HD. These results are noteworthy, as cardiovascular disease is responsible for approximately 40% of deaths in ESKD patients [51], and the prevalence of DM-2, according to global estimations, will increase by 3% to 6% at the end of 2025, with approximately 3 million patients with DM-2 [52]. On the other hand, a limitation of this study is its small sample size, mainly due to the difficulties of recruiting patients for long-term exercise training studies. Secondly, even though patients with coronary heart disease were excluded from the study, different types of silent cardiomyopathies, such as hypertrophic cardiomyopathy (HCM) [53], arrhythmogenic right ventricular cardiomyopathy (ARVC) [54] or idiopathic dilated cardiomyopathy (DCM) [55] that may affect our results, had not included in the study's exclusion criteria. Thirdly, since cardiac implantable electronic devices (CIED) are widely used as a therapeutic measure in cardiac rhythm disorders and heart failure management in the DKD population, the possible impact of exercise intervention in CIED-related complications [56] and quality of life [57] of HD patients should have been taken under consideration. However, in the current randomized clinical trial, we neither include patients with CIEDs nor examined the protentional impact on the study population considering related complications (such as infections) and quality of life. Finally, the long-term prognostic ability of exercise training was not evaluated.

5. Conclusions

In conclusion, a 6-month, home-based combined exercise training program can improve cardiac autonomic function, cardiorespiratory efficiency, and metabolic profile in DKD patients in HD. Encouraging DKD patients to increase their physical activity levels may be the key to improving their daily life and reducing cardiovascular risk and mortality.

Author Contributions: Methodology, V.M., M.A. and E.K.; software, V.M.; validation, V.M, E.K., A.D. and V.L; formal analysis, V.M.; Investigation, V.M., S.R., A.R. and G.T.; resources, V.M.; data curation, V.M.; writing—original draft preparation, V.M.; writing—review and editing, V.M., E.K., A.D., V.L., S.R. and A.R.; visualization, V.M.; supervision, E.K., V.L., A.D., A.P. and P.Z.; project administration, E.K. All authors have read and agreed to the published version of the manuscript.

Funding: This research received no external funding. The APC was funding by Life Journal on MDPI, on behalf of Professor Kouidi Evangelia.

Institutional Review Board Statement: The study was conducted in accordance with the Declaration of Helsinki, and approved by the Ethics Committee of the Aristotle University of Thessaloniki, Greece (Protocol number: 117461/2019).

Informed Consent Statement: Informed consent was obtained from all subjects involved in the study.

Data Availability Statement: The data presented in this study are available on request from the corresponding author. The data are not publicly available due to ethical restrictions.

Conflicts of Interest: The authors of the present research study declare no conflict of interest regarding the present article.

References

1. Pop-Busui, R.; Boulton, A.J.M.; Feldman, E.L.; Bril, V.; Freeman, R.; Malik, R.A.; Sosenko, J.M.; Ziegler, D. Diabetic Neuropathy: A Position Statement by the American Diabetes Association. *Diabetes Care* **2017**, *40*, 136–154. [CrossRef] [PubMed]
2. Yun, J.-S.; Ahn, Y.-B.; Song, K.-H.; Yoo, K.-D.; Kim, H.-W.; Park, Y.-M.; Ko, S.-H. The association between abnormal heart rate variability and new onset of chronic kidney disease in patients with type 2 diabetes: A ten-year follow-up study. *Diabetes Res. Clin. Pract.* **2015**, *108*, 31–37. [CrossRef] [PubMed]
3. Agashe, S.; Petak, S. Cardiac Autonomic Neuropathy in Diabetes Mellitus. *Methodist Debakey Cardiovasc. J.* **2018**, *14*, 251. [CrossRef]
4. Dimitropoulos, G. Cardiac autonomic neuropathy in patients with diabetes mellitus. *World J. Diabetes* **2014**, *5*, 17. [CrossRef] [PubMed]
5. Vinik, A.I.; Erbas, T. Cardiovascular autonomic neuropathy: Diagnosis and management. *Curr. Diabetes Rep.* **2006**, *6*, 424–430. [CrossRef]
6. Laursen, J.C.; Rasmussen, I.K.B.; Zobel, E.H.; Hasbak, P.; von Scholten, B.J.; Holmvang, L.; Hansen, T.W. The Association Between Cardiovascular Autonomic Function and Changes in Kidney and Myocardial Function in Type 2 Diabetes and Healthy Controls. *Front. Endocrinol.* **2021**, *12*, 780679. [CrossRef]

7. Perkovic, V.; Heerspink, H.L.; Chalmers, J.; Woodward, M.; Jun, M.; Li, Q.; MacMahon, S.; Cooper, M.E.; Hamet, P.; Marre, M.; et al. Intensive glucose control improves kidney outcomes in patients with type 2 diabetes. *Kidney Int.* **2013**, *83*, 517–523. [CrossRef]
8. Lim, A. Diabetic nephropathy & complications and treatment. *Int. J. Nephrol. Renov. Dis.* **2014**, *7*, 361.
9. Cai, Z.; Yang, Y.; Zhang, J. Effects of physical activity on the progression of diabetic nephropathy: A meta-analysis. *Biosci. Rep.* **2021**, *41*, BSR20203624. [CrossRef]
10. Neveen, A.R.A.; Nesreen, E.N.G.; Nadia, M.K.; al Shaimaa, G.M. Effect of intradialytic aerobic exercise on patients with diabetic nephropathy. *Bull. Fac. Phys. Ther.* **2019**, *24*, 1–7. [CrossRef]
11. Huppertz, N.; Beetham, K.S.; Howden, E.J.; Leicht, A.S.; Isbel, N.M.; Coombes, J.S. A 12-month lifestyle intervention does not improve cardiac autonomic function in patients with chronic kidney disease. *Auton. Neurosci.* **2020**, *224*, 102642. [CrossRef] [PubMed]
12. Reboredo, M.d.M.; Pinheiro, B.d.V.; Neder, J.A.; Ávila, M.P.W.; Araujo, E.; Ribeiro, M.L.d.B.; de Mendonça, A.F. Effects of aerobic training during hemodialysis on heart rate variability and left ventricular function in end-stage renal disease patients. *J. Br. Nefrol.* **2010**, *32*, 367–373.
13. Charlson, M.E.; Carrozzino, D.; Guidi, J.; Patierno, C. Charlson Comorbidity Index: A Critical Review of Clinimetric Properties. *Psychother. Psychosom.* **2022**, *91*, 8–35. [CrossRef] [PubMed]
14. Fraccaro, P.; Kontopantelis, E.; Sperrin, M.; Peek, N.; Mallen, C.; Urban, P.; Buchan, I.E.; Mamas, M.A. Predicting mortality from change-over-time in the Charlson Comorbidity Index. *Medicine* **2016**, *95*, e4973. [CrossRef]
15. Quan, H.; Li, B.; Couris, C.M.; Fushimi, K.; Graham, P.; Hider, P.; Januel, J.-M.; Sundararajan, V. Updating and Validating the Charlson Comorbidity Index and Score for Risk Adjustment in Hospital Discharge Abstracts Using Data From 6 Countries. *Am. J. Epidemiol.* **2011**, *173*, 676–682. [CrossRef]
16. Huang, Y.-Q.; Gou, R.; Diao, Y.-S.; Yin, Q.-H.; Fan, W.-X.; Liang, Y.-P.; Chen, Y.; Wu, M.; Zang, L.; Li, L.; et al. Charlson comorbidity index helps predict the risk of mortality for patients with type 2 diabetic nephropathy. *J. Zhejiang Univ. Sci. B* **2014**, *15*, 58–66. [CrossRef]
17. Nerlekar, N.; Mulley, W.; Rehmani, H.; Ramkumar, S.; Cheng, K.; Vasanthakumar, S.A.; Mottram, P.M. Feasibility of exercise stress echocardiography for cardiac risk assessment in chronic kidney disease patients prior to renal transplantation. *Clin. Transplant.* **2016**, *30*, 1209–1215. [CrossRef]
18. Bhati, P.; Shenoy, S.; Hussain, M.E. Exercise training and cardiac autonomic function in type 2 diabetes mellitus: A systematic review. *Diabetes Metab. Syndr. Clin. Res. Rev.* **2018**, *12*, 69–78.
19. Chowdhary, S.; Townend, J.N. Role of nitric oxide in the regulation of cardiovascular autonomic control. *Clin. Sci.* **1999**, *97*, 5–17. [CrossRef]
20. Townend, J.N.; Al-Ani, M.; West, J.N.; Littler, W.A.; Coote, J.H. Modulation of Cardiac Autonomic Control in Humans by Angiotensin II. *Hypertension* **1995**, *25*, 1270–1275. [CrossRef]
21. Buch, A.N.; Coote, J.H.; Townend, J.N. Mortality, cardiac vagal control and physical training—What's the link? *Exp. Physiol.* **2002**, *87*, 423–435. [CrossRef] [PubMed]
22. Gourine, A.V.; Ackland, G.L. Cardiac Vagus and Exercise. *Physiology* **2019**, *34*, 71–80. [CrossRef] [PubMed]
23. Cole, C.R.; Blackstone, E.H.; Pashkow, F.J.; Snader, C.E.; Lauer, M.S. Heart-Rate Recovery Immediately after Exercise as a Predictor of Mortality. *N. Engl. J. Med.* **1999**, *341*, 1351–1357. [CrossRef] [PubMed]
24. Cole, C.R.; Foody, J.M.; Blackstone, E.H.; Lauer, M.S. Heart Rate Recovery after Submaximal Exercise Testing as a Predictor of Mortality in a Cardiovascularly Healthy Cohort. *Ann. Intern. Med.* **2000**, *132*, 552. [CrossRef] [PubMed]
25. Myers, J.; Handley, D.; Oswald, U.; Bruner, K.; Kottman, W.; Hsu, L.; Dubach, P. Effects of exercise training on heart rate recovery in patients with chronic heart failure. *Am. Heart J.* **2007**, *153*, 1056–1063. [CrossRef]
26. Pearson, M.J.; Smart, N.A. Exercise therapy and autonomic function in heart failure patients: A systematic review and meta-analysis. *Heart Fail Rev.* **2018**, *23*, 91–108. [CrossRef]
27. Task Force of the European Society of Cardiology and the North American Society of Pacing Electrophysiology. Heart Rate Variability. Standards of Measurement, Physiological Interpretation, and Clinical Use. *Circulation* **1996**, *93*, 1043–1065. [CrossRef]
28. Bonete, G.; Dias, B.A.L.; Leandro, D.A.; Fernandes, A.T.N.; Pereira, C.H.; Ribeiro, C.T.D.; Sousa, A.G.P.; Resqueti, V.; Fregonezi, G.A.F.; Dias, F.A.L. Impaired heart rate variability, Valsalva and 30:15 ratio indexes are associated with reduced submaximal exercise capacity in subjects with diabetes mellitus. *Diabetes Res. Clin. Pract.* **2019**, *155*, 107813. [CrossRef]
29. Kleiger, R.E.; Miller, J.P.; Bigger, J.T.; Moss, A.J. Decreased heart rate variability and its association with increased mortality after acute myocardial infarction. *Am. J. Cardiol.* **1987**, *59*, 256–262. [CrossRef]
30. Pagkalos, M.; Koutlianos, N.; Kouidi, E.; Pagkalos, E.; Mandroukas, K.; Deligiannis, A. Heart rate variability modifications following exercise training in type 2 diabetic patients with definite cardiac autonomic neuropathy. *Br. J. Sports Med.* **2007**, *42*, 47–54. [CrossRef]
31. Piralaiy, E.; Siahkuhian, M.; Nikookheslat, S.D.; Pescatello, L.S.; Sheikhalizadeh, M.; Khani, M. Cardiac Autonomic Modulation in Response to Three Types of Exercise in Patients with Type 2 Diabetic Neuropathy. *J. Diabetes Metab. Disord.* **2021**, *20*, 1469–1478. [CrossRef] [PubMed]
32. Deligiannis, A.; Kouidi, E.; Tourkantonis, A. Effects of physical training on heart rate variability in patients on hemodialysis. *Am. J. Cardiol.* **1999**, *84*, 197–202. [CrossRef] [PubMed]

33. Kouidi, E.J.; Grekas, D.M.; Deligiannis, A.P. Effects of Exercise Training on Noninvasive Cardiac Measures in Patients Undergoing Long-term Hemodialysis: A Randomized Controlled Trial. *Am. J. Kidney Dis.* **2009**, *54*, 511–521. [CrossRef] [PubMed]
34. Kouidi, E.; Karagiannis, V.; Grekas, D.; Iakovides, A.; Kaprinis, G.; Tourkantonis, A.; Deligiannis, A. Depression, heart rate variability, and exercise training in dialysis patients. *Eur. J. Cardiovasc. Prev. Rehabil.* **2010**, *17*, 160–167. [CrossRef]
35. Morais, M.J.d.D.; de Abreu, L.C.; de Oliveira, F.S.; Bezerra, I.M.P.; Raimundo, R.D.; Silva, R.P.M. Is aerobic exercise training during hemodialysis a reliable intervention for autonomic dysfunction in individuals with chronic kidney disease? A prospective longitudinal clinical trial. *J. Multidiscip. Healthc.* **2019**, *12*, 711–718. [CrossRef] [PubMed]
36. Wooton, A.K.; Melchior, L. Diabetes-associated cardiac autonomic neuropathy. *Nurse Pract.* **2020**, *45*, 24–31. [CrossRef]
37. Calvo, C.; Maule, S.; Mecca, F.; Quadri, R.; Martina, G.; Perin, P.C. The influence of autonomic neuropathy on hypotension during hemodialysis. *Clin. Auton. Res.* **2002**, *12*, 84–87. [CrossRef] [PubMed]
38. Watanabe, M.A.; Schmidt, G. Heart rate turbulence: A 5-year review. *Heart Rhythm* **2004**, *1*, 732–738. [CrossRef]
39. Lin, K.; Wei, L.; Huang, Z.; Zeng, Q. Combination of Ewing test, heart rate variability, and heart rate turbulence analysis for early diagnosis of diabetic cardiac autonomic neuropathy. *Medicine* **2017**, *96*, e8296. [CrossRef]
40. Disertori, M.; Masè, M.; Rigoni, M.; Nollo, G.; Ravelli, F. Heart Rate Turbulence Is a Powerful Predictor of Cardiac Death and Ventricular Arrhythmias in Postmyocardial Infarction and Heart Failure Patients. *Circ. Arrhythm. Electrophysiol.* **2016**, *9*, e004610. [CrossRef] [PubMed]
41. Myers, J.; Chan, K.; Chen, Y.; Lit, Y.; Patti, A.; Massaband, P.; Kiratli, B.J.; Tamura, M.; Chertow, G.M.; Rabkin, R. Effect of a Home-Based Exercise Program on Indices of Physical Function and Quality of Life in Elderly Maintenance Hemodialysis Patients. *Kidney Blood Press. Res.* **2021**, *46*, 196–206. [CrossRef] [PubMed]
42. Baggetta, R.; on behalf of the EXCITE Working group; D'Arrigo, G.; Torino, C.; ElHafeez, S.A.; Manfredini, F.; Mallamaci, F.; Zoccali, C.; Tripepi, G. Effect of a home based, low intensity, physical exercise program in older adults dialysis patients: A secondary analysis of the EXCITE trial. *BMC Geriatr.* **2018**, *18*, 248. [CrossRef] [PubMed]
43. De Villar, L.O.P.; Martínez-Olmos, F.J.; Pérez-Domínguez, F.d.B.; Benavent-Caballer, V.; Montañez-Aguilera, F.J.; Mercer, T.; Segura-Orti, E. Comparison of intradialytic versus home-based exercise programs on physical functioning, physical activity level, adherence, and health-related quality of life: Pilot study. *Sci. Rep.* **2020**, *10*, 8302. [CrossRef] [PubMed]
44. Muscella, A.; Stefàno, E.; Marsigliante, S. The effects of exercise training on lipid metabolism and coronary heart disease. *Am. J. Physiol. Heart Circ. Physiol.* **2020**, *319*, H76–H88. [CrossRef]
45. Sanavi, S.; Afshar, R.; Shegarfy, L.; Shavandi, N. Effects of aerobic exercise and resistance training on lipid profiles and inflammation status in patients on maintenance hemodialysis. *Indian J. Nephrol.* **2010**, *20*, 185. [CrossRef] [PubMed]
46. Song, W.J.; Sohng, K.Y. Effects of Progressive Resistance Training on Body Composition, Physical Fitness and Quality of Life of Patients on Hemodialysis. *J. Korean Acad. Nurs.* **2012**, *42*, 947. [CrossRef] [PubMed]
47. Torres, E.; Aragoncillo, I.; Moreno, J.; Vega, A.; Abad, S.; García-Prieto, A.; Macias, N.; Hernandez, A.; Godino, M.T.; Luño, J. Exercise training during hemodialysis sessions: Physical and biochemical benefits. *Ther. Apher. Dial.* **2020**, *24*, 648–654. [CrossRef]
48. Meher, M.; Panda, J.K. Impact of glycemic control over cardiac autonomic neuropathy. *J. Diabetes Metab. Disord.* **2020**, *19*, 1339–1344. [CrossRef]
49. Loimaala, A.; Huikuri, H.V.; Kööbi, T.; Rinne, M.; Nenonen, A.; Vuori, I. Exercise Training Improves Baroreflex Sensitivity in Type 2 Diabetes. *Diabetes* **2003**, *52*, 1837–1842. [CrossRef]
50. Achmad, C.; Lim, N.S.; Pramudyo, M.; Iqbal, M.; Karwiky, G.; Febrianora, M.; Natalia, N. Relation Between Glycemic Control and Cardiac Autonomic Neuropathy in Patients with Diabetes Mellitus Type 2. *Curr. Probl. Cardiol.* **2022**, 101135. [CrossRef]
51. Ng, C.H.; Ong, Z.H.; Sran, H.K.; Wee, T.B. Comparison of cardiovascular mortality in hemodialysis versus peritoneal dialysis. *Int. Urol. Nephrol.* **2021**, *53*, 1363–1371. [CrossRef] [PubMed]
52. Wild, S.; Roglic, G.; Green, A.; Sicree, R.; King, H. Global Prevalence of Diabetes. *Diabetes Care* **2004**, *27*, 1047–1053. [CrossRef]
53. Mascia, G.; Crotti, L.; Groppelli, A.; Canepa, M.; Merlo, A.C.; Benenati, S.; Michele, P. Syncope in hypertrophic cardiomyopathy (part I): An updated systematic review and meta-analysis. *Int. J. Cardiol.* **2022**, *357*, 88–94. [CrossRef]
54. Krahn, A.D.; Wilde, A.A.M.; Calkins, H.; la Gerche, A.; Cadrin-Tourigny, J.; Roberts, J.D. Arrhythmogenic Right Ventricular Cardiomyopathy. *JACC Clin. Electrophysiol.* **2022**, *8*, 533–553. [CrossRef]
55. Mohananey, A.; Tseng, A.S.; Julakanti, R.R.; Gonzalez-Bonilla, H.M.; Kruisselbrink, T.; Prochnow, C.; Rodman, S.; Lin, G.; Redfield, M.M.; Rosenbaum, A.N.; et al. An intervention strategy to improve genetic testing for dilated cardiomyopathy in a heart failure clinic. *Genet. Med.* **2022**, 100341. [CrossRef] [PubMed]
56. Lin, Y.S.; Chen, T.H.; Lin, M.S.; Chen, D.Y.; Mao, C.T.; Hsu, J.T.; Chen, M.C. Impact of Chronic Kidney Disease on Short-Term Cardiac Implantable Electronic Device Related Infection. *Medicine* **2016**, *95*, e2587. [CrossRef] [PubMed]
57. Fumagalli, S.; Pieragnoli, P.; Haugaa, K.H.; Potpara, T.S.; Rasero, L.; Ramacciati, N.; Ricciardi, G.; Solimene, F.; Mascia, G.; Mascioli, G.; et al. The influence of age on the psychological profile of patients with cardiac implantable electronic devices: Results from the Italian population in a multicenter study conducted by the European Heart Rhythm Association. *Aging Clin. Exp. Res.* **2019**, *31*, 1219–1226. [CrossRef]

life

MDPI

Article

Assessment of Countermovement Jump: What Should We Report?

Zdravko Anicic [1], Danica Janicijevic [1], Olivera M. Knezevic [1], Amador Garcia-Ramos [2,3], Milos R. Petrovic [1], Dimitrije Cabarkapa [4] and Dragan M. Mirkov [1,*]

1 Faculty of Sport and Physical Education, The Research Centre, University of Belgrade, 11000 Belgrade, Serbia
2 Department of Physical Education and Sport, Faculty of Sport Sciences, University of Granada, 18011 Granada, Spain
3 Department of Sports Sciences and Physical Conditioning, Faculty of Education, Universidad Católica de la Santísima Concepción, Concepción 4030000, Chile
4 Jayhawk Athletic Performance Laboratory—Wu Tsai Human Performance Alliance, University of Kansas, Lawrence, KS 66045, USA
* Correspondence: dragan.mirkov@fsfv.bg.ac.rs

Abstract: The purpose of the present study was (i) to explore the reliability of the most commonly used countermovement jump (CMJ) metrics, and (ii) to reduce a large pool of metrics with acceptable levels of reliability via principal component analysis to the significant factors capable of providing distinctive aspects of CMJ performance. Seventy-nine physically active participants (thirty-seven females and forty-two males) performed three maximal CMJs while standing on a force platform. Each participant visited the laboratory on two occasions, separated by 24–48 h. The most reliable variables were performance variables (CV = 4.2–11.1%), followed by kinetic variables (CV = 1.6–93.4%), and finally kinematic variables (CV = 1.9–37.4%). From the 45 CMJ computed metrics, only 24 demonstrated acceptable levels of reliability (CV ≤ 10%). These variables were included in the principal component analysis and loaded a total of four factors, explaining 91% of the CMJ variance: performance component (variables responsible for overall jump performance), eccentric component (variables related to the breaking phase), concentric component (variables related to the upward phase), and jump strategy component (variables influencing the jumping style). Overall, the findings revealed important implications for sports scientists and practitioners regarding the CMJ-derived metrics that should be considered to gain a comprehensive insight into the biomechanical parameters related to CMJ performance.

Keywords: force platform; kinematic; kinetic; testing; vertical jump

Citation: Anicic, Z.; Janicijevic, D.; Knezevic, O.M.; Garcia-Ramos, A.; Petrovic, M.R.; Cabarkapa, D.; Mirkov, D.M. Assessment of Countermovement Jump: What Should We Report? *Life* **2023**, *13*, 190. https://doi.org/10.3390/life13010190

Academic Editors: Ines Drenjančević and Helena Lenasi

Received: 9 November 2022
Revised: 24 December 2022
Accepted: 28 December 2022
Published: 9 January 2023

1. Introduction

The countermovement jump (CMJ) is one of the most implemented testing modalities for the assessment of lower body mechanical capacities. It has been primarily used for monitoring sports performance [1], inter-limb asymmetries [2], neuromuscular fatigue [3], and the effectiveness of different training programs [4]. The main reasons for the widespread use of the CMJ may be attributed to the simplicity of the testing protocols and the higher ecological validity when compared to other traditionally used testing methods (e.g., isokinetic or isometric tests). Moreover, the global market offers a variety of devices, from more rudimentary (e.g., meter-scale-based devices) to highly sophisticated ones (e.g., high-frequency motion capture systems), which facilitate the computation of different CMJ metrics. However, despite the exponential growth of performance monitoring technologies, force platforms are still considered the gold standard testing modality for the assessment of CMJ performance, since they allow sports scientists and practitioners to obtain a plethora of biomechanical variables with high levels of reliability [5,6].

The CMJ variables computed using force platforms can be roughly categorized into: (i) performance, (ii) kinetic, and (iii) kinematic. Undoubtedly, performance variables are the most frequently considered in the scientific literature and practical setting [7]. One variable that stands out from all performance metrics is jump height, due to its direct association with the performance on a variety of sport-specific tasks and physical preparedness of the athletes [3,8]. Moreover, some complex metrics, such as leg stiffness (i.e., ratio of vertical ground reaction force to minimal centre of mass displacement) and modified reactive strength index (RSI modified; i.e., quotient of jump height and push-off duration), are gaining massive popularity in terms of being considered as indirect measures of elastic/reactive properties of lower body muscles [9–11]. Kinetic variables (e.g., mean force, peak force, impulse) have been frequently computed by sports scientists and practitioners due to their ability to provide in-depth insight into the mechanics of the CMJ execution [12]. On the other hand, kinematic variables (e.g., jump duration, time to peak force, propulsive phase duration) have been analyzed to quantitatively portray the jump strategy used to achieve a given jump performance. However, they seem to receive less attention in the scientific literature when compared to the performance and kinetic variables [13].

The practical utility of CMJ-derived variables is largely contingent on their reliability. In addition to the natural variability of the human system (i.e., biological variability), the reliability of the CMJ-derived variables is greatly affected by the complexity of the computational methods. In other words, variables that require a greater number of computational steps (i.e., kinematic variables) tend to have lower reliability, while the variables directly calculated from the force–time curve usually display higher levels of reliability (i.e., performance and kinetic variables) [14]. For instance, good-to-excellent reliability was reported for the jump height and RSI modified (performance variables) [15–18], mean and peak force, power and breaking impulse (kinetic variables) [14,16,19], and some kinematic variables (i.e., peak propulsive velocity and jump depth) [16,19,20]. Other kinematic variables, such as the duration of the unloading, breaking, and landing phases, revealed poor levels of reliability [5,14,16,21]. Moreover, it has been generally argued that variables derived from the downward portion of the force–time curve have lower reliability than the variables derived from the upward portion [14,19], while there is no consensus regarding the reliability of the duration of the propulsive phase and peak breaking velocity [14,16,19,21].

Several studies evaluated the possibility and usefulness of reducing the large pool of CMJ-derived variables to a more pragmatical number of essential variables for describing and understanding the overall CMJ performance [22–25]. This has been usually performed by applying factor analysis to all computed CMJ metrics and defining the number of factors, as well as their structure (i.e., corresponding variables). Interestingly, different studies reported between two and four main factors with a certain difference in the structure of the variables that are loading different factors. For example, among the studies that identified two main factors, Laffaye et al. [22,23] identified the rate of force development (RFD) as a force factor, while Kipp et al. [24] identified RFD as a velocity factor. Some authors reported that overall CMJ variance should be explained by more than two factors [26]. Likewise, Merrigan et al. [27] identified three main factors (having military personnel as the group of subjects), where the first factor was loaded by the variables defining potential for reaching jump height (e.g., mean propulsive force, RSI modified), second by the variables responsible for the strategy to reach a high jump (e.g., velocity, jump depth), and third by the variables defining overall jump outcome (e.g., jump height). Moreover, the same group of authors identified four relevant factors when examining elite collegiate athletes as a cohort of targeted participants [25]. In addition to the study sample, the possible explanation for discrepancies in the number of factors and their structure may be attributed to the different number of CMJ-derived variables considered for the factor analysis in the aforementioned studies (usually without considering the variables related to the landing phase of the jump).

To solve this problem, 45 force–time-derived variables were identified as potentially important CMJ metrics, and the aims of this study were (i) to explore their within- and between-day reliability, and (ii) to reduce a large pool of identified CMJ-derived variables with acceptable levels of reliability to the significant factors using principal component analysis. From the reliability standpoint, it has been hypothesized that the CMJ variables will be ranked from most to least reliable as follows: performance variables > kinetic variables > kinematic variables. However, the hypothesis regarding the minimum number of factors that can explain the overall CMJ performance could not be set due to the inconsistent findings previously reported in the scientific literature. The results of the present study are expected to reveal the list of highly reliable metrics that should be used to thoroughly explore the lower body neuromuscular capacities through the CMJ.

2. Materials and Methods

2.1. Participants

Seventy-nine physically active participants (37 women (body mass = 65.7 ± 10.5 kg, height = 1.72 ± 0.01 m, age = 22.2 ± 3.6 years) and 42 men (body mass = 79.3 ± 9.5 kg, height = 1.85 ± 0.07 m, age = 22.4 ± 4.1 years)) volunteered to participate in the present study. All participants had at least one year of lower-body resistance training experience and most of them were actively involved in resistance training at the time of the testing. Moreover, they were free of musculoskeletal injuries and/or pain that could negatively impact CMJ performance, and none of them were taking any supplementation at the time of the study. Participants were familiarized with the research protocol in both written and verbal manner. The study was conducted according to the guidelines of the Declaration of Helsinki and approved by the Institutional Review Board of the University of Belgrade, Faculty of sport and physical education (protocol number: 02-273/21-1, date: 16 March 2021).

2.2. Study Design

The present study aimed to explore the reliability of the most frequently used CMJ variables and to provide a list of reliable metrics that are necessary to be considered when exploring lower body neuromuscular capacities through CMJ. For this purpose, participants completed two identical testing sessions, separated 24–48 h apart. During each experiment, participants performed three maximal CMJs without arm swing (having their hands fixed on the hips). The rest interval between each CMJ was 60 s [18]. The sessions were performed under identical laboratory conditions for all subjects.

2.3. Testing Protocol

All testing procedures were performed in the university research laboratory and at the same time of the day for each subject (±30 min). The participants were required to avoid any intense physical activity 24 h prior to each testing session to ensure their full readiness for the testing. Upon arrival at the laboratory, participants performed a standardized warm-up procedure consisting of 5 min of stationary cycling at a self-selected pace and a set of dynamic exercises (e.g., calf raises, hip hinges, lunges, squats, and hopping), followed by five submaximal CMJs with a 30 s inter-jump rest interval. Two minutes after the completion of the warm-up protocol, participants stepped on a previously calibrated force plate wearing their habitual training shoes during both sessions. They were instructed to stand steadily for 5 s, jump as high as possible after the "go" signal given by the research assistant, and land at approximately the same spot on the force plate. Each participant completed three maximal CMJs using a self-selected countermovement depth. The rest interval between consecutive CMJ was 60 s [28,29]. The same protocol was repeated during the second session. Verbal encouragement was provided systematically for every CMJ. The highest jump from the first and second sessions were recorded and incorporated into the between-day reliability analysis.

2.4. Data Processing

Vertical ground reaction force (vGRF) data were recorded using fixed bilateral three-dimensional force platforms (AMTI BP600400, Watertown, MA, USA) at a sampling frequency of 1000 Hz. The vGRF data obtained from both platforms was summed and processed using custom-made software (MATLAB and Statistics Toolbox Release 2015a, The MathWorks, Inc., Natick, MA, USA). Body weight was established during the 2 s motionless period prior to the beginning of the downward phase of the CMJ motion. The initiation of the downward phase was identified as the moment when the force–time curve trace dropped 10 N below body weight, the take-off when the vGRF was below 5 N, and the landing as the point where vGRF exceeded the 5 N threshold. The centre of mass (COM) acceleration was calculated as the net vGRF (absolute vGRF—body weight) divided by the participant's body mass. In addition, COM velocity was calculated as a numeric integration of acceleration data with respect to time and COM displacement as a double integral of COM acceleration. All force-derived variables were scaled to the subject's body mass. Impulse-momentum and flight time methods were used to calculate jump height [17]. RSI was calculated by dividing jump height by the time duration between initiation of the downward phase and the take-off [20]. Likewise, leg stiffness was calculated as the ratio between peak breaking force and COM displacement during the breaking phase.

Additionally, CMJ force–time curve was calculated for the following five phases (Figure 1): (i) Unloading phase—a portion of the force–time curve portion between jump initiation and minimum vGRF; (ii) Breaking phase—a part of the force–time curve between reaching the minimum vGRF and reaching the lowest downward displacement of the COM; (iii) Propulsive phase—a portion of the force–time curve from reaching the lowest COM position until the take-off; (iv) Flight phase—the time in the air between the take-off and the first contact during landing; and (iv) Landing phase—the portion of the force–time curve beginning with the initiation of the landing and the moment when the COM velocity is 0 m·s^{-1}. The list of all CMJ-derived dependent variables (45 total), categorized into performance, kinetic, and kinematic variables, is presented in Table 1.

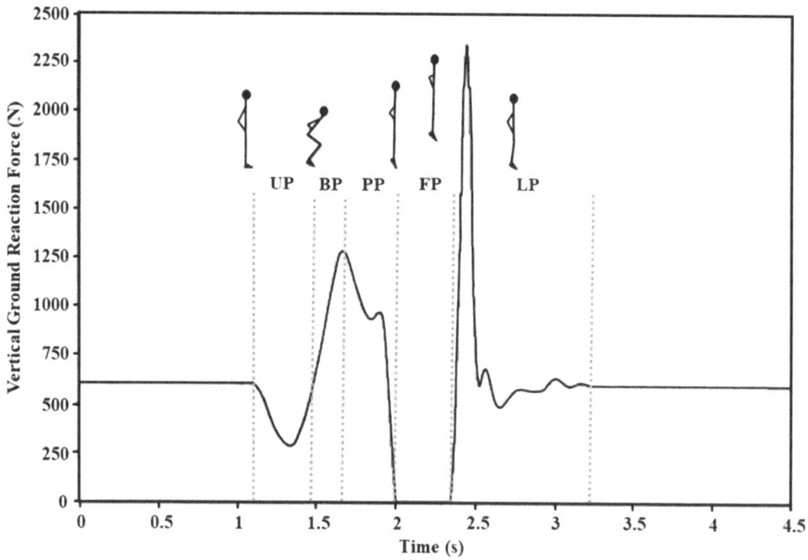

Figure 1. Typical vertical ground reaction force–time signal of the countermovement jump obtained from the force (y axis represents vertical ground reaction force [N], while x axis represents time [ms]). Vertical dotted lines divide typical phases of the countermovement jump (UP, unloading phase; BP, breaking phase; PP, propulsive phase; FP, flight phase, LP, landing phase).

Table 1. List of all the 45 countermovement-jump-derived variables computed in this study along with their units of measurement.

Variable Type	Variable	Unit
Performance variables	Jump height (flight time method)	cm
	Jump height (impulse momentum method)	cm
	Leg stiffness	AU
	Reactive strength index—modified	AU
Kinetic variables	Unloading peak force	N
	Unloading mean force	N
	Breaking peak force	N
	Breaking mean force	N
	Propulsive peak force	N
	Propulsive mean force	N
	Landing first force peak	N
	Landing second force peak	N
	Breaking rate of force development	$N{\cdot}s^{-1}$
	Propulsive peak rate of force development	$N{\cdot}s^{-1}$
	Propulsive mean rate of force development	$N{\cdot}s^{-1}$
	Landing peak rate of force development	$N{\cdot}s^{-1}$
	Landing mean rate of force development	$N{\cdot}s^{-1}$
	Unloading impulse	N·s
	Breaking impulse	N·s
	Propulsive impulse	N·s
	Positive impulse	N·s
	Landing impulse	N·s
	Breaking peak power	W
	Breaking mean power	W
	Propulsive peak power	W
	Propulsive mean power	W
	Landing rate of power development	$W{\cdot}s^{-1}$
Kinematic variables	Jump duration	s
	Unloading phase duration	s
	Breaking phase duration	s
	Propulsive phase duration	s
	Flight phase duration	s
	Landing phase duration	s
	Time to unloading peak force	s
	Time to minimum power	s
	Time to propulsive peak force	s
	Time to first landing force peak	s
	Time to second landing force peak	s
	Peak negative velocity	$m{\cdot}s^{-1}$
	Propulsive peak velocity	$m{\cdot}s^{-1}$
	Take-off velocity	$m{\cdot}s^{-1}$
	Countermovement center of mass depth	cm
	Center of mass at take off	cm
	Flight time:Jump time ratio	AU
	Breaking time:Jump time ratio	AU

Note: AU—arbitrary units.

2.5. Statistical Analysis

Intraclass correlation coefficient (ICC; model 3.1) and coefficient of variation (CV%) were used for assessing within- and between-day reliability for the 45 dependent variables computed in this study. Acceptable reliability was determined as an ICC $\geq$ 0.70 and CV $\leq$ 10% [30]. Paired-sample t-tests were used to detect differences between sessions and uncover the possible learning effect. The magnitude of the change was determined using Cohen's *d* effect size (ES) and interpreted using the criterion proposed by Hopkins [31]: trivial <0.20; small = 0.20–0.59; moderate = 0.60–1.19; large = 1.20–2.00; and extremely large >2.00. Repeated measures ANOVA was used for comparing the magnitude of the three jumps performed within each day. Correlations between the metrics and partial correlations relative to the full correlations of data were assessed using Bartlett's test of sphericity and the Kaiser–Mayer–Olkin (KMO) measurement. Then, principal component analysis (PCA) with Varimax rotation was used to extract principal components, based on Eigenvalue (i.e., the total amount of variance that can be explained by a given principal component) >1 [32]. All statistical analyses were performed using the software package SPSS (Version 25.0; IRB Corp., Armonk, NY, USA) and Microsoft Excel (Microsoft Corp., Redmond, WA, USA).

3. Results

Within- and between-day reliability of the performance, kinetic, and kinematic variables are presented in Figures 2 and 3, respectively. All performance variables presented acceptable within- and between-day reliability (CV $\leq$ 8.57%) with the only exception being leg stiffness, which presented a lower between-day reliability (CV = 11.1%). Regarding kinetic variables, the impulse-related variables always presented an acceptable reliability (CV $\leq$ 8.3%); force- and power-related variables were generally more reliable for the propulsive phase (CV $\leq$ 4.35%) than for the unloading, breaking, and landing phases (CV $\leq$ 44.3%); and RFD-related variables never reached an acceptable level of reliability (CV $\geq$ 22.2%). Finally, only 8 out of 18 kinematic variables reached acceptable levels of reliability: all velocity variables (CV $\leq$ 8.5%), depth of the countermovement and height of the COM at take-off (CV $\leq$ 8.2%), two variables that describe the duration of the propulsive (CV = 1.9%) and flight phases (CV = 2.2%), and the variable that describes the ratio between flight time and jump time (CV = 8.3%).

The differences in the magnitude of the variables between sessions 1 and 2 were significant for all performance variables, for 11 out of 23 kinetic variables, and for 11 out of 18 kinematic variables. However, of all the 26 variables that differed between sessions 1 and 2, only 6 presented a higher magnitude during the second testing session (time to reach maximal force during unloading and propulsive phases, time to reach minimal power, depth of the COM, and the duration of the unloading and propulsive phases). Nevertheless, the ES always ranged from trivial to small (ES = 0.05–0.36; Figure 3).

The KMO measure of sampling (0.799) and Bartlett's sphericity test (χ^2 = 6378; $p < 0.001$) suggested sample adequacy and sufficient correlations between variables. Based on Kaiser's criterion factor analysis, four principal components were extracted which explained 91.8% of overall CMJ variance (Table 2; Figure 4). Generally, the first factor was loaded by the variables related to the propulsive and flight phases (performance component) and explained 59% of the variance. The second factor was loaded by the variables related to the breaking phase (eccentric component) and explained 16% of the variance. The third factor was loaded with the variables related to the propulsive and breaking phases (concentric component) and explained 11% of the variance. The fourth factor was loaded by the kinematic variables, which are generally related to the jump strategy (jump strategy component) and explained 6% of the variance. Interestingly, three variables (mean propulsive power, propulsive impulse, and total positive impulse) overlapped both the first and third factors based on loading values.

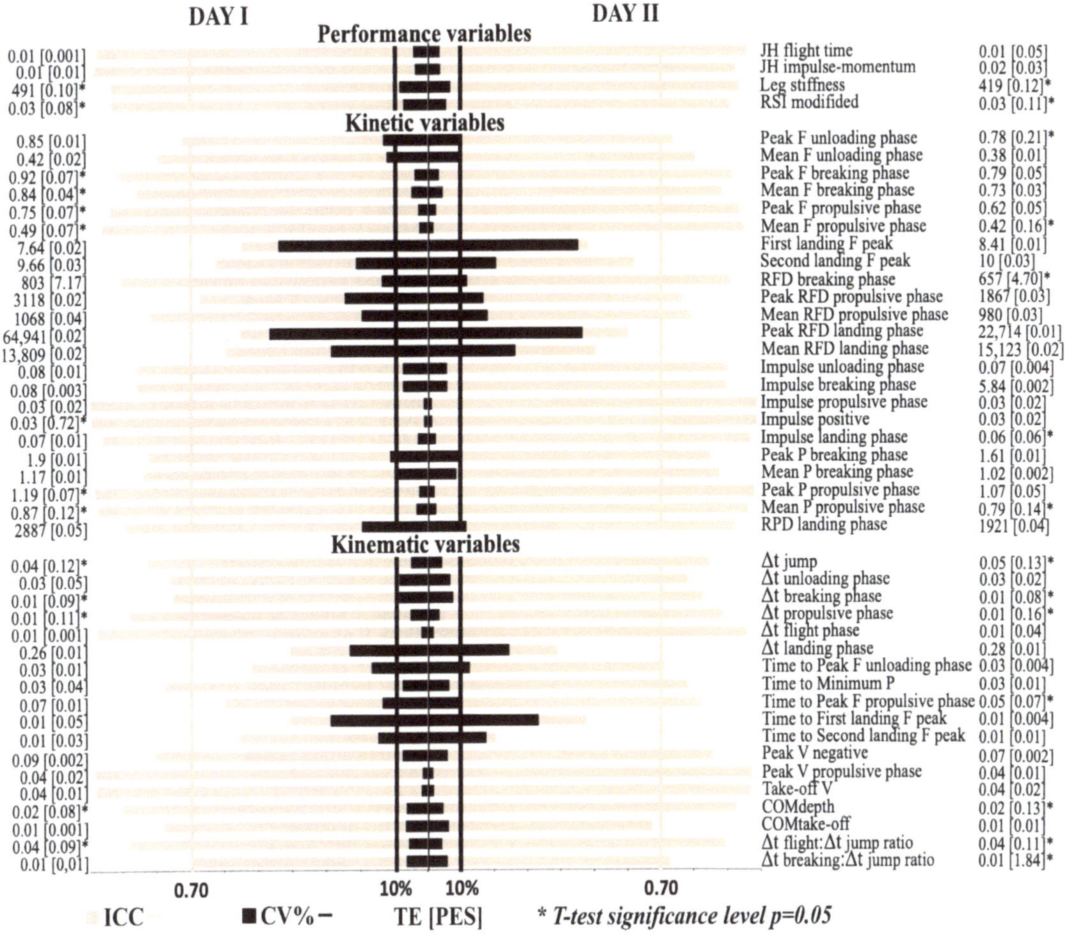

Figure 2. Within-day reliability of the different variables derived from the countermovement jump test. Gray bars indicate the magnitude of the interclass correlation coefficient (ICC), while black bars indicate the coefficient of variation (CV%) for the first (left side of the figure) and the second day (right side of the figure). Force, impulse, and power variables are scaled to body mass. Vertical black lines delimit the area of acceptable reliability based on CV = 10%, while gray delimit acceptable reliability based on ICC = 0.70; PES, partial eta square, * denotes ANOVA significance level ($p \leq 0.05$). JH, jump height; RSI modified, reactive strength index—modified; F, force; RFD, rate of force development; P, power; Δt, duration of the certain phase; V, velocity; COM, center of mass.

4. Discussion

The purpose of the present study was twofold: (i) to assess the reliability of a large number of performance, kinetic, and kinematic CMJ-derived variables and (ii) to provide a reduced list of reliable metrics that should be reported to provide information regarding the distinctive aspects of CMJ performance. The main findings revealed that only 24 out of 45 CMJ-derived variables demonstrated acceptable within- and between-day reliability and that the most reliable metrics were the performance variables (3 out of 4), followed by the kinetic variables (12 out of 23), and finally the kinematic variables (8 out of 18). Four main components were extracted as a result of principal component analysis applied to the 24 reliable CMJ-derived variables and were conveniently addressed as: performance component, eccentric component, concentric component, and jump strategy component, explaining 56%, 16%, 11%, and 6% of the common variance, respectively. These findings present important implications for sports scientists and practitioners regarding the CMJ-derived metrics that should be considered to gain a comprehensive insight into the distinctive aspects of the CMJ performance.

All performance variables were obtained with an acceptable within- and between-day reliability with the only exception being leg stiffness, which presented a between-day reliability lower than the minimal threshold for acceptable reliability (CV > 10%). These results are in line with the findings of previous research reports focused on exploring the reliability of the CMJ-derived performance variables [16,19,21]. For instance, exceptional reliability was obtained in the present investigation for jump height, regardless of the computational method (impulse-momentum or flight time approaches), which is in agreement with several recent studies [16,19,21] that reported CV < 5% for jump height. Moreover, the RSI modified, considered an important indicator of neuromuscular function, showed an acceptable reliability with a CV of 8.6%; despite this, we did not fix the countermovement depth. The acceptable reliability of the RSI modified is in line with the findings of previous studies [18,20,33] (CV $\leq$ 10%). The only performance variable that presented a questionable reliability was the leg stiffness (CV = 11%), a variable that is considered as a quantitative measure of the elastic properties of lower body muscles [9]. Although several studies reported similar findings [17,19,34], Heishman et al. [21] argued that leg stiffness cannot be considered a reliable CMJ-derived variable because they obtained CV values that exceeded 20%. A possible explanation for the discrepancies between studies regarding the reliability of leg stiffness could be the different CMJ strategies implemented by the participants [33,35], as well as the gender and participant's sports background (e.g., familiarity with the CMJ).

Regarding kinetic variables, impulse-related variables were the most reliable (all CV $\leq$ 8.3%), followed by force and power variables (50% of variables with acceptable reliability in each group), while RFD-related variables never reached an acceptable level of reliability. Impulse-related variables were also considered reliable metrics in previous studies [7,17,36], which is an important finding, since they are responsible for the jump height and jump mechanics in general [29,37]. On the other hand, force-related variables, although obtained from the directly recorded force–time curve, were not always obtained with acceptable reliability. For instance, variables related to the breaking and propulsive phase demonstrated acceptable reliability (CV $\leq$ 6.1%), while variables related to the unloading and landing phase did not meet the criteria for acceptable reliability. These results are in line with previous evidence [14,19,21]. Since the computational method was rather direct, the low reliability of the kinetic variables collected during the unloading and landing phases might be explained by the high variability produced when participants are allowed to self-select the countermovement depth. Furthermore, power-related variables collected during the breaking phase showed lower reliability (CV $\geq$ 10.9%) compared to the variables collected during the propulsive phase (CV $\leq$ 4.1%), which is contrary to the findings of Merrigan et al. [19] who reported acceptable levels of reliability for the CMJ-derived power variables collected during the breaking phase of CMJ motion. Finally, although RFD-related variables are considered important metrics for athletes who have limited time to exert force and are frequently considered for monitoring the neuromuscular component of the human movement [14,21,29,38], none of the RFD-related variables reached acceptable reliability in the present study, questioning their use when assessing lower-body neuromuscular performance through the execution of the CMJ.

Confirming our first hypothesis, only 8 out of 18 kinematic variables presented acceptable reliability. Specifically, all velocity- and COM-related variables were reliable, as well as the duration of the propulsive phase, flight phase, and the ratio between flight time and overall jump time. In line with our study, velocity-related variables were shown to be consistently reliable [19,21,39], as well as the COM-related variables (depth of the countermovement and height of the COM at take-off) [14,19,39]. However, the results of previous studies are somewhat contradictory when it comes to the reliability of the duration of the different CMJ phases. Contrary to the findings of the present study, Warr et al. [14] and Heishman et al. [21] reported low reliability for the propulsive phase duration, while the breaking phase duration was sometimes found to be reliable [14,27] but other times was not a reliable metric [21]. The duration of the flight phase seems to be universally accepted as the most reliable metric when it comes to the duration of different CMJ phases [21] because the parameters that are influencing jump height are consistent within the same participants (i.e., similar velocities at take-off and correct jumping technique). On the other hand, the duration of the landing phase and overall duration of the jump were not studied extensively. The variables related to the time necessary to reach certain mechanical peaks on the CMJ curve were generally considered non-reliable, as shown in our study [14].

In summary, a total of 24 (out of 45) CMJ-derived variables demonstrated acceptable reliability and, therefore, were included in the principal component analysis. Considering the number of the included variables and extracted factors, our study is similar to the study of Merrigan et al. [25] who included 19 variables in the principal component analysis and extracted the same number of factors (i.e., four factors). However, the variables that load the four factors in the study of Merrigan et al. [25] differ from our study. Specifically, our 24 variables were grouped into a performance component (loaded by the variables related to the propulsive and flight phases), eccentric component (loaded by the variables related to the breaking phase), concentric component (loaded by the variables related to the concentric phase), and jumps strategy component (loaded by the variables specific for selecting a different jump strategy). Although Merrigan et al. [25] also extracted factors related to the concentric phase and another factor related to overall jump performance, two other factors were loaded with the variables related to the breaking phase. Another important discrepancy between our and the findings of Merrigan et al. [25] is that they did not extract any factor related to the jump strategy. The reason why our findings cannot be directly compared to other studies is due to a lower number of CMJ-related variables (i.e., 11 or 15) included in the analysis procedures [26], as well as due to the CMJ-related variables that were excluded from our principal component analysis because of their lack of reliability (e.g., breaking phase duration, breaking RFD, power during breaking phase). The explanation for why breaking phase-related variables showed low reliability is possibly because they are diverse strategies adopted for performing the downward phase of the CMJ (i.e., different jump depth and lowering velocities).

Several limitations of the present study should be acknowledged. First, the sample consisted of physically active young participants engaged in recreational physical activities, and it is unknown whether the results of the present study could be generalized to other populations (e.g., athletes or sedentary individuals). Second, it is important to acknowledge the low number of subjects included in the study per variable [40], although the total number of participants in our study was larger/similar when compared to other research reports ($n = 79$ vs. $n = 16$–82) [25–27]. Third, it should be noted that the principal component analysis included two variables whose reliability was on the border of the acceptability criteria (breaking mean power [CV = 10.85%] and leg stiffness [CV = 11.01%]).

5. Conclusions

From the large pool of 45 CMJ-derived variables computed in the present study, only 24 demonstrated acceptable within- and between-day levels of reliability. The CMJ-derived variables ranked in the order of highest to lowest reliability magnitude were as follows: performance variables (e.g., jump height, modified reactive strength index), kinetic variables (impulse-related variables were the most reliable), kinematic variables (only 8 out of 18 kinematic variables revealed acceptable reliability). When included into the principal component analysis, these 24 variables loaded four factors, explaining 91% of the variance and were conveniently addressed as performance component (loaded by the variables responsible for overall jump performance), eccentric component (loaded by the variables related to the breaking phase of the CMJ), concentric component (loaded by the variables related to the concentric phase of the CMJ) and jump strategy component (loaded by the variables that are importantly influencing the jumping style, such as the depth of the countermovement). Overall, the findings of the present study reveal important implications for sports scientists and practitioners regarding the CMJ-derived variables that should be considered to gain a comprehensive insight into the mechanics pertaining to CMJ performance.

Author Contributions: Conceptualization, D.M.M., O.M.K. and A.G.-R.; Methodology, Z.A., M.R.P. and D.J.; Investigation, D.C. and A.G.-R.; Data Curation, Z.A.; Writing—Original Draft Preparation, Z.A. and D.J.; Writing—Review and Editing, D.C., O.M.K. and A.G.-R.; Supervision, D.M.M.; Project Administration, M.R.P. All authors have read and agreed to the published version of the manuscript.

Funding: This research received no external funding.

Institutional Review Board Statement: The study was conducted in accordance with the Declaration of Helsinki, and approved by the Institutional Review Board of the Faculty of the Sport and Physical Education (protocol code 02-273/21-1, date: 16 March 2021).

Informed Consent Statement: Informed consent was obtained from all subjects involved in the study.

Data Availability Statement: Data available on request due to privacy restrictions.

Acknowledgments: This work was supported by the grant 415-03-68/2022-14/200154 from the Ministry of Education, Science and Technological Development of the Republic of Serbia.

Conflicts of Interest: The authors declare no conflict of interest.

References

1. McMahon, J.J.; Murphy, S.; Rej, S.J.; Comfort, P. Countermovement-Jump-Phase Characteristics of Senior and Academy Rugby League Players. *Int. J. Sport. Physiol. Perform.* **2017**, *12*, 803–811. [CrossRef] [PubMed]
2. Lake, J.P.; Mundy, P.D.; Comfort, P.; Suchomel, T.J. Do the peak and mean force methods of assessing vertical jump force asymmetry agree? *Sport. Biomech.* **2020**, *19*, 227–234. [CrossRef] [PubMed]
3. Claudino, J.G.; Cronin, J.; Mezêncio, B.; McMaster, D.T.; McGuigan, M.; Tricoli, V.; Amadio, A.C.; Serrão, J.C. The countermovement jump to monitor neuromuscular status: A meta-analysis. *J. Sci. Med. Sport* **2017**, *20*, 397–402. [CrossRef] [PubMed]
4. Berton, R.; Lixandrão, M.E.; Pinto e Silva, C.M.; Tricoli, V. Effects of weightlifting exercise, traditional resistance and plyometric training on countermovement jump performance: A meta-analysis. *J. Sport. Sci.* **2018**, *36*, 2038–2044. [CrossRef]
5. Lake, J.P.; Mundy, P.; Comfort, P.; McMahon, J.J.; Suchomel, T.J.; Carden, P. Concurrent Validity of a Portable Force Plate Using Vertical Jump Force–Time Characteristics. *J. Appl. Biomech.* **2018**, *34*, 410–413. [CrossRef]
6. Wade, L.; Lichtwark, G.A.; Farris, D.J. Comparisons of laboratory-based methods to calculate jump height and improvements to the field-based flight-time method. *Scand. J. Med. Sci. Sport.* **2019**, *30*, 31–37. [CrossRef]
7. Pérez-Castilla, A.; Fernandes, J.F.T.; Rojas, F.J.; García-Ramos, A. Reliability and Magnitude of Countermovement Jump Performance Variables: Influence of the Take-off Threshold. *Meas. Phys. Educ. Exerc. Sci.* **2021**, *25*, 227–235. [CrossRef]
8. Markström, J.L.; Olsson, C.J. Countermovement jump peak force relative to body weight and jump height as predictors for sprint running performances: (In)homogeneity of track and field athletes? *J. Strength Cond. Res.* **2013**, *27*, 944–953. [CrossRef]
9. Struzik, A.; Zawadzki, J.; Rokita, A. Leg stiffness and potential energy in the countermovement phase and the CMJ jump height. *Biomed. Hum. Kinet.* **2016**, *8*, 39–44. [CrossRef]
10. Barker, L.A.; Harry, J.R.; Mercer, J.A. Relationships between countermovement jump ground reaction forces and jump height, reactive strength index, and jump time. *J. Strength Cond. Res.* **2018**, *32*, 248–254. [CrossRef]

11. McMahon, J.; Jones, P.A.; Suchomel, T.J.; Lake, J.; Comfort, P. Influence of reactive strength index modified on force- and power-time curves. *Int. J. Sport. Physiol. Perform.* **2018**, *13*, 220–227. [CrossRef] [PubMed]
12. Moir, G.L.; Garcia, A.; Dwyer, G.B. Intersession Reliability of Kinematic and Kinetic Variables During Vertical Jumps in Men and Women. *Int. J. Sport. Physiol. Perform.* **2009**, *4*, 317–330. [CrossRef] [PubMed]
13. Kirby, T.J.; McBride, J.M.; Haines, T.L.; Dayne, A.M. Relative Net Vertical Impulse Determines Jumping Performance. *J. Appl. Biomech.* **2011**, *27*, 207–214. [CrossRef] [PubMed]
14. Warr, D.M.; Pablos, C.; Sánchez-Alarcos, J.V.; Torres, V.; Izquierdo, J.M.; Redondo, J.C. Reliability of measurements during countermovement jump assessments: Analysis of performance across subphases. *Cogent Soc. Sci.* **2020**, *6*. *in press*. [CrossRef]
15. Chiu, L.Z.F.; Dæhlin, T.E. Comparing Numerical Methods to Estimate Vertical Jump Height Using a Force Platform. *Meas. Phys. Educ. Exerc. Sci.* **2018**, *24*, 25–32. [CrossRef]
16. Harry, J.R.; Barker, L.A.; Tinsley, G.M.; Krzyszkowski, J.; Chowning, L.D.; McMahon, J.J.; Lake, J. Relationships among countermovement vertical jump performance metrics, strategy variables, and inter-limb asymmetry in females. *Sports Biomech.* **2021**, *in press*. [CrossRef]
17. Moir, G.L. Three Different Methods of Calculating Vertical Jump Height from Force Platform Data in Men and Women. *Meas. Phys. Educ. Exerc. Sci.* **2008**, *12*, 207–218. [CrossRef]
18. Carroll, K.M.; Wagle, J.P.; Sole, C.J.; Stone, M.H. Intrasession and Intersession Reliability of Countermovement Jump Testing in Division-I Volleyball Athletes. *J. Strength Cond. Res.* **2019**, *33*, 2932–2935. [CrossRef]
19. Merrigan, J.J.; Stone, J.D.; Hornsby, W.G.; Hagen, J.A. Identifying Reliable and Relatable Force–Time Metrics in Athletes—Considerations for the Isometric Mid-Thigh Pull and Countermovement Jump. *Sports* **2021**, *9*, 4. [CrossRef]
20. McMahon, J.J.; Lake, J.P.; Comfort, P. Reliability of and Relationship between Flight Time to Contraction Time Ratio and Reactive Strength Index Modified. *Sports* **2018**, *6*, 81. [CrossRef]
21. Heishman, A.D.; Daub, B.D.; Miller, R.M.; Freitas, E.D.; Frantz, B.A.; Bemben, M.G. Countermovement Jump Reliability Performed With and Without an Arm Swing in NCAA Division 1 Intercollegiate Basketball Players. *J. Strength Cond. Res.* **2018**, *32*, 546–558. [CrossRef] [PubMed]
22. Laffaye, G.; Bardy, B.G.; Durey, A. Principal Component Structure and Sport-Specific Differences in the Running One-Leg Vertical Jump. *Int. J. Sport. Med.* **2007**, *28*, 420–425. [CrossRef] [PubMed]
23. Laffaye, G.; Wagner, P.P.; Tombleson, T.I.L. Countermovement jump height: Gender and sport-specific differences in the force-time variables. *J. Strength Cond. Res.* **2014**, *28*, 1096–1105. [CrossRef] [PubMed]
24. Kipp, K.; Kiely, M.T.; Geiser, C.F. Reactive Strength Index Modified Is a Valid Measure of Explosiveness in Collegiate Female Volleyball Players. *J. Strength Cond. Res.* **2016**, *30*, 1341–1347. [CrossRef] [PubMed]
25. Merrigan, J.J.; Rentz, L.E.; Hornsby, W.G.; Wagle, J.P.; Stone, J.D.; Smith, H.T.; Galster, S.M.; Joseph, M.; Hagen, J.A. Comparisons of Countermovement Jump Force-Time Characteristics Among National Collegiate Athletic Association Division I American Football Athletes: Use of Principal Component Analysis. *J. Strength Cond. Res.* **2022**, *36*, 411–419. [CrossRef]
26. James, L.P.; Suppiah, H.; McGuigan, M.R.; Carey, D.L. Dimensionality Reduction for Countermovement Jump Metrics. *Int. J. Sport. Physiol. Perform.* **2021**, *16*, 1052–1055. [CrossRef]
27. Merrigan, J.J.; Stone, J.D.; Ramadan, J.; Hagen, J.; Thompson, A. Dimensionality reduction differentiates sensitive force-time characteristics from loaded and unloaded conditions throughout competitive military training. *Sustainability* **2021**, *13*, 6105. [CrossRef]
28. Gathercole, R.J.; Stellingwerff, T.; Sporer, B.C. Effect of acute fatigue and training adaptation on countermovement jump performance in elite snowboard cross athletes. *J. Strength Cond. Res.* **2015**, *29*, 37–46. [CrossRef]
29. Mizuguchi, S.; Sands, W.A.; Wassinger, C.A.; Lamont, H.S.; Stone, M.H. A new approach to determining net impulse and identification of its characteristics in countermovement jumping: Reliability and validity. *Sport. Biomech.* **2015**, *14*, 258–272. [CrossRef]
30. Pérez-Castilla, A.; McMahon, J.J.; Comfort, P.; García-Ramos, A. Assessment of Loaded Squat Jump Height With a Free-Weight Barbell and Smith Machine: Comparison of the Takeoff Velocity and Flight Time Procedures. *J. Strength Cond. Res.* **2017**, *34*, 671–677. [CrossRef]
31. Hopkins, W.G.; Marshall, S.W.; Batterham, A.M.; Hanin, J. Progressive Statistics for Studies in Sports Medicine and Exercise Science. *Med. Sci. Sport. Exerc.* **2009**, *41*, 3–13. [CrossRef]
32. Kollias, I.; Hatzitaki, V.; Papaiakovou, G.; Giatsis, G. Using principal components analysis to identify individual differences in vertical jump performance. *Res. Q. Exerc. Sport* **2001**, *72*, 63–67. [CrossRef] [PubMed]
33. McMahon, J.J.; Rej, S.J.E.; Comfort, P. Sex Differences in Countermovement Jump Phase Characteristics. *Sports* **2017**, *5*, 8. [CrossRef] [PubMed]
34. Choukou, M.A.; Laffaye, G.; Taiar, R. Reliability and validity of an accele-rometric system for assessing vertical jumping performance. *Biol. Sport* **2014**, *31*, 55–62. [CrossRef] [PubMed]
35. Mcmahon, J.J.; Comfort, P.; Cscs, D. Lower limb stiffness: Effect on performance and training considerations. *Strength Cond. J.* **2012**, *34*, 94–101. [CrossRef]
36. Lake, J.P.; Mundy, P.D.; Comfort, P.; McMahon, J.J.; Suchomel, T.J.; Carden, P. Effect of Barbell Load on Vertical Jump Landing Force-Time Characteristics. *J. Strength Cond. Res.* **2021**, *35*, 25–32. [CrossRef] [PubMed]
37. Linthorne, N.P. Analysis of standing vertical jumps using a force platform. *Am. J. Phys.* **2001**, *69*, 1198–1204. [CrossRef]

38. McLellan, C.P.; Lovell, D.I.; Gass, G.C. The Role of Rate of Force Development on Vertical Jump Performance. *J. Strength Cond. Res.* **2011**, *25*, 379–385. [CrossRef]
39. Gordon, D.; Hayward, S.; van Lopik, K.; Philpott, L.; West, A. Reliability of bilateral and shear components in a two-legged counter-movement jump. *Proc. Inst. Mech. Eng. Part P J. Sport. Eng. Technol.* **2021**, *236*, 159–171. [CrossRef]
40. Osborne, J.W.; Costello, A.B. Sample size and subject to item ratio in principal components analysis. *Pract. Assess. Res. Eval.* **2004**, *9*, 11.

life

Article

Prediction of Maximal Oxygen Consumption in Cycle Ergometry in Competitive Cyclists

Iva Jurov [1,*], Janez Toplišek [2,3] and Marta Cvijić [2,3]

1 Clinical Institute of Occupational, Traffic and Sports Medicine, University Medical Centre Ljubljana, 1000 Ljubljana, Slovenia
2 Department of Cardiology, University Medical Centre Ljubljana, 1000 Ljubljana, Slovenia
3 Faculty of Medicine, University of Ljubljana, 1000 Ljubljana, Slovenia
* Correspondence: iva.jurov@kclj.si

Abstract: Models for predicting maximal oxygen consumption (VO_{2max}) in average adults might not be suitable for athletes, especially for competitive cyclists who can have significantly higher VO_{2max} than normally active people. The aim of this study was to develop a clinically applicable equation for predicting VO_{2max} during cycle ergometry in competitive cyclists and to compare its accuracy to the traditional American College of Sports Medicine (ACSM) equation. Maximal cycle ergometry tests were performed in 496 male and 84 female competitive cyclists. Six predictors were initially used to model the prediction equation (power output, body weight, body height, fat mass, fat-free mass and age). Power output and body weight were the most important parameters in the model predicting VO_{2max}. Three new equations were derived: for male ($VO_{2max} = 0.10 \times PO - 0.60 \times BW + 64.21$), female cyclists ($0.13 \times PO - 0.83 \times BW + 64.02$) and the non-gender-specific formula ($0.12 \times PO - 0.65 \times BW + 59.78$). The ACSM underestimated VO_{2max} in men by 7.32 mL/min/kg (11.54%), in women by 8.24 mL/min/kg (15.04%) and in all participants by 7.45 mL/min/kg (11.99%), compared to the new equation that underestimated VO_{2max} in men by 0.12 mL/min/kg (0.19%) and in all participants by 0.65 mL/min/kg (1.04%). In female cyclists, the new equation had no relative bias. We recommend that medicine and sports practitioners adapt our proposed equations when working with competitive cyclists. Our findings demonstrate the need to evaluate prediction models for other athletes with a special focus on disciplines that demand high aerobic capacity.

Keywords: oxygen consumption; aerobic exercise; physiology; physical performance; sports medicine

Citation: Jurov, I.; Toplišek, J.; Cvijić, M. Prediction of Maximal Oxygen Consumption in Cycle Ergometry in Competitive Cyclists. *Life* **2023**, *13*, 160. https://doi.org/10.3390/life13010160

Academic Editors: Helena Lenasi and Ines Drenjančević

Received: 5 December 2022
Revised: 21 December 2022
Accepted: 28 December 2022
Published: 5 January 2023

1. Introduction

In sports physiology, the cardiopulmonary exercise test is utilized for testing exercise capacity, for determining training plans and looking for causes of exercise intolerance. It is also used for monitoring effects of interventions and, in normally active individuals, it provides important information in patient diagnosis and management [1]. Indirect calorimetry in exercise testing is considered as a gold standard to detect maximal oxygen uptake (VO_{2max}), which is a primary indicator of cardiorespiratory fitness during the incremental test. Measuring VO_{2max} needs to be performed by a skilled technician by using standardized exercise treadmill protocols or cycle ergometry [2]. Indirect calorimetry with a gas analyser is required to determine VO_{2max}. Measured VO_{2max} is usually compared to predicted values in order to estimate an individual's suboptimal values. In addition, VO_{2max} also has to be estimated when direct measurement is not feasible [3]. Predicting VO_{2max} can be carried out with several prediction models [1]. Traditionally, the American College of Sports Medicine (ACSM) equation is used, which was determined 40 years ago. In recent years, its widespread use has been questioned by some researchers aiming to optimize its estimation in specific populations.

Endurance athletes, especially competitive and professional, are also a specific population that significantly differs from normally active individuals in their high aerobic capacity. During the incremental test, athletes, in general, achieve higher VO_{2max} values than their non-athletic counterparts, but trained endurance athletes exhibit the highest values of all sport disciplines. This is also the case in competitive cyclists. Faria et al. concluded that VO_{2max} values in successful riders are as high as 74 mL/kg/min [4]. Although a high VO_{2max} is a prerequisite in professional cyclists, it is not the only factor influencing results. In moderately trained cyclists, other tests, such as functional threshold power, are used and can predict performance as well [5].

When selecting reference values for VO_{2max}, it is important that the individual tested matches the population in which the reference values were obtained [1]. Traditional equations for predicting VO_{2max} in cardiopulmonary exercise testing, such as the ACSM equation [6], are based on the non-elite, normally active population [3]. Although they are used for athletes as well, they might not be the most appropriate for predicting VO_{2max} in competitive cyclists. Similar conclusions were found in other specific populations, such as in patients with heart failure [7], coronary artery disease patients [8] and also in healthy adults [9,10], where new and improved prediction models have been presented and show greater accuracy.

In cyclists, cycle ergometry is the most suitable type of incremental testing in the laboratory. To achieve absolutely maximal values, it is advised to use the cyclist's own bike that is attached to the ergometry system. In this way, the problem of the cyclists not being able to configure the ergometer to their normal riding posture is eliminated [4]. The test is terminated based on different termination criteria: an athlete's volitional fatigue, pedal cadence cannot be maintained, a given workload cannot be maintained, a plateau in VO_2 or a difference in metabolic parameters (a rise in respiratory exchange ratio above 1.0, 90% of predicted heart rate). One or more criteria can be used, but a plateau in VO_2 is generally considered as the most important marker of reaching $VO_{2max.}$ Incremental tests in competitive cyclists are usually conducted on a regular basis to evaluate their health and performance. Depressed VO_{2max} may not only be a sign of illness but also of fatigue or overtraining. If cyclists are fatigued, power output during the incremental test is typically also affected [11]. Finally, VO_{2max} can be used for training monitoring during training modifications and adaptations [12,13]

Comparing measured values of VO_{2max} of an athlete to predicted VO_{2max} values of normally active people could consequently lead to misdiagnosis. This is why developing a VO_{2max} equation specific for endurance athletes with typically high VO_{2max} is needed. The aim of this research was to develop a clinically applicable equation for predicting VO_{2max} during cycle ergometry in competitive cyclists and to compare its accuracy to the traditional ACSM equation based on a larger sample of 580 competitive cyclists.

2. Materials and Methods

Healthy competitive road cyclists with a valid license were included in this study as part of an annual cycling performance analysis that is set before the competitive period in the season. One day prior to the test, they refrained from high-intensity exercise. They were well rested and received instructions on how to be well hydrated. Informed consent was obtained before starting the procedures.

Two experienced physiologists performed all procedures in the same exercise laboratory in 5 consecutive years (2014–2019). Results of the test were retrospectively analysed for the purpose of this research. This study conformed to the standards set by the latest version of the Declaration of Helsinki. Informed consent was obtained for experimentation with human subjects. Institutional ethical committee approved this study (0120-202/2020/5).

After a 15 min warm-up, a graded exercise test was performed using cycle ergometry (Cyclus 2, Leipzig, Germany). Ventilatory and gas data (measured with V2 mask, Hans Rudolph, Shawnee, KS, USA of appropriate size) were collected during the incremental test with indirect calorimetry (metabolic cart K5, Cosmed, Albano Laziale, Italy). The workload was constantly increased until volitional exhaustion. Only maximal incremental tests were included in the analysis, defined as respiratory exchange ratio R $\geq$ 1.0, a reached plateau in VO_{2max}, BORG scale $\geq$ 8. The test was also terminated if an athlete could not maintain cycling cadence above 60. Two modified Conconi protocols, commonly used in elite athletes [14], were used based on cyclists' age and body mass. Cyclists under 17 years of age or weighing less than 50 kg started protocol at 60 Watts and increased 15 Watts every minute (the 60 + 15 W protocol) and cyclists above 17 years of age and weighing more than 50 kg started protocol at 100 Watts and increased 20 Watts every minute (the 100 + 20 W protocol). Using breath-by-breath data, the VO_{2max} was determined as the average of the 5 s highest values during the last 30 s in the incremental test. Ambient temperature during all procedures was 21 °C. Metabolic cart was calibrated prior to each of the measurements.

Prior to body composition measurement, participants received instructions on how to be adequately hydrated. Body composition was assessed with bioelectrical impedance (Biospace Inbody 720, Cerritos, CA, USA).

Descriptive statistics (average $\pm$ standard deviation) was used for representing measured and predicted values of VO_{2max} and for all baseline characteristics. Pearson correlation coefficients were used to examine the relationship between predictors before performing multiple regression analysis. The generalization of the model was tested by splitting the data randomly on 70% and 30% of participants in each of the groups (male, female and non-gender specific). The model was derived from data of 70% and was confirmed on the remaining 30% by forcing the model [15]. The model is accurate for the sample and generalizable for the population. Assumptions for multiple regression (non-zero variance, multicollinearity, homoscedasticity, normality of distribution, independence, linearity) were checked. Quantitative variable types were used in multiple regression analysis. The predictors did not have variances of 0. We excluded parameters that had correlation above 0.8 between them to confirm the assumption of no multicollinearity. Variance inflation factor (VIF) (well below 10) and tolerance statistics (well above 0.2) were also checked.

Durbin–Watson test for independent errors was 1.657, 1.854 and 1.536 for males, females and the non-gender-specific group, respectively. As seen in Figure A1 (Appendix A), the points are randomly and evenly dispersed throughout the plot, which means that assumptions of linearity and homoscedasticity were met. To test the normality of residuals, we present the histogram and normal P-P plot of regression-standardized residual (Figure A2, Appendix A).

Bland–Altman plots were constructed to graphically illustrate the variance between measured VO_{2max} and VO_{2max} generated from the new and the ACSM equations (Figure 1. Constant error (bias) and standard error of the estimate (SEE) were calculated.

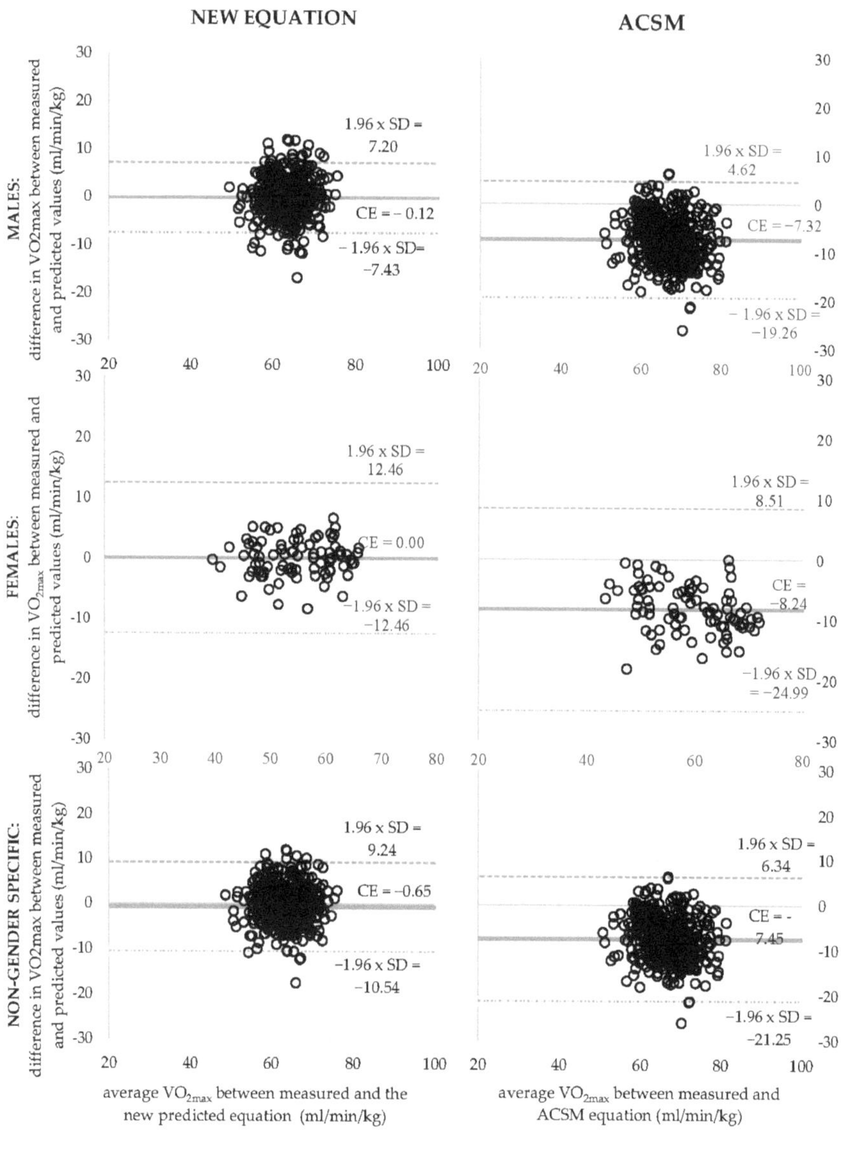

Figure 1. Bland–Altman plots comparing measured VO_{2max} and predicted VO_{2max} using the new gender—specific equations and the American College of Sports Medicine (ACSM) equation.

3. Results

The baseline characteristics of participants are presented in Appendix B (Table A1). Cyclist were recruited from 21 competitive female and male road-cycling teams. On average, male cyclists were younger than female cyclists, they were higher and heavier, they had less body fat and a greater fat-free mass. Their maximal power output was greater, both when measured absolutely (W) and relatively (W/kg). The VO_{2max} in males was greater than in females.

The stepwise method was used in the multiple-regression analysis to indicate which predictors are the best for predicting the measured VO_{2max}. Multiple regression was performed on males, females and on the entire sample to develop the gender non-specific equation. Initially, six predictors were considered based on previously used variables [1]: power output (W), body weight (kg), body height (cm), skeletal muscle mass (kg), fat mas (%) and age.

To test the six predictors, at least 60 participants were required for the analysis [15]. Since skeletal muscle mass and fat mass were highly correlated, fat mass was excluded from further analysis. Age predictor was also excluded since it over- and underestimated predicted VO_{2max} based on confidence intervals. The stepwise technique was repeated with the remaining four predictors. Considering average squared error (average standard error of the estimate) and R square as "goodness-of-fit parameters", model 2 was the most appropriate since adding variables in models 3 and 4 did not change the model significantly (Figure 2).

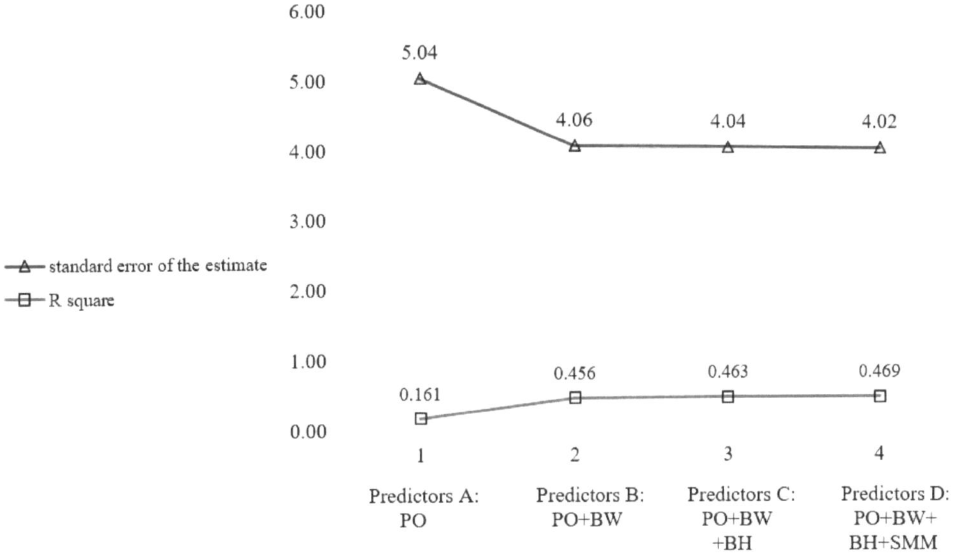

Figure 2. Average standard error of the estimate with variables considered in the multiple regression models (predictors in the models A–D) for male cyclists. Lower values of standard error of the estimate and higher of R square indicate greater predictive value of the model (PO = power output, BW = body weight, BH = body height, SMM = skeletal muscle mass).

In Appendix B (Table A2), we present the influence of parameters on the measured VO_{2max}. Results revealed that the power output has the biggest influence on VO_{2max} followed by body weight. However, the other two predictors (body height and skeletal muscle mass) have a lower impact on the outcome parameter (Table A2).

The model using power output and body weight was chosen in all three groups based on having a high R square (other models with additional parameters do not increase R square appreciably). In Table 1, we present gender-specific and non-gender-specific formulas. R square was 0.456 in male, 0.671 in female cyclists and 0.589 in the non-gender-specific group.

A comparison of relative bias in the new equations and ACSM equation is presented in Figure 1. ANOVA was also performed to compare the ACSM equation and the new equation (Appendix B, Table A3). The ACSM underestimated VO_{2max} in men by 7.32 mL/min/kg (11.54%), in women by 8.24 mL/min/kg (15.04%) and in all participants by 7.45 mL/min/kg (11.99%), compared to the new equation that underestimated VO_{2max} in men by 0.12 mL/min/kg (0.19%) and in all participants by 0.65 mL/min/kg (1.04%). In female cyclists, the new equation had no relative bias (0.00%).

Table 1. New prediction equations for VO_{2max} in competitive cyclists.

	VO_{2max} (mL/min/kg)
male cyclists	$0.10 \times PO - 0.60 \times BW + 64.21$
female cyclists	$0.13 \times PO - 0.83 \times BW + 64.02$
gender non-specific	$0.12 \times PO - 0.65 \times BW + 59.78$

VO_{2max} = maximal oxygen uptake, PO = maximal power output in Watts, BW = body weight in kg.

4. Discussion

The aim of this study was to develop a clinically applicable equation for predicting VO_{2max} during cycle ergometry in competitive cyclists and to compare its accuracy to the traditional ACSM equation. Based on a large database of competitive cyclists, we present our new equations (gender-specific and -non-specific), which proved to be more accurate for competitive cyclists that the traditional ACSM equation.

The new prediction models require the same parameters as the ACSM equation [6] (power output and body weight of the participant) but are improved based on a big database of cyclists included. Similarly to the ACSM, they are simple to use. Based on the results, we would suggest using gender-specific formulas, since the number of both female and male cyclists included in this research was high. The gender-non-specific formula was derived to present the differences if gender is not considered. In addition, the gender-non-specific formula can be used in cases where software programs do not enable usage of two (but only one) equations for predicting VO_{2max} before initiating the incremental test. In this case and as supported by this paper, it is better to use the gender-non-specific formula than the ACSM equation.

Average VO_{2max} 62.2 mL/min/kg confirmed that cyclists included in this study have superior physical fitness based on Cooper Institute for Aerobics Research to classify the degree of fitness compared to normally active individuals [16]. Based on their average maximal power output (5.90 W/kg in male and 5.19 in female cyclists), they are considered top-level competitive cyclists [17]. This supports the prediction that an equation that is based on normally active individuals (the ACSM) might not be appropriate for competitive cyclists.

The applicability of the ACSM equation was tested in specific populations in recent years. Commonly used equations were improved by newly derived equations on the normally active general population [9] and specific populations, such as coronary artery disease patients [8] and heart-failure patients [7]. The same was carried out in this paper as well. Our results suggest that the previously used ACSM equation might not be the most suitable in competitive cyclists. In our sample, the ACSM equation was less accurate than our new prediction model. The ACSM underestimated VO_{2max} in men by 11.54%, in women by 15.04% and in all participants by 11.99%. Our equation had a smaller relative bias. On average, it underestimated VO_{2max} in men by only 0.19% and in all participants by 1.04%. In female cyclists, it had no relative bias (0.00%). Since ACSM is based on a different population aged 20 years or more [3], the findings are not surprising. ACSM is also not derived from competitive cyclists.

The results of this study are in line with research by Koutilanos et al. [18] on male competitive athletes. They concluded that ACSM's equation is not capable of accurately predicting VO_{2max} in athletes aged 18–37 years. In contrast to Koutilanos et al., our study includes female athletes as well. Since female cyclists are much less studied than their male counterparts [19], we believe this is an important aspect of this research. The ACSM equation is certainly useful in adult persons that are normally active. However, based on these findings, its use might be limited in competitive cyclists. Comparing measured values of cyclists during cycle ergometry to more accurately predicted values can lead to better disease diagnosis, disease management, better training optimization, et cetera. If values are compared to the ACSM-predicted values, suboptimal VO_{2max} can easily be unrecognized.

In addition, female athletes do not exhibit the same VO_{2max} values as male athletes and using specific formulas is vital for better athlete management in a sport medicine setting.

The new equation needs only two predictors and is easy to use when performing the incremental test without indirect calorimetry. Malek et al. [20] cross-validated VO_{2max} prediction equations on samples of aerobically trained males and females. They concluded that the equation by Storer et al. was the most accurate. Like our model, this equation calculated VO_{2max} by using power output, body weight, but it also requires age as a predictor. Since it was designed based on 115 males and 116 females, aged 20–70, from the general population, it is not surprising that age had an important impact on the outcome [21]. In our sample, however, the sample was more homogenous (age 17.6 ± 3.5). The average age of our sample is also lower since most competitive cyclists included in this research were tested at the time when they just started competing not only nationally, but also at the international level. In addition, sample size in our research was much larger (580 participants compared to 231). Only accurate predicted VO_{2max} values are useful in the clinical setting when dealing with competitive athletes. Therefore, we recommend that medicine and sports practitioners adapt our proposed equation when working with competitive cyclists.

Limitations

Some limitations of this study should be noted. Cyclists in this study are Caucasian so suggested prediction models might not be attributed to other ethnic groups. More research is needed to confirm if any differences exist. When we calculated the gender-non-specific equation, the group included far more male than female cyclists. To determine a more accurate prediction model, the number of subjects should be more balanced. The reason for less women in our sample is that female competitive cyclists are less common than male cyclists. We assume that the gender-non-specific equation might be different if genders were equally represented in the multiple-regression analysis. Still, we believe the new prediction equation could be more accurate than the ACSM equation. Finally, average age in the male group is lower than in the female group. This is a result of the fact that more "younger" cyclists come to exercise physiology centres where measurements for this article were performed, and some "older" cyclists train and get tested abroad with their pro-cycling teams. The number of young male cyclists is also much higher than female cyclists in the country of measurement and this resulted in slightly different age groups. Nevertheless, since the participants in our paper compete nationally and internationally, we believe they are a good representation of competitive cyclists. Additional studies are needed to explore whether this equation could be used on other specific groups, such as off-road cyclists, mountain bikers or in triathletes.

5. Conclusions

We present two gender-specific equations and a non-gender-specific equation that are based on a large database of competitive cyclists. The current results demonstrate that new prediction equations are more accurate for predicting VO_{2max} in competitive cyclists than the traditional ACSM equation. The equations require only power output and body-weight values, which makes them easy to use. The new equations are more accurate, they have lower bias and have a very low standard error in the estimate. They can be used in a setting where indirect calorimetry to measure VO_{2max} is not possible and to evaluate measured values to predicted values for evaluation. This can serve in the clinical setting for disease detection, disease management and also in the monitoring of performance and training adaptations. Finally, based on the results of this study, the traditionally used ACSM is not suitable for competitive cyclists based on great underestimation, both in male and female cyclists.

Future studies are needed to verify if the proposed prediction model could also be suitable for other disciplines with similar physiological demands, such as triathlon or mountain-bike cycling. There is also a need to develop new prediction models for disciplines with different physiological characteristics. This might be especially useful for cardiovascular preparticipation screening examination in disciplines where cardiovascular risk is increased.

Author Contributions: Conceptualization, I.J. and M.C.; methodology, I.J. and M.C.; formal analysis, I.J.; investigation, I.J.; data curation, I.J.; writing—original draft preparation, I.J.; writing—review and editing, I.J., J.T. and M.C.; supervision J.T. and M.C. All authors have read and agreed to the published version of the manuscript.

Funding: This research received no external funding.

Institutional Review Board Statement: The study was conducted in accordance with the Declaration of Helsinki and approved by the Institutional Review Board (Komisija Republike Slovenije za medicinsko etiko, 0120-202/2020/5).

Informed Consent Statement: Informed consent was obtained from all subjects involved in the study.

Data Availability Statement: The datasets used and analysed during the current study are available from the corresponding author on reasonable request.

Acknowledgments: We thank all cyclists, coaches and their teams for participation. We also thank all participants involved in gathering the data over the years, especially S.R. and R.M.

Conflicts of Interest: The authors declare no conflict of interest.

Abbreviations

VO_{2max}	Maximal oxygen uptake
ACSM	American College of Sports Medicine
SEE	Standard error of the estimate
PO	Maximal power output
BW	Body weight
BH	Body height
SMM	Skeletal muscle mass

Appendix A

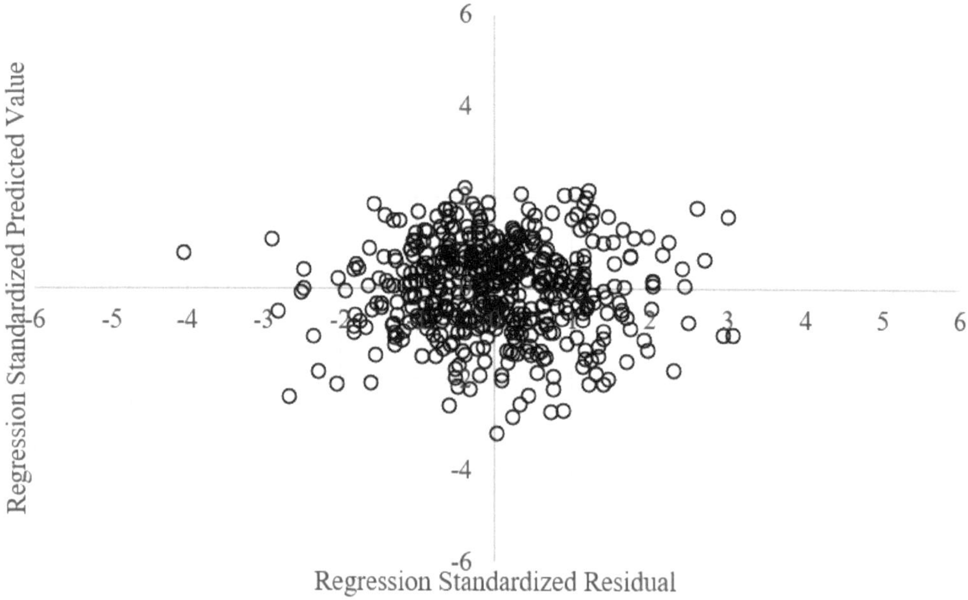

Figure A1. The plot of standardized residuals against standardized predicted values.

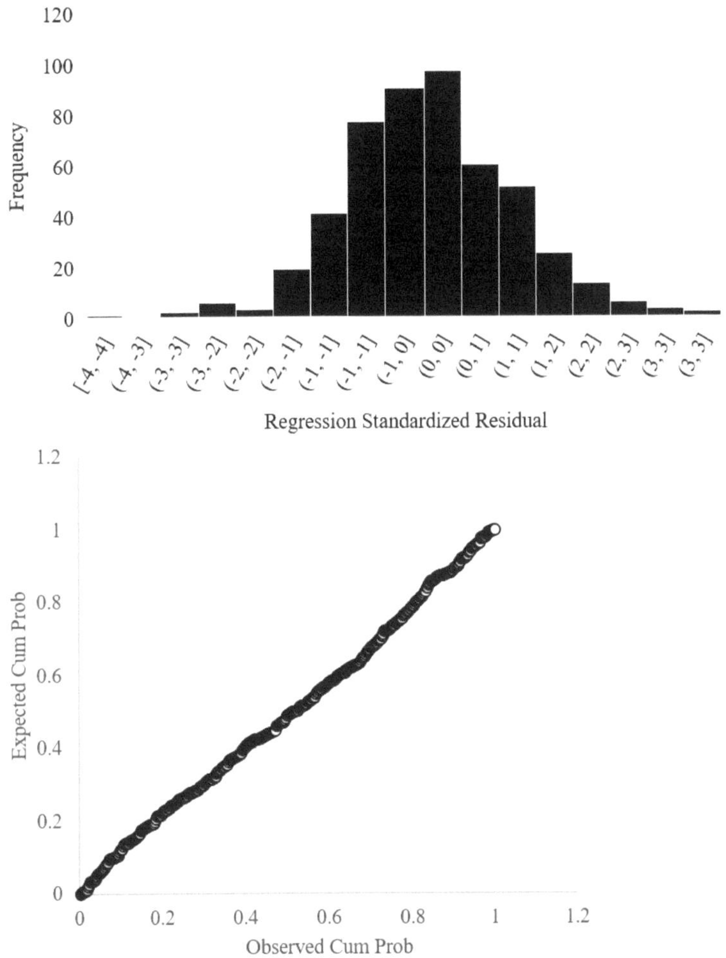

Figure A2. Normality of residuals presented with a histogram and normal probability plot of the residuals.

Appendix B

Table A1. Baseline characteristics of cyclists.

	Males (n = 496) Mean + SD		**Females (n = 84) Mean + SD**	
age (years)	**17.14**	2.72	20.18	5.59
height (cm)	178.67	6.73	166.04	5.57
body mass (kg)	66.72	7.58	57.45	6.46
fat mass (%)	9.29	3.87	17.34	4.67
fat free mass (%)	51.43	1.68	45.98	2.8
maximal power output (W)	393.74	56.48	296.05	45.05
maximal power output (W/kg)	5.90	0.56	5.19	0.79
maximal oxygen consumption (ml/min/kg)	63.43	5.49	54.82	7.02

Table A2. Parameters used in the stepwise multiple-regression analysis and their influence on the VO_{2max} in male cyclists.

	Unstandardized Coefficients	Standardized Coefficients			95% Confidence Interval for B	
	Std. Error	Beta	t	Sig.	Lower	Upper
constant	7.186		9.874	0.000	56.840	85.078
PO	0.005	1.023	20.087	0.000	0.090	0.109
BW	0.047	−0.698	−10.837	0.000	−0.598	−0.415
BH	0.048	−0.184	−3.151	0.002	−0.244	−0.057
SMM	11.628	0.084	2.324	0.021	4.178	49.872

(VO_{2max} = maximal oxygen uptake, PO = maximal power output, BW = body weight, BH = body height, SMM = skeletal muscle mass).

Table A3. ANOVA testing effect of different equations used—for all groups (male, female and gender-non-specific) there are statistically significant differences between the new equation and the ACSM aquation.

		Sum of Squares	df	Mean Square	F	Sig.
male cyclists	new equation	4657.287	190	24.512	3.339	0.000
	ACSM	12,573.567	190	66.177	3.479	0.000
female cyclists	new equation	3319.783	72	46.108	14.312	0.000
	ACSM	6022.488	72	83.646	23.317	0.000
gender non-specific	new equation	11,173.397	221	50.558	5.071	0.000
	ACSM	22,144.828	221	100.203	5.491	0.000

References

1. Palange, P.; Laveneziana, P.; Neder, J.A.; Ward, S.A. (Eds.) *Clinical Exercise Testing*; European Respiratory Society: Sheffield, UK, 2018.
2. Balady, G.J.; Arena, R.; Sietsema, K.; Myers, J.; Coke, L.; Fletcher, G.F.; Forman, D.; Franklin, B.; Guazzi, M.; Gulati, M.; et al. Clinician's Guide to Cardiopulmonary Exercise Testing in Adults: A Scientific Statement from the American Heart Association. *Circulation* **2010**, *122*, 191–225. [CrossRef] [PubMed]
3. American College of Sports Medicine. *ACSM's Guidelines for Exercise Testing and Prescription*, 8th ed.; Thompson, W.R., Gordon, N.F., Pescatello, L.S., Eds.; Wolters Kluwer Health: Philadelphia, USA, 2010; ISBN 978-0-7817-6903-7.
4. Faria, E.W.; Parker, D.L.; Faria, I.E. The Science of Cycling. *Sport. Med.* **2005**, *35*, 285–312. [CrossRef] [PubMed]
5. Sørensen, A.; Aune, T.K.; Rangul, V.; Dalen, T. The Validity of Functional Threshold Power and Maximal Oxygen Uptake for Cycling Performance in Moderately Trained Cyclists. *Sports* **2019**, *7*, 217. [CrossRef] [PubMed]
6. American College of Sports Medicine; Liguori, G.; Feito, Y.; Fountaine, C. *ACSM's Guidelines for Exercise Testing and Prescription*, 11th ed.; Charles, J., Roy, B., Eds.; Wolters Kluwer: Philadelphia, PA, USA, 2021; ISBN 978-1975150181.
7. Kokkinos, P.; Kaminsky, L.A.; Arena, R.; Zhang, J.; Franklin, B.; Kraus, W.; Triantafyllidi, H.; Benas, D.; Whellan, D.J.; Myers, J. New Equations for Predicting Maximum Oxygen Uptake in Patients with Heart Failure. *Am. J. Cardiol.* **2020**, *128*, 7–11. [CrossRef] [PubMed]
8. Jang, W.Y.; Kang, D.O.; Park, Y.; Lee, J.; Kim, W.-S.W.; Choi, J.Y.; Roh, S.-Y.; Jang, Y.; Park, S.-H.; Kim, W.-S.W.; et al. Validation of FRIEND and ACSM Equations for Cardiorespiratory Fitness: Comparison to Direct Measurement in CAD Patients. *J. Clin. Med.* **2020**, *9*, 1889. [CrossRef] [PubMed]
9. Myers, J.; Kaminsky, L.A.; Lima, R.; Christle, J.W.; Ashley, E.; Arena, R. A Reference Equation for Normal Standards for VO2 Max: Analysis from the Fitness Registry and the Importance of Exercise National Database (FRIEND Registry). *Prog. Cardiovasc. Dis.* **2017**, *60*, 21–29. [CrossRef] [PubMed]
10. Kokkinos, P.; Kaminsky, L.A.; Arena, R.; Zhang, J.; Myers, J. A New Generalized Cycle Ergometry Equation for Predicting Maximal Oxygen Uptake: The Fitness Registry and the Importance of Exercise National Database (FRIEND). *Eur. J. Prev. Cardiol.* **2018**, *25*, 1077–1082. [CrossRef] [PubMed]
11. Spragg, J.; Leo, P.; Swart, J. The Relationship between Training Characteristics and Durability in Professional Cyclists across a Competitive Season. *Eur. J. Sport Sci.* **2022**, *1*, 1–10. [CrossRef]
12. Duc, S.; Urianstad, T.; Rønnestad, B.R. Adding Vibration During Varied-Intensity Work Intervals Increases Time Spent Near Maximal Oxygen Uptake in Well-Trained Cyclists. *Int. J. Sports Physiol. Perform.* **2022**, *17*, 1565–1573. [CrossRef] [PubMed]
13. Hebisz, R.; Hebisz, P.; Danek, N.; Michalik, K.; Zatoń, M. Predicting Changes in Maximal Oxygen Uptake in Response to Polarized Training (Sprint Interval Training, High-Intensity Interval Training, and Endurance Training) in Mountain Bike Cyclists. *J. Strength Cond Res.* **2022**, *36*, 1726–1730. [CrossRef] [PubMed]
14. Lucia, A.; Hoyos, J.; Chicharro, J.L. Physiology of Professional Road Cycling. *Sport. Med.* **2001**, *31*, 325–337. [CrossRef] [PubMed]
15. Field, A.P. *Discovering Statistics Using IBM SPSS Statistics*, 5th ed.; SAGE Publications, Inc.: Thousand Oaks, CA, USA, 2018; p. 368.
16. Cooper Institute. *Principles of Health and Fitness for Fitness Professionals*; Cooper Institute: Dallas, TX, USA, 2017.

17. Faria, E.W.; Parker, D.L.; Faria, I.E. The Science of Cycling: Factors Affecting Performance—Part 2. *Sports* **2005**, *35*, 313–337. [CrossRef] [PubMed]
18. Koutlianos, N.; Dimitros, E.; Metaxas, T.; Deligiannis, A.S.; Kouidi, E. Indirect Estimation of VO_{2max} in Athletes by ACSM's Equation: Valid or Not? *Hippokratia* **2013**, *17*, 136. [PubMed]
19. Van Erp, T. The Development of Women's Professional Cycling. *J. Sci. Cycl.* **2019**, *8*, 1–2. [CrossRef]
20. Malek, M.H.; Berger, D.E.; Housh, T.J.; Coburn, J.W.; Beck, T.W. Validity of VO_{2max} Equations for Aerobically Trained Males and Females. *Med. Sci. Sports Exerc.* **2004**, *36*, 1427–1432. [CrossRef] [PubMed]
21. Storer, T.W.; Davis, J.A.; Caiozzo, V.J. Accurate Prediction of VO_2(Max) in Cycle Ergometry. *Med. Sci. Sports Exerc.* **1990**, *22*, 704–712. [CrossRef] [PubMed]

life

Article

The Energetic Costs of Uphill Locomotion in Trail Running: Physiological Consequences Due to Uphill Locomotion Pattern—A Feasibility Study

Paul Zimmermann [1,2,3,*], Nico Müller [2], Volker Schöffl [2,4,5,6,7], Benedikt Ehrlich [2], Othmar Moser [3,8] and Isabelle Schöffl [2,9]

1 Department of Cardiology, Klinikum Bamberg, 96049 Bamberg, Germany
2 Interdisciplinary Center of Sportsmedicine Bamberg, Klinikum Bamberg, 96049 Bamberg, Germany
3 Division of Exercise Physiology and Metabolism, Department of Sport Science, University of Bayreuth, 95447 Bayreuth, Germany
4 Department of Orthopedic and Trauma Surgery, Klinikum Bamberg, 96049 Bamberg, Germany
5 Department of Orthopedic and Trauma Surgery, Friedrich-Alexander University Erlangen-Nurnberg, 91054 Erlangen, Germany
6 School of Clinical and Applied Sciences, Leeds Beckett University, Leeds LS1 3HE, UK
7 Section of Wilderness Medicine, Department of Emergency Medicine, University of Colorado School of Medicine, Denver, CO 80045, USA
8 Interdisciplinary Metabolic Medicine Research Group, Division of Endocrinology and Diabetology, Medical University of Graz, 8036 Graz, Austria
9 Department of Pediatric Cardiology, Friedrich-Alexander-University Erlangen-Nurnberg, 91054 Erlangen, Germany
* Correspondence: paul.zimmermann@arcormail.de

Citation: Zimmermann, P.; Müller, N.; Schöffl, V.; Ehrlich, B.; Moser, O.; Schöffl, I. The Energetic Costs of Uphill Locomotion in Trail Running: Physiological Consequences Due to Uphill Locomotion Pattern—A Feasibility Study. *Life* **2022**, *12*, 2070. https://doi.org/10.3390/life12122070

Academic Editors: Helena Lenasi, Herbert Loellgen and Ines Drenjančević

Received: 23 October 2022
Accepted: 7 December 2022
Published: 9 December 2022

Abstract: The primary aim of our feasibility reporting was to define physiological differences in trail running (TR) athletes due to different uphill locomotion patterns, uphill running versus uphill walking. In this context, a feasibility analysis of TR athletes' cardiopulmonary exercise testing (CPET) data, which were obtained in summer 2020 at the accompanying sports medicine performance center, was performed. Fourteen TR athletes (n = 14, male = 10, female = 4, age: 36.8 ± 8.0 years) were evaluated for specific physiological demands by outdoor CPET during a short uphill TR performance. The obtained data of the participating TR athletes were compared for anthropometric data, CPET parameters, such as $\dot{V}Emaximum$, $\dot{V}O_2maximum$, maximal breath frequency (BF_{max}) and peak oxygen pulse as well as energetic demands, i.e., the energy cost of running (C_r). All participating TR athletes showed excellent performance data, whereby across both different uphill locomotion strategies, significant differences were solely revealed for $\dot{V}Emaximum$ ($p = 0.033$) and time to reach mountain peak ($p = 0.008$). These results provide new insights and might contribute to a comprehensive understanding of cardiorespiratory consequences to short uphill locomotion strategy in TR athletes and might strengthen further scientific research in this field.

Keywords: trail running; cardiopulmonary exercise testing; uphill running; uphill walking; energetic demands; short trail running performance

1. Introduction

Mountain endurance running, especially trail running (TR), has increased its popularity in the recent years [1]. The International Trail Running Association defined TR as a "pedestrian off road race in a natural environment (e.g., mountain) with minimal possible paved or asphalt road (<20% of the total duration race)" and TR profiles start with short distances (<42 km) and may be extended to ultralong distances (>100 km) [2]. The growing popularity of this sport has led to new scientific research fields in sports science referring to physiological consequences influencing an athlete's performance determinants and factors influencing an individual athlete's neuromuscular fatigue [1,3–6]. As TR represents a complex sport in terms of cardiorespiratory and biomechanical demands, individual athlete's race performance prediction is challenging [7,8]. Therefore, evaluating clinical and sports performance, incremental running tests have been developed to elucidate maximal cardiorespiratory capabilities [9–11].

As previously reported, an athlete's performance in level running depends on several influencing factors, such as the energy cost of running (Cr), the individual maximal oxygen uptake ($\dot{V}O_2maximum$), and the fraction of $\dot{V}O_2$peak that can be kept up during an athlete's race performance [12].

In the recent decades, the growing popularity of TR has attracted interest in specific cardiorespiratory performance characterization of these athletes, including maximum oxygen uptake values ($\dot{V}O_2maximum$) [13–15]. However, to date, few studies have evaluated the impact of a slope on specific maximal physiological values, such as $\dot{V}O_2maximum$, during incremental running tests [10]. The existing literature reveals contrasting results and shows the lack of consensus caused by different athlete's population and/or testing protocols [10]. Therefore, on the one hand, previous research revealed similar results for $\dot{V}O_2maximum$ in level and uphill running [1,11,16,17], while, on the other hand, contrary findings for the physiological demands during uphill sections in comparison to level conditions were elucidated previously in this scientific field [10,18–20]. Considering downhill running, previous research revealed that $\dot{V}O_2maximum$ can not be reached and that the $\dot{V}O_2maximum$ is estimated to be 16–18% lower than in level and uphill maximal incremental running tests [21]. Previous research in this specific scientific area by Schöffel et al., and Balducci et al., could not reveal significant changes in $\dot{V}O_2maximum$ with an increasing slope for TR athletes, but they did find a progressive increase in ventilation ($\dot{V}E$) [1,22].

In contrast, other studies revealed a correlation between real-world uphill TR performance and certain parameters, such as running economy, maximal strength, local endurance assessment by fatigue index (FI), and the athlete's characteristics, such as body fat percentage and athlete's age [2,13,23]. The classical physiological variables of endurance running, such as $\dot{V}O_2maximum$ and/or percentage of $\dot{V}O_2maximum$ at ventilatory threshold (VT), did not allow meaningful prediction of short TR performance [2].

Next to the cardiorespiratory demands, the bioenergetic demands and biomechanical work for level and sloped surface running have been studied before [23–27]. As previously reported, the Cr in general depends on the characteristics of the terrain, on the incline as well as on the biomechanics of uphill locomotion and is independent of speed [12,27–30]. Several influencing factors on Cr might lead to interindividual athlete performance variability, such as higher resting metabolism and leg architecture, such as muscle strength of the plantar flexor muscle and triceps surae muscle, as well as a combination of eccentric and concentric actions forming the stretch-shortening muscle function resulting in elastic energy storage and reuse [31]. In fact, factors influencing the Cr of running are rather well identified in uphill versus level running, whereby additional influencing cardiorespiratory, metabolic, and biomechanical factors in highly trained or elite runners have been reported in previous research [32,33]. In this context, metabolic adaptations within the muscle, such as increased mitochondria and oxidative enzymes, and more efficient individual athlete's mechanics requiring less energy are the main putative factors [32]. Considering that many factors are influencing Cr [31,32], it might be assumed that an individual runner's characteristics contribute to different adaptive strategies to uphill locomotion with variable Cr levels in level and uphill running [1].

TR performance is multifaceted, and comparative studies on this scientific topic are rare up to now. While certain variable circumstances influence an athlete's performance, the execution of comparable studies is hindered. In this context, variable determinants such as uphill locomotion strategy, the usage of poles to minimize energy expenditure, testing profiles and procedures as well as comparable homogeneous cohorts of athletes have to be taken into consideration when conducting a study and to better characterize short TR performance and the energy expenditure during uphill locomotion [2,22,34,35].

Hence, the aim of the presented feasibility study was to compare the cardiorespiratory variables in TR athletes in a short outdoor field-testing protocol focusing different uphill locomotion strategies, such as uphill running vs. uphill walking. It was hypothesized that there would be no differences in the TR CPET performance data between both analyzed uphill locomotion strategies. In detail, we evaluated the specific interindividual CPET performance data values due to the uphill locomotion strategy, such as $\dot{V}O_2maximum$, maximal ventilation ($\dot{V}Emaximum$), time to reach anaerobic VT, breath frequency (BF), peak oxygen pulse (peak $O_{2\,pulse}$), athlete's blood lactate level, and athlete's specific Cr data. Since the sport-specific cardiorespiratory demands in TR differ immensely from road running, the presented novel data allow a better understanding of the sport-specific physiological demands in short TR performance determined by the variable uphill locomotion strategy [22].

2. Materials and Methods

2.1. Study Design

This was a single-center feasibility study of outdoor field uphill CPET performance data from 14 participating TR athletes, which were obtained in 2020 during short outdoor field testing. After inclusion in the study, participating trail runners were assigned to ascending numbers and afterwards were allocated to the order in which the testing protocols (uphill running vs. uphill walking) were conducted in a cross-over randomized fashion with the software Research Randomizer 4.0 (Social Psychology Network®, Lancaster, PA, USA) (1:1) [36].

The obtained data of participating TR athletes (n = 14) were compared for the physiological consequences measured by CPET performance data and Cr parameters for short uphill running versus uphill walking locomotion.

2.2. Ethical Consideration

The study protocol (228_20 B) was approved by the local ethics committee of the University of Nurnberg-Erlangen. The research was conducted in conformity with the declaration of Helsinki and Good Clinical Practice [37]. Prior to any trial-related activities and data acquirement, our participating TR athletes gave their written informed consent and were informed about the study protocol and the following data measurements.

2.3. Participating Trail Running Athletes

Eligibility criteria included male or female TR athletes aged between 25 to 50 years with a body mass index (BMI) between 19 and 25 kg/m^2. The characteristics of the included participants displayed an age of 36.8 ± 8.0 years, with a total height of 179 ± 8.4 cm, a body mass of 70.4 ± 10.0 kg, and BMI of 21.8 ± 1.8 kg/m^2. Regular attendance to TR competitions with a minimum of 21 km distance was a precondition to take part in our study. Our participating athletes did not use poles during uphill locomotion testing.

During the study protocol, 2 participants of the initially recruited 16 athletes had to be withdrawn from the analysis due to technical problems during the outdoor field CPET data assessment. Hence, 14 data sets were included for analysis and are presented in the results section. Anthropometric data, training, and race performance parameters of participating athletes are displayed in Tables 1 and 2.

Table 1. Anthropometric data of trail running athletes.

	Male n = 10	Female n = 4	*p*-Value
Age (y)	36.2 ± 9.2	38.3 ± 4.0	n.s.
Height (cm)	183.3 + 5.4	168.3 ± 2.5	0.0002
Body mass (kg)	75.3 ± 7.1	58.3 ± 2.2	0.0006
BMI (kg/m^2)	22.5 ± 1.8	20.8 ± 1.3	n.s.

Data are presented as a mean with standard deviation; Abbreviations: n, number; y, years, cm, centimeter; kg, kilogram; m^2, square-meter, n.s., not significant.

Table 2. Training and Race Performance Parameter of participating Tral Rummers (n = 14).

Parameter	Meant ± SD
Favorite TR race distance	43.69 ± 26.56 km
Race participation per annum	5.38± 4.41
Current training period	Race: 7.69% Tapering 23.08 Recreation: 69.23
Denivelation running per training week	1200.00 ± 769.58 m
Training distance per week	60.41 ± 26.15 km
Competition in road level running	Yes: 69.24% No: 30.76
Best time in 10 km official race	44.6 ± 0.10 min
Best time in 1000 m denivelation official race	63.83 ± 3.00 min
Years of specific TR training	4.27 ± 3.99
Uphill locomotion strategy - Preferring uphill running - Preferring uphill Walking - Both combined	Running 57.14% Walking 35.71 Both combined: 7.15
Severe Injury break during TR career	Yes: 21.43% No: 78.57

Abbreviations: n, number of athletes; standard deviation; TR, trail running; % percentage.

By an individual questionnaire, each athlete was evaluated for the displayed training and race information in Table 2. In this context, our tapering period was defined by the following aspects: cutting back training up to 3–4 weeks prior to TR competition, reduced training volume (by roughly 30%), maintaining training frequency with upholding intensity (fewer repetitions, less miles), reduced high risk injury training sessions, and speed workouts. So, the tapering was estimated to display the right balance in our athletes between training volume, intensity, and frequency. These points were evaluated to discriminate between race and tapering period in our athletes. Referring to their best performance in an official 1000 m denivelation race, no slope, race length, or ground surfaces were defined or evaluated in our athletes. The purpose of the evaluation was to estimate roughly the individual athlete's performance and the homogeneity of the athletic cohort; no specific evaluation of their race performance took place. Our participating athletes, who all took part in several TR competitions, were asked for their preferred locomotion strategy due to their race experience depending on individual TR profiles, ground surfaces, and TR distances. In this context, the athlete's answers were based on their race assessment and personal experiences.

As a test site, the Wiesenttal mountain in the upper Franconian Switzerland was chosen. The topographical profile with a length of 375 m and a maximal incline of 29.3% (mean incline overall 22.3 ± 7%) represents an optimal testing area for short TR, and the mountain is part of an actual TR race course. Further information and the topographic profile of this TR race course are provided at https://www.outdooractive.com/de/route/trailrunning/fraenkische-schweiz/wiesenttal-trail-neideck-1000/105762273/#dm=1 (accessed on 23 October 2022). Individuals were excluded if they were enrolled in a different study, had a history of acute infection, and/or CPET testing was contraindicated. A medical investigator assessed inclusion and exclusion criteria before enrolment in the study. In this context, an infection-free interval of at least 4 weeks and no musculoskeletal injuries within the last 4 weeks were a precondition to be involved in the study.

2.4. Outdoor Uphill Field Measurements

Participants were instructed about all study-related procedures during the first visit. After assessing the anthropometric data and training and race performance data, the participating TR athletes were obligated to proceed with a warm-up for 10 min prior to uphill locomotion testing to be in comparable physiological readiness before the testing. A maximum of 5 min between the end of the warm-up and the beginning of testing was permitted. The outdoor field-testing data acquirement was scheduled during the weekend after having a small, not prescribed breakfast two hours prior to testing, and the athletes were asked to avoid intensive training units for two days prior to testing. Each athlete completed two trail runs at the test site according to the allocated testing protocol, one in an uphill running locomotion and the other in an uphill walking locomotion. During the outdoor uphill field CPET, participants received a chest belt for continuous heart rate (HR) monitoring via a Bluetooth smart HR sensor (chest belt Dual ANT+/Bluetooth smart; Kalenji, Decathlon®, S.A., Lille, France) for safety reasons and assessment of peak and post-exercise maximum HR levels (HR measured in beats per minute (bpm)). We acquired our CPET data during the outdoor sport-specific field testing by using the mobile field test spiroergometry (MetaMax 3B-R2, Cortex medical®, Leipzig, Germany). The calibration procedure was performed at the laboratory site prior to testing in the morning. It has to be stated that the air humidity and the temperature levels in the testing site varied from the laboratory conditions. Prior to outdoor testing, no further specific calibration was performed. The Metamax system and the mask were placed on the participating subject at the testing site with a maximum delay of 30 min from laboratory calibration to maintain the stability and sensitivity of the instrumentation.

During the testing days, the environmental conditions showed comparable conditions with similar outdoor temperature, dry TR track, and no rain showers. Immediately after reaching the summit of the test site, the individual time for the test track (recorded time in minutes) and one point lactate concentration to determine peak-exercise lactate level (measured in $mmol{\cdot}L^{-1}$) with a capillary blood analysis from the earlobe were obtained within the first minute post-exercise in each participating athlete (Lactate Scout 4 and EKF Diagnostics, EKF-Germany®, Cardiff, Wales, UK). In between the two uphill locomotion test tracks, one hour of recovery was granted for each participating athlete. The athletes were allowed to recover by resting in a sitting or lying position, using recovery techniques such as stretching, and refueling energy by hydration. For each TR athlete, individual CPET performance was analyzed by breath-by-breath analysis for this outdoor sport-specific field testing. As we were able to define a plateau in our TR athlete's cardiorespiratory response and taking additional secondary criteria for a maximal effort into consideration [38], we identified the following TR athletes' maximal cardiorespiratory responses during CPET: individual $\dot{V}O_2maximum$, individual $\dot{V}Emaximum$, time to reach anaerobic threshold, maximal breath frequency (BF_{max}), peak $O_{2\ pulse}$, athlete's peak-exercise lactate level, and specific Cr data (Cr locomotion $_{mean}$). The time of uphill locomotion was also acquired and analyzed. The ventilatory thresholds VT1 and VT2 during the outdoor field testing were determined as described before in our previous research on outdoor uphill testing in TR athletes [22]. According to di Prampero et al., 1986, and Vernillo et al., 2017, "the energy cost of running (C_r), is defined as the amount of energy spent to transport the subject's body a given distance" [12,15]. In our study, we calculated the Cr based on the formula described by Balducci et al. [1]:

$$\mathrm{Cr} = \frac{(\mathrm{VO_2\,peak} - 0.083)}{\mathrm{m} \times \mathrm{v}};\ \text{with v } \left(\mathrm{m/s^{-1}}\right),\ \text{m (kg), and the absolute term } 0.083 \left(\mathrm{mL/O_2/s^{-1}}\right).$$

Furthermore, the individual TR athlete's time to reach the respiratory exchange ratio equal to 1.01 (RER = 1.01) was measured, pointing out the maximal lactate steady state (MLSS) during the race performance ($Time_{RER1.01}$ (min)) [39,40].

2.5. Statistical Analysis

Data were analyzed with SPSS software version 21.0 (IBM, SPSS® software, Ehningen, Germany). Firstly, all acquired data were assessed for normal distribution by analyzing the data by means of Shapiro–Wilk testing and, secondly, the homogeneity of variances was asserted by Levene's testing, which showed that equal variances could be assumed.

Afterwards, a *t*-test for paired samples was used as the statistical test for hypothesis testing and to compare the means of the two samples. Therefore, the means of the two samples were compared to determine whether the two samples were different from one another. In calculating the *t*-test, the following three fundamental data points were essential: values including the difference between mean values from each data set, the number of data values, and the standard deviation of each group. Results are presented as mean $\pm$ standard deviation. $p \leq 0.05$ was accepted as statistically significant. Afterwards, a gender-specific analysis for the interesting parameters was utilized equally.

3. Results

As described in the methods section, all athletes performed the TR testing under similar environmental conditions. During the testing days, which were conducted over the summer months, the outdoor air temperature ranged from 12 to 25 °C and the air humidity was 50 $\pm$ 10%.

The cardiorespiratory and metabolic performance parameters of the outdoor uphill performance testing in our TR athletes are presented in Table 3.

Table 3. Cardiorespiratory and metabolic performance parameters of uphill TR exercise testing.

Parameter	Uphill Running (n = 14) Male	Uphill Running (n = 14) Female	Uphill Walking (n = 14) Male	Uphill Walking (n = 14) Female	*p*-Value Male	*p*-Value Female	Overall *p*-Value
$\dot{V}O_2$*maximum* ($mL{\cdot}kg^{-1}{\cdot}min^{-1}$)	57.61 ± 37.0	49.3 ± 3.4	56.5 ± 6.6	49.0 ± 2.2	ns	ns	**0.362**
	55.2 ± 7.2		**54.4 ± 6.6**				
Peak $O_{2\ pulse}$ (mL/bpm)	26.3 ± 2.4	17.5 ± 1.7	24.8 ± 2.0	17.3 ± 1.0	ns	ns	**0.154**
	23.7 ± 4.7		**22.6 ± 4.0**				
Breath frequency $_{peak}$ (Hz)	55.5 ± 7.7	54.5 ± 8.6	54.9 ± 7.4	49.0 ± 3.5	ns	ns	**0.191**
	55.2 ± 7.6		**53.2 ± 7.0**				
Time $_{RER1.01}$ (min)	2.0 ± 1.6	3.2 ± 2.8	2.5 ± 2.3	3.3 ± 3.2	ns	ns	**0.212**
	1.83 + 1.44		**2.73 ± 2.47**				
VE maximum (L·min)	166.4 ± 23.0	105.9 ± 17.0	158.0 ± 23.7	97.9 ± 7.8	0.096	0.2330	**0.033 ***
	149.2 + 35.2		**140.8 ± 34.6**				
Peak heart rate (bpm)	176.2 ± 12.4	171.8 ± 3.0	177.5 ± 11.7	172.3 ± 2.6	ns	ns	**0.297**
	175 ± 11		**176 ± 10**				
Lacdate $_{peak\ exercise}$ ($mmol{\cdot}L^{-1}$)	9.4 ± 3.8	9.7 ± 3.2	9.6 ± 3.3	8.0 ± 2.1	ns	ns	**0.752**
	8.7 ± 4.1		**9.1 ± 4.3**				
Cr locomotion $_{mean}$ ($J{\cdot}kg^{-1}{\cdot}m^{-1}$)	7.1 ± 2.0	6.5 ± 2.1	7.2 ± 2.3	6.6 ± 2.2	ns	ns	**0.581**
	6.87 ± 2.25		**7.02 ± 2.51**				
Time $_{uphill}$ (min)	4.7 ± 1.1	6.1 ± 0.4	5.1 ± 0.6	6.4 ± 0.5	0.0217	0.312	**0.009 ***
	5.13 ± 0.86		**5.42 ± 0.86**				

Data are presented as mean with standard deviation. *p* value *, statistically significant ($p < 0.05$). Abbreviations: SD, standard deviation; Cr locomotion, cost of locomotion; ns, not significant.

No significant differences could be revealed in the CPET data analyses for the $\dot{V}O_2$*maximum* ($mL{\cdot}kg^{-1}{\cdot}min^{-1}$) between both analyzed uphill locomotion strategies in the TR athletes (p = 0.362, data presented in Table 3). Furthermore, no significant differences were elucidated for the peak $O_{2\ pulse}$ (mL/bpm) in the TR athletes (p = 0.154, presented in Table 3) nor for the maximum breath frequency during exercise (p = 0.191, results presented in Table 3). Additionally, $Time_{RER1.01}$ (min) did not significantly differ in between the two analyzed TR motion performances ($p = 0.212$, data presented in Table 3). The only significant difference for CPET variables between the two locomotion strategies was in $\dot{V}Emaximum$ (L·min^{-1}) ($p = 0.033$, results presented in Table 3 and Figure 1A). By analyzing the individual TR athletes' time to reach the peak finishing line in the uphill outdoor test track (recorded time in minutes), significant differences could be proven, whereby the TR athletes were significantly faster performing the running test motion (Time $_{uphill}$, $p = 0.009$, data presented in Table 3 and Figure 1B).

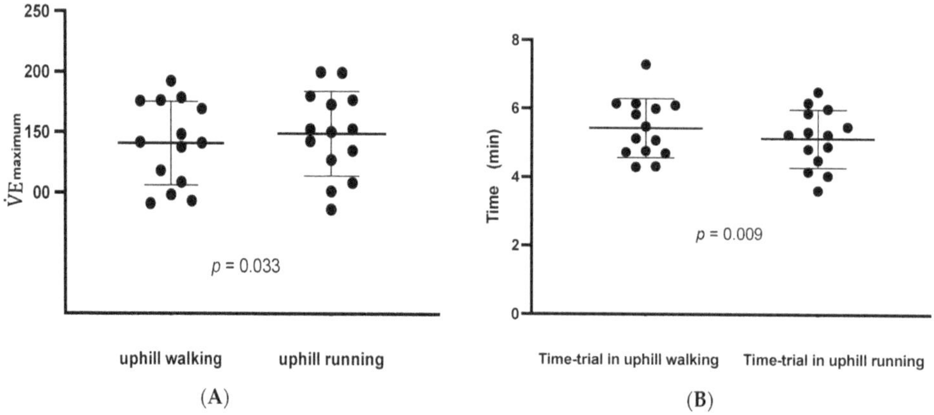

Figure 1. Panel (**A**) Differences in $\dot{V}Emaximum$ in uphill walking versus uphill running. Panel (**B**) Time trials in uphill walking vs. running.

The gender-specific subgroup analysis, as stated in the limitation section, might point out interesting gender-specific insights in the presented CPET parameters. In this context, no significant (ns) differences between both uphill locomotion strategies were revealed, except the significantly faster uphill running time in male TR athletes (results shown in Table 3).

4. Discussion

This feasibility outdoor uphill study was undertaken to determine physiological differences in TR athletes regarding different uphill locomotion strategies, uphill running versus uphill walking. The aim of the presented work was to provide new insights into sport-specific cardiorespiratory demands due to uphill locomotion strategy.

Previous research revealed correlations between performance in endurance running and anthropometric characteristics, such as body fat percentage, body mass, height, and BMI [41–43]. The most able runners were shorter and lighter than the other competing athletes [41]. Next to these conditions, evaluating the performance of TR athletes seems to include additional multifaceted aspects, especially environmental conditions, such as topographic uphill running profiles and sport-specific demands on the muscle composition, especially the lower limbs [24].

To date, research has mainly focused on CPET parameters and lactate thresholds concepts to predict an athlete's performance in road running [44,45], but little is known about performance prediction and LT in TR athletes [46]. As key physiological parameters characterizing an athlete's running performance, $\dot{V}O_2$peak, $\dot{V}O_2$ at lactate threshold, and running economy are known [22,47–50], whereby the determining physiological factors in uphill running are mainly metabolic, biochemical—such as factors of energy cost—and cardiovascular [33]. Taking these multifaceted aspects into consideration, we were not able to provide significant different cardiorespiratory performance parameters in our TR athletes due to uphill locomotion strategy, except $\dot{V}Emaximum$. These results emphasize the importance of having a good running economy to provide good race performance by the ability to maintain a high intensity level for as long as possible in addition to having a high $\dot{V}O_{2max}$ [51]. In our previous research, we could elucidate comparable results in CPET parameters for TR athletes compared to road running athletes with regards to $\dot{V}O_2$peak, $\dot{V}O_2$ at LT, and peak $O_{2\,pulse}$ [22]. Considering these parameters, the question arises whether there are various multifaceted aspects for uphill TR which might influence an athlete's performance to be more effective and less energy demanding next to the physiological demands.

In our TR athletes, who seemed to be on a comparable fitness level and who are famous for their enhanced aerobic and anaerobic capacity in uphill locomotion [22,52], we solely could elucidate significant differences due to cardiorespiratory performance parameters for $\dot{V}Emaximum$. Our findings might be influenced by various notable performance parameters: firstly, previous research showed that $\dot{V}O_2$peak is not estimated to be a systematically reliable predictor of running performance and generally a low variability of $\dot{V}O_2$peak in these trained TR athletes is observed [2,49]. However, next to $\dot{V}O_2$peak, the term "velocity at $\dot{V}O_2maximum$" was introduced in 1984 to combine $\dot{V}O_2maximum$ and economy to identify aerobic differences between runners [53]. Secondly, the likely uphill locomotion velocity in our participating athletes on the short TR course might result in our comparable CPET measurements. Ortiz et al., revealed previously that oxygen consumption and metabolic running power (indexed W/kg) increased linearly with velocity in vertical kilometer (VK) race athletes, whereby at speeds slower than 0.7 $m{\cdot}s^{-1}$, walking required less metabolic power than running and, at speeds of 0.8 $m{\cdot}s^{-1}$, there were no metabolic cost differences, suggesting that running likely costs less energy than walking in a laboratory setting [54]. Taken together, slower athletes in VK races should walk uphill and faster racers should run to minimize their specific metabolic locomotion power needs and to optimize energetic savings [54]. These findings are supported by Giovanelli et al., who revealed a range of optimal inclines (steeper than 15.8°) during uphill walking and running to reduce energy expenditure [55]. Additionally, the vertical ascent rate might be maximized up to slopes between 15° and 25° by using poles in uphill locomotion to delay fatigue effects [35].

No significant differences could be proven for the parameter lactate peak exercise due to uphill locomotion strategy. Blood lactate levels are known to be influenced by age, sex, training status, and the athlete's overall effort [9], and previous research revealed higher blood lactate concentrations in uphill running than in level running with a certain response in metabolic variables to increasing slopes [15,22,56]. In this context, experienced runners on uphill grades are able to provide a certain running economy reflecting both intrinsic physiological demands and skill [14]. Additionally, Lemire et al., could reveal that experienced endurance athletes showed a variable cardiorespiratory response to uphill and downhill running with regard to maximal oxygen uptake, heart rate, and ventilation response, and TR athletes did not reach $\dot{V}O_2maximum$ during maximal incremental downhill testing [11]. Therefore, multifaceted variables, such as different interindividual athlete's running economy or higher recruitment of muscle mass during uphill running, might result in an individual athlete's lactate levels and RER response [9,14,22,57]. An individual selective recruitment of type II glycolytic and type I oxidative muscle fiber activation during uphill locomotion and a variable glycogen depletion during uphill running, especially in the gastrocnemius, soleus, and vastus lateralis muscle—revealed in previous research in humans and rats—entail a certain individual variability in lactate levels and might finally result in comparable results in our athletes [13,58,59]. Referring to these variables, metabolic sport-specific physiological adaptions might result in probably compensatory higher $\dot{V}Emaximum$ during uphill running locomotion in our TR athletes to compensate higher CO_2 production in the involved muscles and to buffer the accumulating acid [22].

Another aspect to be focused is that previous research revealed an association between reductions in thoraco-abdominal coordination during uphill running and reduced breathing efficiency with a less efficient ventilatory pattern [60]. Both conditions are known for determining ventilatory efficiency and represent a key tool for an athlete's performance evaluation [60]. The reduction in thoraco-abdominal coordination during increased slope running displayed by greater forward inclination of the trunk and displacement of the center of mass might influence the obtained changes in ventilatory patterns in our uphill running athletes. The decreased efficient ventilatory pattern represented by an increased respiratory rate for the same amount of ventilation, such as higher $\dot{V}Emaximum$, during uphill running locomotion might be the cardiophysiological consequence. Similar findings regarding a higher VE/VT ratio in previous research, which ratio was not analyzed in our study, indicate the same cardiophysiological adaption [60].

There were no significant differences due to uphill locomotion strategy for athletes' peak heart rate and peak $O_{2\,pulse}$. The previously studied muscle mechanoreceptor responding to muscle stretch with its inhibition of cardiac vagal activity, and subsequent increased heart rate response and enhanced cardiac output, might contribute to our athlete's comparable exhausting CPET data [61]. Nevertheless, analyzing our obtained results, we were not able to elucidate significant differences in cardiac output due to uphill locomotion strategy in our TR athletes. Nonetheless, trained athletes in general are predisposed for functional and structural cardiac remodeling and enhanced cardiac output during exercise, as described before [62]. Our observed comparable dynamic data might be influenced by the following facts: firstly, the dynamic cardiac output data remain unchanged due to the locomotion strategy because of short TR race duration; secondly, the CPET data display equally exhausting performance parameters because TR athletes provide excellent oxygen extraction during uphill locomotion [22].

In our study, we could not prove significant differences in Cr locomotion $_{mean}$ between different uphill locomotion in TR athletes. Previous research in this scientific area indicated that uphill running in well-trained athletes—with greater maximal power of the lower limbs—showed small changes in running mechanics and subsequently lower fatigue-induced alterations in Cr [2]. Minetti et al., revealed that for both walking and running strategy at a given steep incline—up to 24.2%—metabolic power linearly increases with treadmill velocity [26]. In our TR athletes, the Cr locomotion $_{mean}$ did not significantly differ in between the two uphill locomotion patterns, whereby uphill running locomotion is characterized by a higher step frequency, shorter swing/aerial phase duration, increased mechanical work due to the off-road ground, and a progressive adoption of a mid-to-forefoot strike pattern [2,15]. These effects on muscle contraction pattern and on biomechanics, as well as increased working demands during uphill running locomotion, particularly the hip, might contribute to the varying physiological response as we obtained a significantly higher peak $\dot{V}E$ during uphill running in comparison to uphill walking strategy [2,15]. The Cr locomotion $_{mean}$ did not differ in between our athletes' performance, whereby this fact is confirmed by previous research that showed that greater energetic demands in uphill locomotion were not compensated by lower metabolic demands in downhill running [63].

In summary, different uphill locomotion strategies in TR seem not to be associated with variable physiological CPET parameters—except $\dot{V}Emaximum$. Our findings are supported by previous research, whereby $\dot{V}O_2$ at a specified velocity and rate of $\dot{V}O_2$peak did not allow to predict an athlete's vertical race performance based on classical physiological parameters [55]. Other predictive variables seem to be more appropriate to better describe TR athletes' running performance, such as the specificity of the running course profile and individual mechanics of running [2,64–67]. In the end, it has to be stated that predicting TR race performance due to different uphill locomotion modalities is difficult, as the short TR race performance is related to multifaceted variables. In this context, prolonged concentric and eccentric muscle actions during uphill and downhill sections, the individual athlete's physiological and biomechanical determinants as well as the topographic trail's characteristics play an important role [2,15,68].

Our feasibility study is not without limitations. First of all, the sample size of analyzed TR athletes is relatively small, which is due to the fact that a selective local recruitment was performed. This resulted in a heterogeneous distribution of male and female TR athletes with a certain mixture of young and experienced athletes, entailing an interindividual variability in relation to lifetime training hours and training schedule variability, race experience, and age. These characteristics might contribute to a certain standard deviation in our data assessment. Due to the relatively small number of enrolled TR athletes, we are not able to draw statistically reliable conclusions, but we might point out interesting trends in TR athletes' physiological responses due to uphill locomotion in our feasibility trial. Although the number of participating athletes is small, we did perform sex-specific data analyses to provide novel insights in this scientific field, but we are aware that we would not be able derive statistically reliable conclusions. Secondly, our data assessment was performed in variable athletes' training periods and under "real world" weather conditions. These conditions might contribute to a certain data variability. Due to the small number of TR athletes, we did not run a control arm for this feasibility study. Additionally, as a limitation, it might be stated that one hour resting and recreational time in between two uphill races might be not enough considering the physiological consequences of a maximal intensity eccentric muscle action exercise. This fact has to be taken into consideration for the design of further studies. Nevertheless, the uniqueness of our feasibility study described characteristic physiological responses in short uphill TR due to locomotion pattern. Furthermore, the topographical profile of the Wiesenttal mountain with its short length and single-incline distance solely represents a testing area for short TR, but does not allow conclusions for long-distance TR. Although our new insights might be regarded as an interesting descriptive feasibility preliminary report, they might open the door for investigating this locomotion strategy deeply with more TR athletes including additive locomotion characteristics, such as step frequency and breathing pattern.

5. Conclusions

In conclusion, the findings of our feasibility study on sport-specific physiological energetic demands in TR athletes provide new evidence that significant differences in short outdoor uphill CPET assessment can be identified for different uphill locomotion strategies. For short TR performance, significantly higher $\dot{V}Emaximum$ ($L{\cdot}min^{-1}$) and faster race performance time during the uphill running strategy could be elucidated. However, this preliminary reporting study indicates that short TR physiological race performance cannot successfully be explained by the classical physiological model of endurance running. Due to multifaceted TR performance variables, our findings provide new insights and might contribute to a comprehensive understanding for individual TR athletes' race performance and uphill locomotion strategy. Further scientific research might be warranted in order to identify the physiological predictors of short TR performance with a homogeneous group of trained TR athletes and to strengthen the scientific evidence of our reported novel insights in outdoor real-world TR performance assessment.

Author Contributions: Study design, I.S. and V.S.; Data curation, N.M., I.S. and B.E.; formal analysis, I.S. and N.M.; investigation, P.Z., I.S., B.E. and N.M.; methodology, N.M. and I.S.; project administration, I.S.; supervision, I.S.; writing—original draft, P.Z.; writing—review and editing, V.S., I.S., N.M., B.E. and O.M. All authors have read and agreed to the published version of the manuscript.

Funding: This research received no external funding.

Institutional Review Board Statement: The study was conducted according to the guidelines of the Declaration of Helsinki and approved by the Ethics Committee of the University of Nurnberg-Erlangen (study protocol 228_20 B).

Informed Consent Statement: Written informed consent has been obtained from the patient(s) to publish this paper.

Data Availability Statement: Individual anonymized data supporting the analyses of this study contained in this manuscript will be made available upon reasonable written request from researchers whose proposed use of data for a specific purpose has been approved.

Acknowledgments: We want to thank the participants, who were examined by Interdisciplinary Center of Sportsmedicine, Klinikum Bamberg and Saludis, Klinikum Bamberg.

Conflicts of Interest: The authors declare no conflict of interest.

References

1. Balducci, P.; Clémençon, M.; Morel, B.; Quiniou, G.; Saboul, D.; Hautier, C.A. Comparison of Level and Graded Treadmill Tests to Evaluate Endurance Mountain Runners. *J. Sports Sci. Med.* **2016**, *15*, 239–246. [PubMed]
2. Ehrström, S.; Tartaruga, M.P.; Easthope, C.S.; Brisswalter, J.; Morin, J.-B.; Vercruyssen, F. Short Trail Running Race: Beyond the Classic Model for Endurance Running Performance. *Med. Sci. Sports Exerc.* **2018**, *50*, 580–588. [CrossRef] [PubMed]
3. Saugy, J.; Place, N.; Millet, G.Y.; Degache, F.; Schena, F.; Millet, G.P. Alterations of Neuromuscular Function after the World's Most Challenging Mountain Ultra-Marathon. *PLoS ONE* **2013**, *8*, e65596. [CrossRef] [PubMed]
4. Millet, G.P.; Millet, G.Y. Ultramarathon Is an Outstanding Model for the Study of Adaptive Responses to Extreme Load and Stress. *BMC Med.* **2012**, *10*, 77. [CrossRef] [PubMed]
5. Millet, G.Y.; Tomazin, K.; Verges, S.; Vincent, C.; Bonnefoy, R.; Boisson, R.-C.; Gergelé, L.; Féasson, L.; Martin, V. Neuromuscular Consequences of an Extreme Mountain Ultra-Marathon. *PLoS ONE* **2011**, *6*, e17059. [CrossRef]
6. Hoffman, M.D.; Ingwerson, J.L.; Rogers, I.R.; Hew-Butler, T.; Stuempfle, K.J. Increasing Creatine Kinase Concentrations at the 161-Km Western States Endurance Run. *Wilderness Environ. Med.* **2012**, *23*, 56–60. [CrossRef]
7. Scheer, V.; Janssen, T.I.; Vieluf, S.; Heitkamp, H.-C. Predicting Trail Running Performance With Laboratory Exercise Tests and Field Based Results. *Int. J. Sports Physiol. Perform.* **2018**, *14*, 30–133. [CrossRef]
8. Lemire, M.; Remetter, R.; Hureau, T.J.; Kouassi, B.Y.L.; Lonsdorfer, E.; Geny, B.; Isner-Horobeti, M.-E.; Favret, F.; Dufour, S.P. High-Intensity Downhill Running Exacerbates Heart Rate and Muscular Fatigue in Trail Runners. *J. Sports Sci.* **2021**, *39*, 815–825. [CrossRef]
9. Beltz, N.M.; Gibson, A.L.; Janot, J.M.; Kravitz, L.; Mermier, C.M.; Dalleck, L.C. Graded Exercise Testing Protocols for the Determination of VO_2 Max: Historical Perspectives, Progress, and Future Considerations. *J. Sport. Med.* **2016**, *2016*, 1–12. [CrossRef]
10. Cassirame, J.; Godin, A.; Chamoux, M.; Doucende, G.; Mourot, L. Physiological Implication of Slope Gradient during Incremental Running Test. *Int. J. Environ. Res. Public Health* **2022**, *19*, 12210. [CrossRef]
11. Lemire, M.; Hureau, T.J.; Remetter, R.; Geny, B.; Kouassi, B.Y.L.; Lonsdorfer, E.; Isner-Horobeti, M.-E.; Favret, F.; Dufour, S.P. Trail Runners Cannot Reach V˙O2max during a Maximal Incremental Downhill Test. *Med. Sci. Sports Exerc.* **2020**, *52*, 1135–1143. [CrossRef]
12. di Prampero, P.E.; Atchou, G.; Brückner, J.C.; Moia, C. The Energetics of Endurance Running. *Eur. J. Appl. Physiol. Occup. Physiol.* **1986**, *55*, 259–266. [CrossRef]
13. Lemire, M.; Hureau, T.J.; Favret, F.; Geny, B.; Kouassi, B.Y.L.; Boukhari, M.; Lonsdorfer, E.; Remetter, R.; Dufour, S.P. Physiological Factors Determining Downhill vs. Uphill Running Endurance Performance. *J. Sci. Med. Sport* **2021**, *24*, 85–91. [CrossRef]
14. Breiner, T.J.; Ortiz, A.L.R.; Kram, R. Level, Uphill and Downhill Running Economy Values Are Strongly Inter-Correlated. *Eur. J. Appl. Physiol.* **2019**, *119*, 257–264. [CrossRef]
15. Vernillo, G.; Giandolini, M.; Edwards, W.B.; Morin, J.-B.; Samozino, P.; Horvais, N.; Millet, G.Y. Biomechanics and Physiology of Uphill and Downhill Running. *Sports Med.* **2017**, *47*, 615–629. [CrossRef]
16. Kasch, F.W.; Wallace, J.P.; Huhn, R.R.; Krogh, L.A.; Hurl, P.M. VO_2max during Horizontal and Inclined Treadmill Running. *J. Appl. Physiol.* **1976**, *40*, 982–983. [CrossRef]
17. Davis, J.A.; Whipp, B.J.; Lamarra, N.; Huntsman, D.J.; Frank, M.H.; Wasserman, K. Effect of Ramp Slope on Determination of Aerobic Parameters from the Ramp Exercise Test. *Med. Sci. Sports Exerc.* **1982**, *14*, 339–343. [CrossRef]
18. Paavolainen, L.; Nummela, A.; Rusko, H. Muscle Power Factors and VO2max as Determinants of Horizontal and Uphill Running Performance. *Scand. J. Med. Sci. Sports* **2000**, *10*, 286–291. [CrossRef] [PubMed]
19. Scheer, V.; Ramme, K.; Reinsberger, C.; Heitkamp, H.-C. VO2max Testing in Trail Runners: Is There a Specific Exercise Test Protocol? *Int. J. Sports Med.* **2018**, *39*, 456–461. [CrossRef] [PubMed]
20. Pringle, J.S.M.; Carter, H.; Doust, J.H.; Jones, A.M. Oxygen Uptake Kinetics during Horizontal and Uphill Treadmill Running in Humans. *Eur. J. Appl. Physiol.* **2002**, *88*, 163–169. [CrossRef] [PubMed]
21. Taylor, H.L.; Buskirk, E.; Henschel, A. Maximal Oxygen Intake as an Objective Measure of Cardio-Respiratory Performance. *J. Appl. Physiol.* **1955**, *8*, 73–80. [CrossRef]
22. Schöffl, I.; Jasinski, D.; Ehrlich, B.; Dittrich, S.; Schöffl, V. Outdoor Uphill Exercise Testing for Trail Runners, a More Suitable Method? *J. Hum. Kinet.* **2021**, *79*, 123–133. [CrossRef]
23. Snyder, K.L.; Farley, C.T. Energetically Optimal Stride Frequency in Running: The Effects of Incline and Decline. *J. Exp. Biol.* **2011**, *214*, 2089–2095. [CrossRef]
24. Padulo, J.; Powell, D.; Milia, R.; Ardigò, L.P. A Paradigm of Uphill Running. *PLoS ONE* **2013**, *8*, e69006. [CrossRef]
25. Padulo, J.; Annino, G.; Migliaccio, G.M.; D'ottavio, S.; Tihanyi, J. Kinematics of Running at Different Slopes and Speeds. *J. Strength Cond. Res.* **2012**, *26*, 1331–1339. [CrossRef]

26. Minetti, A.E.; Moia, C.; Roi, G.S.; Susta, D.; Ferretti, G. Energy Cost of Walking and Running at Extreme Uphill and Downhill Slopes. *J. Appl. Physiol.* **2002**, *93*, 1039–1046. [CrossRef]
27. Minetti, A.E.; Ardigò, L.P.; Saibene, F. Mechanical Determinants of the Minimum Energy Cost of Gradient Running in Humans. *J. Exp. Biol.* **1994**, *195*, 211–225. [CrossRef]
28. Hoogkamer, W.; Taboga, P.; Kram, R. Applying the Cost of Generating Force Hypothesis to Uphill Running. *PeerJ* **2014**, *2*, e482. [CrossRef]
29. Lussiana, T.; Fabre, N.; Hébert-Losier, K.; Mourot, L. Effect of Slope and Footwear on Running Economy and Kinematics. *Scand. J. Med. Sci. Sports* **2013**, *23*, e246–e253. [CrossRef]
30. Margaria, R.; Cerretelli, P.; Aghemo, P.; Sassi, G. Energy Cost of Running. *J. Appl. Physiol.* **1963**, *18*, 367–370. [CrossRef]
31. Lacour, J.-R.; Bourdin, M. Factors Affecting the Energy Cost of Level Running at Submaximal Speed. *Eur. J. Appl. Physiol.* **2015**, *115*, 651–673. [CrossRef] [PubMed]
32. Saunders, P.U.; Pyne, D.B.; Telford, R.D.; Hawley, J.A. Factors Affecting Running Economy in Trained Distance Runners. *Sports Med.* **2004**, *34*, 465–485. [CrossRef] [PubMed]
33. Lemire, M.; Remetter, R.; Hureau, T.J.; Geny, B.; Lonsdorfer, E.; Favret, F.; Dufour, S.P. Energy Cost of Running in Well-Trained Athletes: Toward Slope-Dependent Factors. *Int. J. Sports Physiol. Perform.* **2022**, *17*, 423–431. [CrossRef] [PubMed]
34. de Waal, S.J.; Gomez-Ezeiza, J.; Venter, R.E.; Lamberts, R.P. Physiological Indicators of Trail Running Performance: A Systematic Review. *Int. J. Sports Physiol. Perform.* **2021**, *16*, 325–332. [CrossRef] [PubMed]
35. Giovanelli, N.; Sulli, M.; Kram, R.; Lazzer, S. Do Poles Save Energy during Steep Uphill Walking? *Eur. J. Appl. Physiol.* **2019**, *119*, 1557–1563. [CrossRef] [PubMed]
36. Urbaniak, G.C.; Plous, S. Research Randomizer (Version 4.0) [Computer Software]. Available online: https://Randomizer.Org/about/ (accessed on 27 September 2021).
37. Harriss, D.J.; MacSween, A.; Atkinson, G. Ethical Standards in Sport and Exercise Science Research: 2020 Update. *Int. J. Sports Med.* **2019**, *40*, 813–817. [CrossRef]
38. Schaun, G.Z. The maximal Oxygen uptake verification phase: A light at the end of the tunnel? *Sports Med. Open* **2017**, *3*, 44. [CrossRef]
39. Shulman, R.G.; Rothman, D.L. The "Glycogen Shunt" in Exercising Muscle: A Role for Glycogen in Muscle Energetics and Fatigue. *Proc. Natl. Acad. Sci. USA* **2001**, *98*, 457–461. [CrossRef]
40. Leti, T.; Mendelson, M.; Laplaud, D.; Flore, P. Prediction of Maximal Lactate Steady State in Runners with an Incremental Test on the Field. *J. Sports Sci.* **2012**, *30*, 609–616. [CrossRef]
41. Bale, P.; Bradbury, D.; Colley, E. Anthropometric and Training Variables Related to 10km Running Performance. *Br. J. Sports Med.* **1986**, *20*, 170–173. [CrossRef]
42. Knechtle, B.; Barandun, U.; Knechtle, P.; Zingg, M.A.; Rosemann, T.; Rüst, C.A. Prediction of Half-Marathon Race Time in Recreational Female and Male Runners. *Springerplus* **2014**, *3*, 248. [CrossRef]
43. Alvero-Cruz, J.; Carnero, E.; García, M.; Alacid, F.; Correas-Gómez, L.; Rosemann, T.; Nikolaidis, P.; Knechtle, B. Predictive Performance Models in Long-Distance Runners: A Narrative Review. *Int. J. Environ. Res. Public Health* **2020**, *17*, 8289. [CrossRef]
44. Svedahl, K.; MacIntosh, B.R. Anaerobic Threshold: The Concept and Methods of Measurement. *Can. J. Appl. Physiol.* **2003**, *28*, 299–323. [CrossRef]
45. Sales, M.M.; Sousa, C.V.; da Silva Aguiar, S.; Knechtle, B.; Nikolaidis, P.T.; Alves, P.M.; Simões, H.G. An Integrative Perspective of the Anaerobic Threshold. *Physiol. Behav.* **2019**, *205*, 29–32. [CrossRef]
46. Scheer, V.; Vieluf, S.; Janssen, T.I.; Heitkamp, H.-C. Predicting Competition Performance in Short Trail Running Races with Lactate Thresholds. *J. Hum. Kinet.* **2019**, *69*, 159–167. [CrossRef]
47. Foster, C.; Lucia, A. Running Economy: The Forgotten Factor in Elite Performance. *Sports Med.* **2007**, *37*, 316–319. [CrossRef]
48. Joyner, M.J.; Coyle, E.F. Endurance Exercise Performance: The Physiology of Champions. *J. Physiol.* **2008**, *586*, 35–44. [CrossRef]
49. McLaughlin, J.E.; Howley, E.T.; Bassett, D.R.; Thompson, D.L.; Fitzhugh, E.C. Test of the Classic Model for Predicting Endurance Running Performance. *Med. Sci. Sports Exerc.* **2010**, *42*, 991–997. [CrossRef]
50. Daniels, J.; Daniels, N. Running Economy of Elite Male and Elite Female Runners. *Med. Sci. Sports Exerc.* **1992**, *24*, 483–489. [CrossRef]
51. Larsen, H.B. Kenyan Dominance in Distance Running. *Comp. Biochem. Physiol. A Mol. Integr. Physiol.* **2003**, *136*, 161–170. [CrossRef]
52. Willis, S.J.; Gellaerts, J.; Mariani, B.; Basset, P.; Borrani, F.; Millet, G.P. Level Versus Uphill Economy and Mechanical Responses in Elite Ultra-Trail Runners. *Int. J. Sports Physiol. Perform.* **2019**, *14*, 1001–1005. [CrossRef] [PubMed]
53. Billat, L.V.; Koralsztein, J.P. Significance of the Velocity at VO2max and Time to Exhaustion at This Velocity. *Sports Med.* **1996**, *22*, 90–108. [CrossRef] [PubMed]
54. Ortiz, A.L.R.; Giovanelli, N.; Kram, R. The Metabolic Costs of Walking and Running up a 30-Degree Incline: Implications for Vertical Kilometer Foot Races. *Eur. J. Appl. Physiol.* **2017**, *117*, 1869–1876. [CrossRef] [PubMed]
55. Giovanelli, N.; Ortiz, A.L.R.; Henninger, K.; Kram, R. Energetics of Vertical Kilometer Foot Races; Is Steeper Cheaper? *J. Appl. Physiol.* **2016**, *120*, 370–375. [CrossRef]
56. Itoh, H.; Ohkuwa, T.; Yamazaki, Y.; Miyamura, M. Human Blood Lactate and Ammonia Levels after Supramaximal Uphill and Downhill Running. *Nagoya J. Med. Sci.* **1996**, *59*, 135–142. [PubMed]

57. Midgley, A.W.; McNaughton, L.R.; Polman, R.; Marchant, D. Criteria for Determination of Maximal Oxygen Uptake. *Sport. Med.* **2007**, *37*, 1019–1028. [CrossRef]
58. Costill, D.L.; Jansson, E.; Gollnick, P.D.; Saltin, B. Glycogen Utilization in Leg Muscles of Men during Level and Uphill Running. *Acta Physiol. Scand.* **1974**, *91*, 475–481. [CrossRef]
59. Armstrong, R.B.; Laughlin, M.H.; Rome, L.; Taylor, C.R. Metabolism of Rats Running up and down an Incline. *J. Appl. Physiol. Respir. Environ. Exerc. Physiol.* **1983**, *55*, 518–521. [CrossRef]
60. Bernardi, E.; Pratali, L.; Mandolesi, G.; Spiridonova, M.; Roi, G.S.; Cogo, A. Thoraco-Abdominal Coordination and Performance during Uphill Running at Altitude. *PLoS ONE* **2017**, *12*, e0174927. [CrossRef]
61. Gladwell, V.F.; Coote, J.H. Heart rate at the onset of muscle contraction and during passive muscle stretch in humans: A role for mechanoreceptors. *J. Physiol.* **2002**, *540*, 1095–1102. [CrossRef]
62. Zimmermann, P.; Schöffl, I.; Schöffl, V.; Zimmermann, L.; Eckstein, M.L.; Moser, O.; Wüstenfeld, J. Physiological Effects of Training in Elite German Winter Sport Athletes: Sport Specific Remodeling Determined Using Echocardiographic Data and CPET Performance Parameters. *J. Cardiovasc. Dev. Dis.* **2022**, *9*, 235. [CrossRef]
63. Gregor, R.J.; Costill, D.L. A Comparison of the Energy Expenditure during Positive and Negative Grade Running. *J. Sports Med. Phys. Fit.* **1973**, *13*, 248–252.
64. Abe, D.; Yanagawa, K.; Yamanobe, K.; Tamura, K. Assessment of Middle-Distance Running Performance in Sub-Elite Young Runners Using Energy Cost of Running. *Eur. J. Appl. Physiol. Occup. Physiol.* **1998**, *77*, 320–325. [CrossRef]
65. Mooses, M.; Mooses, K.; Haile, D.W.; Durussel, J.; Kaasik, P.; Pitsiladis, Y.P. Dissociation between Running Economy and Running Performance in Elite Kenyan Distance Runners. *J. Sports Sci.* **2015**, *33*, 136–144. [CrossRef]
66. Giovanelli, N.; Taboga, P.; Rejc, E.; Lazzer, S. Effects of Strength, Explosive and Plyometric Training on Energy Cost of Running in Ultra-Endurance Athletes. *Eur. J. Sport Sci.* **2017**, *17*, 805–813. [CrossRef]
67. Lazzer, S.; Salvadego, D.; Taboga, P.; Rejc, E.; Giovanelli, N.; di Prampero, P.E. Effects of the Etna Uphill Ultramarathon on Energy Cost and Mechanics of Running. *Int. J. Sports Physiol. Perform.* **2015**, *10*, 238–247. [CrossRef]
68. Giandolini, M.; Vernillo, G.; Samozino, P.; Horvais, N.; Edwards, W.B.; Morin, J.-B.; Millet, G.Y. Fatigue Associated with Prolonged Graded Running. *Eur. J. Appl. Physiol.* **2016**, *116*, 1859–1873. [CrossRef]

life

Review

Role of Exercise Stress Echocardiography in Pulmonary Hypertension

Mojca Škafar [1,2], Jana Ambrožič [1], Janez Toplišek [1] and Marta Cvijić [1,2,*]

[1] Department of Cardiology, University Medical Centre Ljubljana, Zaloška 2, 1000 Ljubljana, Slovenia; mojca.skafar@gmail.com (M.Š.); jana.ambrozic@gmail.com (J.A.); jtoplisek@gmail.com (J.T.)
[2] Faculty of Medicine, University of Ljubljana, Vrazov trg 2, 1000 Ljubljana, Slovenia
* Correspondence: marta.cvijic@gmail.com; Tel.: +386-1-522-4467

Abstract: Resting and exercise right heart catheterisation is the gold standard method to diagnose and differentiate types of pulmonary hypertension (PH). As it carries technical challenges, the question arises if non-invasive exercise stress echocardiography may be used as an alternative. Exercise echocardiography can unmask exercise PH, detect the early stages of left ventricular diastolic dysfunction, and, therefore, differentiate between pre- and post-capillary PH. Regardless of the underlying aetiology, a developed PH is associated with increased mortality. Parameters of overt right ventricle (RV) dysfunction, including RV dilation, reduced RV ejection fraction, and elevated right-sided filling pressures, are detectable with resting echocardiography and are associated with worse outcome. However, these measures all fail to identify occult RV dysfunction. Echocardiographic measures of RV contractile reserve during exercise echocardiography are very promising and provide incremental prognostic information on clinical outcome. In this paper, we review pulmonary haemodynamic response to exercise, briefly describe the modalities for assessing pulmonary haemodynamics, and discuss in depth the contemporary key clinical application of exercise stress echocardiography in patients with PH.

Keywords: pulmonary hypertension; exercise stress echocardiography; pulmonary arterial pressure; haemodynamics; diagnosis; prognosis

Citation: Škafar, M.; Ambrožič, J.; Toplišek, J.; Cvijić, M. Role of Exercise Stress Echocardiography in Pulmonary Hypertension. *Life* **2023**, *13*, 1385. https://doi.org/10.3390/life13061385

Academic Editor: Emilio Nardi

Received: 1 May 2023
Revised: 31 May 2023
Accepted: 12 June 2023
Published: 14 June 2023

1. Introduction

Pulmonary hypertension (PH) is a pathophysiological disorder that may involve various clinical conditions and may be associated with numerous respiratory and/or cardiovascular diseases. Regardless of the underlying pathobiological process and the site of the functional changes, PH is defined as a resting mean pulmonary arterial pressure (mPAP) of >20 mmHg measured by right heart catheterisation (RHC) [1].

However, the elevated mPAP is not sufficient to define the subgroups of PH since it could be due to increased pulmonary vascular resistance (PVR) or increased pulmonary artery wedge pressure (PAWP), or just a consequence of alteration in cardiac output (CO) or intrathoracic pressure [1,2]. Taking mPAP, PAWP, and PVR into consideration, we differentiate two main phenotypes of PH. Pre-capillary PH is the consequence of pulmonary vascular disease (e.g., pulmonary artery hypertension, lung diseases, and/or hypoxia, pulmonary artery obstructions, …) and is characterised by the presence of elevated PVR (>2 mmHg/L/min) and normal PAWP ($\leq$15 mmHg). Post-capillary PH is due to left heart disease and is haemodynamically defined as mPAP > 20 mmHg and PAWP > 15 mmHg. The detection and risk stratification of PH patients in everyday clinical practice is often challenging due to the complexity of PH phenotypes; however, it is of paramount importance for adequate treatment.

RHC is the gold standard method to assess pulmonary haemodynamics. It enables diagnosing and classifying PH [1]. On the other hand, echocardiography takes a central role in detecting the consequences of right ventricle (RV) pressure overload, such as RV dysfunction, as well as in detecting the causes of PH, particularly in PH associated with left heart disease. Recent studies have demonstrated that exercise could reveal abnormal pulmonary haemodynamic response in patients with PH, which might provide important prognostic and functional information [3]. According to the current consensus statement, an exercise stress test during RHC is the recommended method to assess cardiopulmonary haemodynamics. However, exercise RHC carries some technical challenges, requiring experienced personnel, special equipment, and additional procedural time. This raises the question of whether non-invasive exercise stress echocardiography may be used as an alternative technique.

The purpose of this manuscript is to review the basic concept of the pulmonary haemodynamic response to exercise in health and disease and to discuss in depth the role of exercise stress echocardiography with a particular emphasis on its diagnostic and prognostic implication in patients with PH.

2. How Does Pulmonary Circulation Response to Exercise?

In healthy subjects: An increase in CO during exercise stresses the pulmonary circulation, which results in a physiological increase in pulmonary arterial pressure (PAP). The increase in PAP is strongly dependent on the level of the exercise, and at high exercise levels, it can frequently exceed the mPAP > 20 mmHg [4]. Changes in PAP during exercise are determined by the interplay between CO, pulmonary artery compliance, PVR, and PAWP and can be presented with the following simplified equitation:

$$mPAP = PVR \times CO + PAWP.$$

In a normal individual, mPAP and PAWP increase significantly during exercise, while PVR slightly but not significantly decreases [3,5]. The physiological decrease in PVR during exercise is due to the recruitment and distension of the pulmonary resistance vessels with increased flow [6]. The increase in PAP is flow-dependent and needs to be reported in relation to the increase in CO (Figure 1) [7].

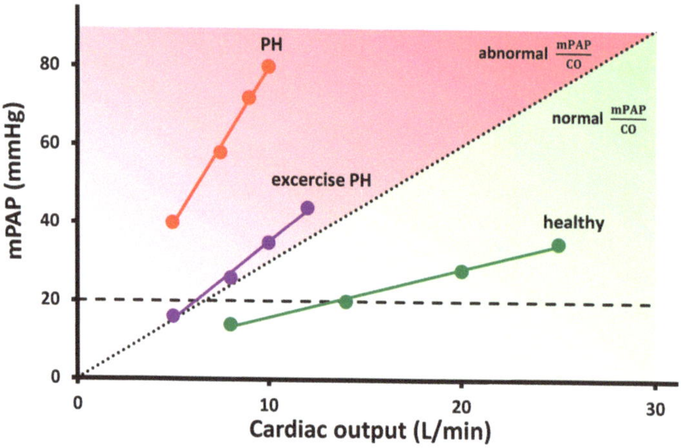

Figure 1. Mean pulmonary arterial pressure (mPAP)–cardiac output (CO) relationship for characterisation of pulmonary haemodynamics during exercise in three representative cases: a patient with pulmonary hypertension (PH), a patient with exercise PH, and a healthy subject. The patient with PH has elevated mPAP at rest and demonstrates disproportionate increase in mPAP during exercise (mPAP/CO = 8 mmHg/L/min), while the patient with exercise, PH has normal resting pulmonary haemodynamics and abnormal response during exercise (mPAP/CO slope = 4 mmHg/L/min). Healthy subject has flow-dependent increase in mPAP, which exceeds 20 mmHg at high CO; however, mPAP/CO slope remains very low (1 mmHg/L/min).

In contrast, the mPAP/CO slope is largely unaffected by workload, but it is strongly age dependent. The age dependency of the mPAP/CO slope is mainly determined by the age dependency of the PAWP/CO slope [5]. Higher PAWP/CO slope with ageing is likely due to a decrease in left ventricular (LV) compliance and relaxation during exercise, which is part of the physiological ageing process [8,9]. On the other hand, the distensibility of the pulmonary vessels may remain largely unaffected by age.

In pathological conditions: Unrelated to the pathology, a disproportionate increase in PAP during low levels of exercise is observed, and this contributes to abnormally high mPAP/CO slope (>3 mmHg/L/min) (Figure 1).

In patients with heart failure (post-capillary PH), impaired early diastolic relaxation, reduced increments in suction, and poor LV contractility and/or LV compliance led to inadequate increases in stroke volume and CO with exercise and increased LV filling pressures and PAWP. The high pressure from the left heart is transmitted into the pulmonary circulation, consequently resulting in an increase in mPAP [10–12].

On the other hand, patients with pulmonary vascular disease have primarily an increased pulmonary vascular tone and remodelling of the small pulmonary arteries. An abnormal exercise-induced increase in PVR due to pulmonary vascular abnormalities is the most important factor contributing to exercise PAP in this patient's population [6]. These patients often have reduced CO already at rest and/or fail to adequately augment CO through exercise [13], which results in a steeper mPAP/CO slope during exercise (Figure 1).

A small group of patients have normal pulmonary haemodynamics at rest, but during exercise, a pathological increase in PAP can be detected, evident by an mPAP/CO slope >3 mmHg/L/min between rest and exercise (Figure 1) [5]. The latest evidence demonstrated that this pathological increase in PAP during exercise is associated with a worse prognosis in patients with exercise dyspnoea [14] and several cardiovascular conditions [15–18]. This entity of exercise PH was recently reintroduced in the 2022 ESC/ERS Guidelines for the diagnosis and treatment of PH as a new phenotype of PH [1].

3. Methods for Assessing Pulmonary Haemodynamics

3.1. RHC

RHC is the gold standard method to assess pulmonary haemodynamics at rest and during exercise [1]. After confirming PH with mPAP > 20 mmHg at rest, the haemodynamic definition further distinguishes PH into two main phenotypes of PH, pre-capillary and post-capillary PH. PVR is used to differentiate between patients with post-capillary PH who have a significant pre-capillary component (PVR > 2 mmHg/L/min; combined post- and pre-capillary PH) and those who do not (PVR $\leq$ 2 mmHg/L/min; isolated post-capillary PH) [1]. However, discrimination between subgroups can be challenging if only resting haemodynamics is available, such as when PAWP is close to 15 mmHg or if the clinical characteristics of the patient primarily suggest heart failure with preserved ejection fraction (HFpEF) despite normal resting PAWP [3]. Therefore, the diagnostic work-up requires the use of provocative manoeuvres (e.g., exercise and fluid challenge) during RHC to elicit a dynamic response of the PAWP that may identify the occult post-capillary PH. The latest guidelines recommend using a PAWP cut-off of >25 mmHg during supine exercise for diagnosing post-capillary PH [9]. Although an increased mPAP/CO slope defines an abnormal haemodynamic response to exercise, it does not allow for differentiation between pre- and post-capillary subgroups. However, the PAWP/CO slope with a threshold > 2 mmHg/L/min may best differentiate between pre- and post-capillary causes of exercise PH [19,20].

While the use of resting RHC is widespread, the exercise RHC is a technically demanding diagnostic method as the exercise causes movement artefacts and is available only in specialised centres.

3.2. Echocardiography

Echocardiography is a non-invasive, inexpensive, and widely available imaging modality for estimating the systolic pulmonary arterial pressure (sPAP) and detecting additional signs suggestive of PH at rest. However, conventional echocardiography alone is insufficient to confirm a diagnosis of PH as the correlation of sPAP by echocardiography compared with sPAP by RHC was modest, with a correlation coefficient of 0.70 (95% CI 0.67 to 0.73) and with the trends toward underestimation of sPAP by echocardiography [21]. Furthermore, the diagnostic accuracy of echocardiography for PH was also modest, with a sensitivity and specificity of around 80% and 70%, respectively [21,22]. There is no good conventional echocardiographic method that can reliably discriminate between pre- and post-capillary PH at rest. Early studies showed only a modest correlation between the echocardiographic Doppler parameter for the estimation of LV filling pressure (E/e′) at rest and invasively measured PAWP [11]. Similar to the concept of exercise RHC, introducing an exercise test to standard Doppler echocardiography might improve the diagnostic accuracy for detecting post-capillary PH. Furthermore, exercise stress echocardiography has an added value in the management of PH patients, as it also allows us to assess complex RV mechanics (e.g., contractile reserve) and the ability of the pulmonary vasculature to accommodate the increased flow.

Ideally, a semi-supine bicycle test or an upright bicycle exercise protocol with imaging is used [11,12], but there are no universally adopted protocols (Table 1). None of these protocols have been shown to be superior to others [9]. The European Association of Cardiovascular Imaging and the American Society of Echocardiography recommend a stepped protocol on a semi-supine bicycle until the patient reaches his maximal predicted workload and/or maximal predicted heart rate (220—age in years) and/or develops limiting symptoms [12]. However, some patients cannot perform that protocol, and a ramped exercise test has also been proposed for a submaximal target heart rate of 100–110/min or until the patient develops limiting symptoms [23]. There is a paucity of data on isometric exercise, which has little or no effect on CO and may change intrathoracic and systemic arterial pressure and systemic vascular resistance. Therefore, isometric exercise might not be suitable for challenging pulmonary circulation [24].

Table 1. Commonly used exercise stress echocardiography protocols currently employed in clinical practice.

Author	Type	Protocol
Ha, 2020 [25]	semi-supine bicycle	stepped protocol: cycling at a cadence of 60 r.p.m. starting at 25 W and increasing in increments of 25 W every 3 min
Erdei, 2014 [23]	semi-supine bicycle	ramp protocol: cycling at a cadence of 60 r.p.m. starting at 15 W with 5 W increments every minute
Motram, 2004 [26]	treadmill	Bruce protocol
Jake Samuel, 2017 [27]	isometric handgrip	holding the dynamometer at 40% of MVC for 3 min
Pongpaopattanakul, 2022 [28]	dynamic handgrip	squeezing the dynamometer at 2 kg at a cadence of 30 r.p.m. for 3 min

Legend: MVC—maximal voluntary contraction; r.p.m.—revolutions per minute.

4. Clinical Application of Exercise Stress Echocardiography

4.1. Diagnostic Role

In patients with exertional dyspnoea and suspected PH due to HFpEF, there is a possibility to unmask early stages of LV diastolic dysfunction to detect increased LV filling pressures during exercise and, therefore, to differentiate between pre- and post-capillary PH.

As in resting echocardiography, the increase in the E/e′ ratio during exercise is suggestive of elevated LV filling pressures. However, studies comparing haemodynamic data acquired by echocardiography and by RHC during exercise are limited [3,25]. Even though the E/e′ ratio during exercise had only a moderate correlation with directly invasively measured PAWP ($r = 0.57$; $p < 0.001$), adding the peak exercise E/e′ ratio to the ESC proposed algorithm of diastolic dysfunction improved sensitivity (up to 90%) and can be used to rule out post-capillary PH [10]. Using low-level exercise (20 W) seems to be a good alternative, as E/e′ at 20 W could reliably predict normal PAWP during exercise (AUC: 0.77; $p < 0.01$) [29]. The authors proposed a cut-off of 12.4 for E/e′ at 20 W (specificity 83%, sensitivity 75%). However, these studies comprised only healthy controls and patients with HFpEF. There is only one study that tested the echocardiographic mPAP/CO ratio to identify patients with abnormal pulmonary vascular response to exercise [30]. In a study group of healthy subjects and mainly patients with chronic thromboembolic pulmonary hypertension, mPAP/CO via exercise stress echocardiography of 3.2 mmHg/L/min was identified as the most favourable threshold. Of note, this cut-off is perfectly in line with the proposed cut-off obtained by RHC.

In spite of the above-mentioned data, the diagnostic value of the stress tests during echocardiography to distinguish between PH subtypes is currently uncertain due to the lack of prospective data, especially regarding its use to identify cases of combined post- and pre-capillary PH. Based on the data from the literature, exercise echocardiography is considered abnormal if the average E/e′ ratio at peak stress increases to ≥15, with or without a peak TR velocity >3.4 m/s (Figure 2) [10–12]. An increase in TR velocity only should not be used to diagnose post-capillary PH because it might be a normal hyperdynamic response to exercise with increased pulmonary blood flow in the absence of LV diastolic dysfunction [31].

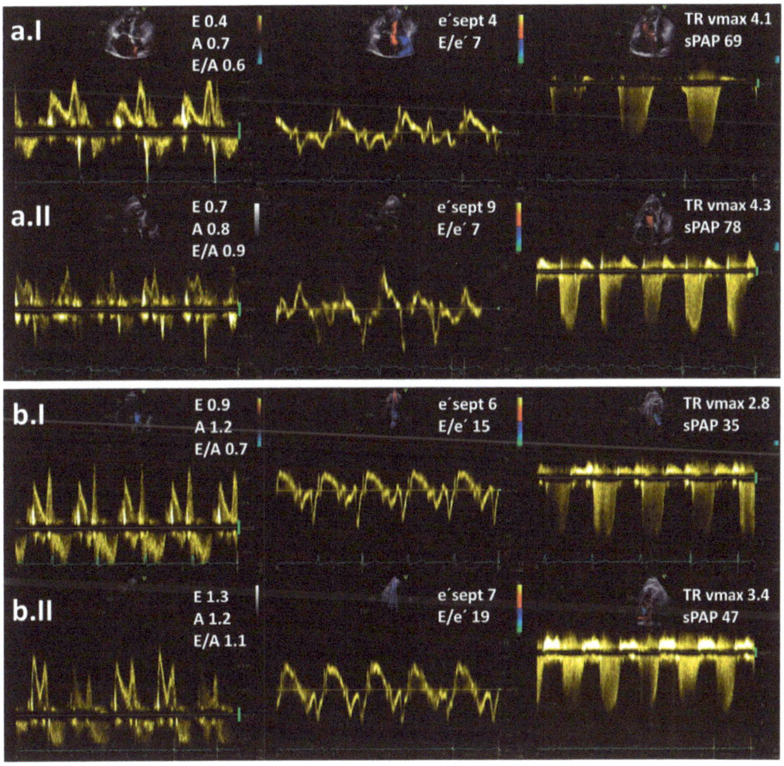

Figure 2. Resting (**I**) and exercise stress echocardiography (**II**) in two patients with pulmonary hypertension (PH): (**a**) a patient with pre-capillary PH and (**b**) a patient with post-capillary PH. Maximal tricuspid regurgitation velocity (TR vmax) is elevated in both patients. Note an increase in E/e′ ratio at low-level exercise from 15 to 19 in a patient with post-capillary PH but no increase in a patient with pre-capillary PH. Legend: sPAP—systolic pulmonary arterial pressure.

Exercise stress echocardiography could be extremely useful as an effective gatekeeper to the RHC for patients with exertional dyspnoea of unknown aetiology and normal resting echocardiographic results and also for identifying patients with a high risk for developing PH. It has been demonstrated that exercise stress echocardiography can distinguish between noncardiac and cardiac causes of unexplained dyspnoea [10,32]. The diagnostic value of exercise stress echocardiography was also evaluated in asymptomatic relatives of patients with idiopathic and familial PAH [33]. Hypertensive response to exercise, defined by TR velocity > 3.1 m/s, was more often present in relatives of PAH patients than in control subjects. Additionally, exercise stress echocardiography is considered to be reasonable, especially in patients with connective tissue disease [34]. It has been reported that up to 50% of this patient population with normal resting mPAP had an abnormal increase in mPAP during exercise. In patients with systemic sclerosis, PH was confirmed by RHC in 81% of patients with positive exercise stress echocardiography [35]. Moreover, exercise stress echocardiography could unmask exercise PH in patients with systemic sclerosis and baseline echocardiographic PAP within the grey zone [36]. It is important to note that in both studies, authors used a definition of exercise PH, which is not in line with nowadays valid definition (an increase of 20 mmHg over the resting sPAP or sPAP > 50 mmHg was considered as a positive test result). However, the clinical value of exercise PH identified by exercise stress echocardiography remains uncertain because of the lack of validated criteria and prospective confirmatory data [1]. Therefore, data from exercise stress echocardiography are not sufficient to be a substitute for invasive haemodynamic data under all circumstances, especially if a therapeutic decision depends on the results [9].

4.2. Prognostic Role

Regardless of the underlying aetiology, the developed PH is associated with worsening symptoms and substantially increased mortality [37]. Even though the detection of exercise PH via exercise stress echocardiography is considered an early and mild phase of PAH [38], patients with exercise PH already had worse outcomes than subjects without exercise PH [39].

The survival of PH patients depends on the capability of the RV to adapt to chronically elevated PAP [40]. Over time, adaptive concentric RV hypertrophy with preserved RV function can evolve into RV dilatation and systolic dysfunction [41,42]. RV function is a major determinant of functional capacity and prognosis when RV afterload is elevated [43–45]. Echocardiographic measures of RV function that are independent predictors of mortality in PH include the tricuspid annular plane systolic excursion (TAPSE < 18 mm [46–48]), RV fractional area change (FAC < 35% [49,50]), peak systolic tricuspid lateral annular velocity (S′ < 9.7 cm/s [51]) and Tei index (>0.40 by pulse Doppler or >0.55 by tissue Doppler [52]). Conventional 2-dimensional echocardiographic evaluation of the RV is difficult due to the complex 3-dimensional (3D) anatomical shape of the RV. This limitation can be overcome with 3D echocardiography and/or cardiac magnetic resonance [53,54] and recently, an increased 3D RVESVi has been shown to correlate with increased mortality [55].

However, these parameters all fail to identify occult RV dysfunction in patients with PH [56] as they reflect already established RV dysfunction. Subtle RV dysfunction could possibly be recognised by the use of advanced echocardiographic techniques, such as strain/myocardial deformation and myocardial work [57–59]. Previous studies demonstrated that RV longitudinal strain was a powerful predictor of survival in patients with PH and provided incremental prognostic value over conventional clinical and echocardiographic variables [60,61].

Additionally, the assessment of RV contractile reserve via RV–pulmonary arterial (PA) coupling shows promising results in detecting subclinical RV systolic dysfunction [56,62–66]. Gold standard measurement of RV–PA coupling involves conductance catheter measurement of "multi-beat" RV end-systolic elastance (Ees), a method that remains costly, impractical and clinically challenging [62]. However, new echocardiographic indices, such as the TAPSE/sPAP ratio [45,67–69] and RV free wall longitudinal strain/sPAP [59], are tightly linked to RV–PA coupling and are associated with outcomes in patients with PH.

A possible non-invasive measure of the RV contractile reserve using exercise stress echocardiography was first proposed by Grünig et al. They demonstrated that an exercise-induced increase in sPAP was a measure of the RV contractile reserve and was an independent prognostic factor in patients with pre-capillary PH (Table 1) [70]. A lower sPAP increase may reveal an impaired ability of the RV to adapt to pulmonary load and exercise and to further increase pressure and pulmonary blood flow. Similarly, an initial steep increment in PAP during exercise followed by a plateau with a linear pattern was associated with decreased exercise capacity and survival in patients with heart failure [17]. Echocardiographic studies focused only on the peak exercise sPAP or the peak change in sPAP [17,70]; however, it would be preferable to interpret exercise PAP pattern relative to the increase in blood flow (PAP/CO ratio). Invasively obtained haemodynamic data clearly showed that high mPAP/CO during exercise was associated with impaired survival in a heterogeneous group of different PH phenotypes [5,14]. Echocardiographic studies analysing mPAP/CO are limited, but initial results are very promising. A disproportionate increase in mPAP/CO slope during exercise was independently associated with adverse clinical outcomes in patients with HFpEF (Table 2) [71], and this parameter had an incremental value even in patients with preserved RV-PA coupling at rest.

Other authors assessed RV contractile reserve based on echocardiographic parameters of RV systolic function (e.g., change in TAPSE, change in RV FAC and change in S') (Table 2) [72,73]. The magnitude of the increase in all three parameters was significantly lower in patients with pre-capillary PH than in healthy controls [72,73]; however, no prognostic data have been available for these parameters. Ireland et al. prospectively studied RV contractile reserve in PH patients who underwent cardiac magnetic resonance, echocardiography, and supine invasive cardiopulmonary exercise testing with concomitant RV pressure-volume catheterisation. RV contractile reserve during exercise, measured by Ees during exertion, was associated with an elevation in PAP but the preservation of RV volumes. The lack of RV reserve, on the other hand, was tightly coupled with acute RV dilation during exercise [62]. RV ejection fraction during exercise was shown to be a robust surrogate for RV contractile reserve (Table 2), and it best predicted occult RV dysfunction among a variety of resting and exercise RV measures and was also associated with clinical worsening [62]. Therefore, echocardiographic parameters of RV contractile reserve and exercise stress echocardiography could be useful for follow-up assessment, especially to identify PH patients at high risk [70].

Table 2. Non-invasive measures of the right ventricle (RV) contractile reserve during exercise stress echocardiography.

Author	Subjects (n)	Echocardiographic Parameters	Most Relevant Findings
Grünig, 2013 [70]	124 PH patients (PAH, CTEPH) and impaired RV systolic function	ΔsPAP	Exercise-induced sPAP increase $\leq$ 30 mmHg was related to the worst outcome (HR 2.84, 95% CI 1.92–6.78; $p = 0.018$).
Almeida, 2014 [73]	14 subjects (7 controls, 7 patients with PH)	ΔS′, ΔTAPSE and ΔFAC	The magnitude of increase in ΔS′, ΔTAPSE and ΔFAC in healthy controls was higher than in patients (all $p < 0.05$.
Guo, 2019 [72]	46 subjects (31 patients with pre-capillary PH, 15 controls)	ΔS′, ΔTAPSE and ΔFAC	Significant increase in ΔS′ ($p = 0.002$), ΔTAPSE ($p < 0.001$) and ΔFAC ($p < 0.001$) was noted only in healthy controls.

Table 2. *Cont.*

Author	Subjects (n)	Echocardiographic Parameters	Most Relevant Findings
Ireland, 2021 [62]	35 subjects with known or suspected PH	Exercise RVEF	Exercise RVEF can detect occult RV dysfunction (AUC = 0.81, cut off of exercise RVEF = 38%). Patients with exercise RVEF < 38% had an increased propensity for clinical worsening over 4 years compared to patients with RVEF > 38% (p = 0.014).
Saito, 2023 [71]	345 patients (1666 HFpEF, 179 controls)	mPAP/CO slope	Patients with mPAP/CO slope > 5.2 mmHg/L/min had a higher rate of adverse events (all-cause mortality, HF events) compared to those with mPAP/CO slope < 5.2 mmHg/L/min (p = 0.0002).

Legend: CO—cardiac output, CTEPH—chronic thromboembolic pulmonary hypertension, FAC—fractional area change, HFpEF—heart failure with preserved ejection fraction, mPAP—mean pulmonary arterial pressure, PAH—pulmonary arterial hypertension, PH—pulmonary hypertension, RVEF—right ventricular ejection fraction, sPAP—systolic pulmonary arterial pressure, S′ -Doppler-derived tricuspid lateral annular peak systolic velocity, and TAPSE—tricuspid annular plane systolic excursion.

4.3. Practical Approach to Exercise Stress Echocardiography

Detailed practical guidelines on stress echocardiography have already been published [74]. For the purpose of this review, only some important factors are emphasised:

- In the case of a step protocol, haemodynamic measurements are performed towards the end of each exercise level when a steady state in oxygen consumption on a given exercise level is achieved (usually in 3–5 min). For practical reasons, shorter time intervals can be chosen (e.g., 2 min steps aiming for a duration of the exercise time of∼10 min), which appear to be a good compromise [3].
- The feasibility of obtaining diagnostic-quality measurements of TR velocity decreases with increasing exercise load, with 54% at low exercise (20 W) and 49% at peak exercise [10]. The administration of agitated colloids enhances a continuous Doppler tricuspid regurgitation signal and allows reliable estimation of sPAP during exercise [30].
- At higher heart rates, the fusion of the mitral E and A waves prevents the estimation of the LV filling pressures. It was reported that E/e′ could not be measured in about 10% of subjects during submaximal exercise (20 W) and in about 25% of patients during peak exercise [29]. Therefore, the acquisition of images during the submaximal phase before the fusion of E and A waves is advised (heart rate 100–110 bpm) [75].
- Acquisition during early recovery is not optimal, as haemodynamics change very rapidly after cessation of exercise, and previous invasive studies demonstrated that PAWP returned to the baseline levels already 1 min post-exercise [76].

5. Future Perspectives

Many different protocols of exercise stress echocardiography are used in everyday clinical practice, and further research is needed to compare these protocols and possibly universally adopt one of them. The diagnosis of exercise PH, which was reintroduced in the newest ESC Guidelines for the diagnosis and treatment of PH, is challenging [1]. It requires exercise RHC, which is not readily available. Further research is needed to determine non-invasive cut-offs that would reveal possibly abnormal exercise pulmonary haemodynamics. Moreover, there is no reliable non-invasive method for discrimination between pre-capillary and post-capillary PH in dubious cases. There are some non-invasive stress tests that could be used in such cases to avoid invasive procedures, but the data about their usability are limited. RV dysfunction is the most important determinant of survival in patients with PH [43,44]. However, the manifestations of RV dysfunction not only include changes in global RV systolic function but also abnormalities in the pattern of contraction and synchrony that can be analysed using RV strain and RV myocardial work.

The use of these novel parameters for prognostic assessment of PH patients is relatively new and mainly used for research purposes [57–59]. Further data are needed to implement them into exercise stress echocardiography in PH patients.

6. Conclusions

Exercise stress echocardiography is a promising non-invasive tool for the assessment of pulmonary haemodynamics, as it provides clinically relevant diagnostic and prognostic information in patients with PH. However, studies comparing echocardiography and RHC data during exercise are limited, and there is no single echocardiographic parameter that reliably identifies the underlying aetiology of PH. Therefore, RHC remains the gold standard method for the diagnosis and classification of different types of PH. In contrast to RHC, echocardiography has the ability to identify RV remodelling and systolic dysfunction, which are important determinants of survival in patients with PH. However, more research is needed since there are still many different protocols of exercise stress echocardiography used in everyday clinical practice, and many new promising echocardiographic parameters for the prognostic assessment of PH patients are emerging.

Author Contributions: Conception and design of the manuscript, or both (M.Š. and M.C.); Manuscript drafting or critical revision for important intellectual content (M.Š., J.A., J.T. and M.C.); Final approval of the manuscript submitted (M.Š., J.A., J.T. and M.C.). All authors have read and agreed to the published version of the manuscript.

Funding: This research received no external funding.

Institutional Review Board Statement: Not applicable.

Informed Consent Statement: Not applicable.

Data Availability Statement: Data sharing not applicable.

Conflicts of Interest: The authors declare no conflict of interest.

References

1. Humbert, M.; Kovacs, G.; Hoeper, M.M.; Badagliacca, R.; Berger, R.M.F.; Brida, M.; Carlsen, J.; Coats, A.J.S.; Escribano-Subias, P.; Ferrari, P.; et al. 2022 ESC/ERS Guidelines for the diagnosis and treatment of pulmonary hypertension. *Eur. Heart J.* **2022**, *43*, 3618–3731. [CrossRef] [PubMed]
2. Galiè, N.; McLaughlin, V.V.; Rubin, L.J.; Simonneau, G. An overview of the 6th World Symposium on Pulmonary Hypertension. *Eur. Respir. J.* **2019**, *53*, 1802148. [CrossRef]
3. Kovacs, G.; Herve, P.; Barbera, J.A.; Chaouat, A.; Chemla, D.; Condliffe, R.; Garcia, G.; Grünig, E.; Howard, L.; Humbert, M.; et al. An official European Respiratory Society statement: Pulmonary haemodynamics during exercise. *Eur. Respir. J.* **2017**, *50*, 1700578. [CrossRef]
4. Kovacs, G.; Berghold, A.; Scheidl, S.; Olschewski, H. Pulmonary arterial pressure during rest and exercise in healthy subjects: A systematic review. *Eur. Respir. J.* **2009**, *34*, 888–894. [CrossRef]
5. Zeder, K.; Banfi, C.; Steinrisser-Allex, G.; Maron, B.A.; Humbert, M.; Lewis, G.D.; Berghold, A.; Olschewski, H.; Kovacs, G. Diagnostic, prognostic and differential-diagnostic relevance of pulmonary haemodynamic parameters during exercise: A systematic review. *Eur. Respir. J.* **2022**, *60*, 2103181. [CrossRef] [PubMed]
6. Lewis, G.D.; Bossone, E.; Naeije, R.; Grünig, E.; Saggar, R.; Lancellotti, P.; Ghio, S.; Varga, J.; Rajagopalan, S.; Oudiz, R.J.; et al. Pulmonary vascular hemodynamic response to exercise in cardiopulmonary diseases. *Circulation* **2013**, *128*, 1470–1479. [CrossRef]
7. Naeije, R.; Vanderpool, R.; Dhakal, B.P.; Saggar, R.; Saggar, R.; Vachiery, J.-L.; Lewis, G.D. Exercise-induced pulmonary hypertension: Physiological basis and methodological concerns. *Am. J. Respir. Crit. Care Med.* **2013**, *178*, 576–583. [CrossRef]
8. Wolsk, E.; Bakkestrøm, R.; Thomsen, J.H.; Balling, L.; Andersen, M.J.; Dahl, J.S.; Hassager, C.; Møller, J.E.; Gustafsson, F. The Influence of Age on Hemodynamic Parameters During Rest and Exercise in Healthy Individuals. *JACC Heart Fail.* **2017**, *5*, 337–346. [CrossRef] [PubMed]
9. Pieske, B.; Tschöpe, C.; de Boer, A.R.; Fraser, A.G.; Anker, S.D.; Donal, E.; Edelmann, F.; Fu, M.; Guazzi, M.; Lam, C.S.P.; et al. How to diagnose heart failure with preserved ejection fraction: The HFA-PEFF diagnostic algorithm: A consensus recommendation from the Heart Failure Association (HFA) of the European Society of Cardiology (ESC). *Eur. Heart J.* **2019**, *40*, 3297–3317. [CrossRef]
10. Obokata, M.; Kane, G.C.; Reddy, Y.N.V.; Olson, T.P.; Melenovsky, V.; Borlaug, B.A. Role of Diastolic Stress Testing in the Evaluation for Heart Failure with Preserved Ejection Fraction: A Simultaneous Invasive-Echocardiographic Study. *Circulation* **2017**, *135*, 825–838. [CrossRef]

11. Nagueh, S.F.; Smiseth, O.A.; Appleton, C.P.; Byrd, B.F., 3rd; Dokainish, H.; Edvardsen, T.; Flachskampf, F.A.; Gillebert, T.C.; Klein, A.L.; Lancellotti, P.; et al. Recommendations for the Evaluation of Left Ventricular Diastolic Function by Echocardiography: An Update from the American Society of Echocardiography and the European Association of Cardiovascular Imaging. *J. Am. Soc. Echocardiogr.* **2016**, *29*, 277–314. [CrossRef] [PubMed]
12. Lancellotti, P.; Pellikka, P.A.; Budts, W.; Chaudhry, F.A.; Donal, E.; Dulgheru, R.; Edvardsen, T.; Garbi, M.; Ha, J.-W.; Kane, G.C.; et al. The clinical use of stress echocardiography in non-ischaemic heart disease: Recommendations from the European Association of Cardiovascular Imaging and the American Society of Echocardiography. *Eur. Heart J. Cardiovasc. Imaging* **2016**, *17*, 1191–1229. [CrossRef] [PubMed]
13. Claeys, M.; Claessen, G.; La Gerche, A.; Petit, T.; Belge, C.; Meyns, B.; Bogaert, J.; Willems, R.; Claus, P.; Delcroix, M. Impaired Cardiac Reserve and Abnormal Vascular Load Limit Exercise Capacity in Chronic Thromboembolic Disease. *JACC Cardiovasc. Imaging* **2019**, *12*, 1444–1456. [CrossRef] [PubMed]
14. Ho, J.E.; Zern, E.K.; Lau, E.S.; Wooster, L.; Bailey, C.S.; Cunningham, T.; Eisman, A.S.; Hardin, K.M.; Farrell, R.; Sbarbaro, J.A.; et al. Exercise Pulmonary Hypertension Predicts Clinical Outcomes in Patients With Dyspnea on Effort. *J. Am. Coll. Cardiol.* **2020**, *75*, 17–26. [CrossRef]
15. Stamm, A.; Saxer, S.; Lichtblau, M.; Hasler, E.D.; Jordan, S.; Huber, L.C.; Bloch, K.E.; Distler, O.; Ulrich, S. Exercise pulmonary haemodynamics predict outcome in patients with systemic sclerosis. *Eur. Respir. J.* **2016**, *48*, 1658–1667. [CrossRef]
16. Zeder, K.; Avian, A.; Bachmaier, G.; Douschan, P.; Foris, V.; Sassmann, T.; Moazedi-Fuerst, F.C.; Graninger, W.B.; Hafner, F.; Brodmann, M.; et al. Exercise Pulmonary Resistances Predict Long-Term Survival in Systemic Sclerosis. *Chest* **2021**, *159*, 781–790. [CrossRef]
17. Lewis, G.D.; Murphy, R.M.; Shah, R.V.; Pappagianopoulos, P.P.; Malhotra, R.; Bloch, K.D.; Systrom, D.M.; Semigran, M.J. Pulmonary vascular response patterns during exercise in left ventricular systolic dysfunction predict exercise capacity and outcomes. *Circ. Heart Fail.* **2011**, *4*, 276–285. [CrossRef]
18. Hasler, E.D.; Müller-Mottet, S.; Furian, M.; Saxer, S.; Huber, L.C.; Maggiorini, M.; Speich, R.; Bloch, K.E.; Ulrich, S. Pressure-Flow During Exercise Catheterization Predicts Survival in Pulmonary Hypertension. *Chest* **2016**, *150*, 57–67. [CrossRef]
19. Bentley, R.F.; Barker, M.; Esfandiari, S.; Wright, S.P.; Valle, F.H.; Granton, J.T.; Mak, S. Normal and abnormal relationships of pulmonary artery to wedge pressure during exercise. *J. Am. Heart Assoc.* **2020**, *9*, e016339. [CrossRef]
20. Eisman, A.S.; Shah, R.V.; Dhakal, B.P.; Pappagianopoulos, P.P.; Wooster, L.; Bailey, C.; Cunningham, T.F.; Hardin, K.M.; Baggish, A.L.; Ho, J.E.; et al. Pulmonary Capillary Wedge Pressure Patterns During Exercise Predict Exercise Capacity and Incident Heart Failure. *Circ. Heart Fail.* **2018**, *11*, e004750. [CrossRef]
21. Janda, S.; Shahidi, N.; Gin, K.; Swiston, J. Diagnostic accuracy of echocardiography for pulmonary hypertension: A systematic review and meta-analysis. *Heart* **2011**, *97*, 612–622. [CrossRef]
22. Ni, J.-R.; Yan, P.-J.; Liu, S.-D.; Hu, Y.; Yang, K.-H.; Song, B.; Lei, J.-Q. Diagnostic accuracy of transthoracic echocardiography for pulmonary hypertension: A systematic review and meta-analysis. *BMJ Open* **2019**, *9*, e033084. [CrossRef]
23. Erdei, T.; Smiseth, O.A.; Marino, P.; Fraser, A.G. A systematic review of diastolic stress tests in heart failure with preserved ejection fraction, with proposals from the EU-FP7 MEDIA study group. *Eur. J. Heart Fail.* **2014**, *16*, 1345–1361. [CrossRef]
24. Motoji, Y.; Forton, K.; Pezzuto, B.; Faoro, V.; Naeije, R. Resistive or dynamic exercise stress testing of the pulmonary circulation and the right heart. *Eur. Respir. J.* **2017**, *50*, 1700151. [CrossRef]
25. Ha, J.W.; Andersen, O.S.; Smiseth, O.A. Diastolic Stress Test: Invasive and Noninvasive Testing. *JACC Cardiovasc. Imaging* **2020**, *13*, 272–282. [CrossRef]
26. Mottram, P.M.; Haluska, B.A.; Marwick, T.H. Response of B-type natriuretic peptide to exercise in hypertensive patients with suspected diastolic heart failure: Correlation with cardiac function, hemodynamics, and workload. *Am. Heart J.* **2004**, *148*, 365–370. [CrossRef]
27. Samuel, T.J.; Beaudry, R.; Haykowsky, M.J.; Sarma, S.; Park, S.; Dombrowsky, T.; Bhella, P.S.; Nelson, M.D. Isometric handgrip echocardiography: A noninvasive stress test to assess left ventricular diastolic function. *Clin. Cardiol.* **2017**, *40*, 1247–1255. [CrossRef]
28. Pongpaopattanakul, P.; Imerbtham, T.; Kitimala, J.; Phatthanawayu, S.; Kaewkong, P. Effects of a dynamic handgrip exercise on left ventricular diastolic functions in diabetes mellitus patients: A preliminary clinical data. *Chula. Med. J.* **2022**, *66*, 283–291. [CrossRef]
29. Effects of a dynamic handgrip exercise on left ventricular diastolic functions in diabetes mellitus patients: A preliminary clinical data Utility of E/e′ Ratio During Low-Level Exercise to Diagnose Heart Failure With Preserved. *JACC Cardiovasc. Imaging* **2023**, *16*, 145–155. [CrossRef] [PubMed]
30. Claessen, G.; La Gerche, A.; Voigt, J.-U.; Dymarkowski, S.; Schnell, F.; Petit, T.; Willems, R.; Claus, P.; Delcroix, M.; Heidbuchel, H. Accuracy of Echocardiography to Evaluate Pulmonary Vascular and RV Function During Exercise. *JACC Cardiovasc. Imaging* **2016**, *9*, 532–543. [CrossRef] [PubMed]
31. Nagueh, S.F.; Chang, S.M.; Nabi, F.; Shah, D.J.; Estep, J.D. Cardiac imaging in patients with heart failure and preserved ejection fraction. *Circ Cardiovasc. Imaging* **2017**, *10*, e006547. [CrossRef]
32. Talreja, D.R.; Nishimura, R.A.; Oh, J.K. Estimation of Left Ventricular Filling Pressure with Exercise by Doppler Echocardiography in Patients with Normal Systolic Function: A Simultaneous Echocardiographic-Cardiac Catheterization Study. *J. Am. Soc. Echocardiogr.* **2007**, *20*, 477–479. [CrossRef]

33. Grünig, E.; Weissmann, S.; Ehlken, N.; Fijalkowska, A.; Fischer, C.; Fourme, T.; Galié, N.; Ghofrani, A.; Harrison, R.E.; Huez, S.; et al. Stress doppler echocardiography in relatives of patients with idiopathic and familial pulmonary arterial hypertension results of a multicenter european analysis of pulmonary artery pressure response to exercise and hypoxia. *Circulation* **2009**, *119*, 1747–1757. [CrossRef]
34. Shaikh, F.; Anklesaria, Z.; Shagroni, T.; Saggar, R.; Gargani, L.; Bossone, E.; Ryan, M.; Channick, R.; Saggar, R. A review of exercise pulmonary hypertension in systemic sclerosis. *J. Scleroderma Relat. Disord.* **2019**, *4*, 225–237. [CrossRef]
35. Steen, V.; Chou, M.; Shanmugam, V.; Mathias, M.; Kuru, T.; Morrissey, R. Exercise-induced pulmonary arterial hypertension in patients with systemic sclerosis. *Chest* **2008**, *134*, 146–151. [CrossRef]
36. Rallidis, L.S.; Papangelopoulou, K.; Anthi, A.; Tsangaris, I.; Varounis, C.; Makavos, G.; Konstantonis, D.; Vlachoyiannopoulos, P.; Orfanos, S.E.; Iliodromitis, E.K. The role of exercise doppler echocardiography to unmask pulmonary arterial hypertension in selected patients with systemic sclerosis and equivocal baseline echocardiographic values for pulmonary hypertension. *Diagnostics* **2021**, *11*, 1200. [CrossRef]
37. Hoeper, M.M.; Humbert, M.; Souza, R.; Idrees, M.; Kawut, S.M.; Sliwa-Hahnle, K.; Jing, Z.-C.; Gibbs, J.S.R. A global view of pulmonary hypertension. *Lancet Respir. Med.* **2016**, *4*, 306–322. [CrossRef]
38. Tolle, J.J.; Waxman, A.B.; Van Horn, T.L.; Pappagianopoulos, P.P.; Systrom, D.M. Exercise-induced pulmonary arterial hypertension. *Circulation* **2008**, *118*, 2183–2189. [CrossRef]
39. Shim, C.Y.; Kim, S.-A.; Choi, D.; Yang, W.-I.; Kim, J.-M.; Moon, S.-H.; Lee, H.-J.; Park, S.; Choi, E.-Y.; Chung, N.; et al. Clinical outcomes of exercise-induced pulmonary hypertension in subjects with preserved left ventricular ejection fraction: Implication of an increase in left ventricular filling pressure during exercise. *Heart* **2011**, *97*, 1417–1424. [CrossRef]
40. Ghio, S.; Klersy, C.; Magrini, G.; D'Armini, A.M.; Scelsi, L.; Raineri, C.; Pasotti, M.; Serio, A.; Campana, C.; Viganò, M. Prognostic relevance of the echocardiographic assessment of right ventricular function in patients with idiopathic pulmonary arterial hypertension. *Int. J. Cardiol.* **2010**, *140*, 272–278. [CrossRef]
41. Vonk-Noordegraaf, A.; Haddad, F.; Chin, K.M.; Forfia, P.R.; Kawut, S.M.; Lumens, J.; Naeije, R.; Newman, J.; Oudiz, R.J.; Provencher, S.; et al. Right Heart Adaptation to Pulmonary Arterial Hypertension. *J Am Coll Cardiol.* **2013**, *62*, D22–D33. [CrossRef] [PubMed]
42. Vonk Noordegraaf, A.; Westerhof, B.E.; Westerhof, N. The Relationship Between the Right Ventricle and its Load in Pulmonary Hypertension. *J. Am. Coll Cardiol.* **2017**, *69*, 236–243. [CrossRef] [PubMed]
43. Writing Committee Members; McLaughlin, V.V.; Archer, S.L.; Badesch, D.B.; Barst, R.J.; Farber, H.W.; Lindner, J.R.; Mathier, M.A.; McGoon, M.D.; Park, M.H.; et al. A report of the american college of cardiology foundation task force on expert consensus documents and the american heart association. *Circulation* **2009**, *119*, 2250–2294. [CrossRef]
44. D'Alonzo, G.E. Survival in Patients with Primary Pulmonary Hypertension. *Ann. Intern. Med.* **1991**, *115*, 343–349. [CrossRef] [PubMed]
45. Tello, K.; Dalmer, A.; Axmann, J.; Vanderpool, R.; Ghofrani, H.A.; Naeije, R.; Roller, F.; Seeger, W.; Sommer, N.; Wilhelm, J.; et al. Reserve of Right Ventricular-Arterial Coupling in the Setting of Chronic Overload. *Circ. Heart Fail.* **2019**, *12*, e005512. [CrossRef] [PubMed]
46. Forfia, P.R.; Fisher, M.R.; Mathai, S.C.; Housten-Harris, T.; Hemnes, A.R.; Borlaug, B.A.; Chamera, E.; Corretti, M.C.; Champion, H.C.; Abraham, T.P.; et al. Tricuspid Annular Displacement Predicts Survival in Pulmonary Hypertension. *Am. J. Respir. Crit. Care Med.* **2006**, *174*, 1034–1041. [CrossRef]
47. Burke, M.A.; Katz, D.H.; Beussink, L.; Selvaraj, S.; Gupta, D.K.; Fox, J.; Chakrabarti, S.; Sauer, A.J.; Rich, J.D.; Freed, B.H.; et al. Prognostic Importance of Pathophysiologic Markers in Patients With Heart Failure and Preserved Ejection Fraction. *Circ. Heart Fail.* **2014**, *7*, 288–299. [CrossRef]
48. Mohammed, S.F.; Hussain, I.; AbouEzzeddine, O.F.; Takahama, H.; Kwon, S.H.; Forfia, P.; Roger, V.L.; Redfield, M.M. Right Ventricular Function in Heart Failure With Preserved Ejection Fraction. *Circulation* **2014**, *130*, 2310–2320. [CrossRef]
49. Mauritz, G.-J.; Kind, T.; Marcus, J.T.; Bogaard, H.-J.; van de Veerdonk, M.; Postmus, P.E.; Boonstra, A.; Westerhof, N.; Vonk-Noordegraaf, A. Progressive Changes in Right Ventricular Geometric Shortening and Long-term Survival in Pulmonary Arterial Hypertension. *Chest* **2012**, *141*, 935–943. [CrossRef]
50. Melenovsky, V.; Hwang, S.J.; Lin, G.; Redfield, M.M.; Borlaug, B.A. Right heart dysfunction in heart failure with preserved ejection fraction. *Eur. Heart J.* **2014**, *35*, 3452–3462. [CrossRef]
51. Champion, H.C.; Michelakis, E.D.; Hassoun, P.M. Comprehensive Invasive and Noninvasive Approach to the Right Ventricle–Pulmonary Circulation Unit. *Circulation* **2009**, *120*, 992–1007. [CrossRef]
52. Yeo, T.C.; Dujardin, K.S.; Tei, C.; Mahoney, D.W.; McGoon, M.D.; Seward, J.B. Value of a Doppler-Derived Index Combining Systolic and Diastolic Time Intervals in Predicting Outcome in Primary Pulmonary Hypertension. *Am. J. Cardiol.* **1998**, *81*, 1157–1161. [CrossRef] [PubMed]
53. Sanz, J.; Conroy, J.; Narula, J. Imaging of the Right Ventricle. *Cardiol Clin.* **2012**, *30*, 189–203. [CrossRef] [PubMed]
54. Addetia, K.; Bhave, N.M.; Tabit, C.E.; Gomberg-Maitland, M.; Freed, B.H.; Dill, K.E.; Lang, R.M.; Mor-Avi, V.; Patel, A.R. Sample Size and Cost Analysis for Pulmonary Arterial Hypertension Drug Trials Using Various Imaging Modalities to Assess Right Ventricular Size and Function End Points. *Circ. Cardiovasc. Imaging* **2014**, *7*, 115–124. [CrossRef] [PubMed]

55. Ryo, K.; Goda, A.; Onishi, T.; Delgado-Montero, A.; Tayal, B.; Champion, H.C.; Simon, M.A.; Mathier, M.A.; Gladwin, M.T.; Gorcsan, J., 3rd. Characterization of Right Ventricular Remodeling in Pulmonary Hypertension Associated With Patient Outcomes by 3-Dimensional Wall Motion Tracking Echocardiography. *Circ. Cardiovasc. Imaging* **2015**, *8*, e003176. [CrossRef]
56. Richter, M.J.; Peters, D.; Ghofrani, H.A.; Naeije, R.; Roller, F.; Sommer, N.; Gall, H.; Grimminger, F.; Seeger, W.; Tello, K. Evaluation and prognostic relevance of right ventricular-arterial coupling in pulmonary hypertension. *Am. J. Respir. Crit. Care Med.* **2020**, *201*, 116–119. [CrossRef]
57. Carluccio, E.; Biagioli, P.; Alunni, G.; Murrone, A.; Zuchi, C.; Coiro, S.; Riccini, C.; Mengoni, A.; D'antonio, A.; Ambrosio, G. Prognostic value of right ventricular dysfunction in heart failure with reduced ejection fraction: Superiority of longitudinal strain over tricuspid annular plane systolic excursion. *Circ. Cardiovasc. Imaging* **2018**, *11*, e006894. [CrossRef]
58. Butcher, S.C.; Fortuni, F.; Montero-Cabezas, J.M.; Abou, R.; El Mahdiui, M.; Van Der Bijl, P.; Van Der Velde, E.T.; Marsan, N.A.; Bax, J.J.; Delgado, V. Right ventricular myocardial work: Proof-of-concept for non-invasive assessment of right ventricular function. *Eur. Heart J. Cardiovasc. Imaging* **2021**, *22*, 142–152. [CrossRef]
59. Ünlü, S.; Bézy, S.; Cvijic, M.; Duchenne, J.; Delcroix, M.; Voigt, J.-U. Right ventricular strain related to pulmonary artery pressure predicts clinical outcome in patients with pulmonary arterial hypertension. *Eur. Heart J. Cardiovasc. Imaging* **2023**, *24*, 635–642. [CrossRef]
60. Haeck, M.L.; Scherptong, R.W.; Marsan, N.A.; Holman, E.R.; Schalij, M.J.; Bax, J.J.; Vliegen, H.W.; Delgado, V. Prognostic Value of Right Ventricular Longitudinal Peak Systolic Strain in Patients With Pulmonary Hypertension. *Circ. Cardiovasc. Imaging* **2012**, *5*, 628–636. [CrossRef]
61. Fine, N.M.; Chen, L.; Bastiansen, P.M.; Frantz, R.P.; Pellikka, P.A.; Oh, J.K.; Kane, G.C. Outcome Prediction by Quantitative Right Ventricular Function Assessment in 575 Subjects Evaluated for Pulmonary Hypertension. *Circ. Cardiovasc. Imaging* **2013**, *6*, 711–721. [CrossRef] [PubMed]
62. Ireland, C.G.; Damico, R.L.; Kolb, T.M.; Mathai, S.C.; Mukherjee, M.; Zimmerman, S.L.; Shah, A.A.; Wigley, F.M.; Houston, B.A.; Hassoun, P.M.; et al. Exercise right ventricular ejection fraction predicts right ventricular contractile reserve. *J. Heart Lung. Transplant.* **2021**, *40*, 504–512. [CrossRef]
63. Hsu, S.; Houston, B.A.; Tampakakis, E.; Bacher, A.C.; Rhodes, P.S.; Mathai, S.C.; Damico, R.L.; Kolb, T.M.; Hummers, L.K.; Shah, A.A.; et al. Right ventricular functional reserve in pulmonary arterial hypertension. *Circulation* **2016**, *133*, 2413–2422. [CrossRef] [PubMed]
64. Hsu, S.; Kokkonen-Simon, K.M.; Kirk, J.A.; Kolb, T.M.; Damico, R.L.; Mathai, S.C.; Mukherjee, M.; Shah, A.A.; Wigley, F.M.; Margulies, K.B.; et al. Right ventricular myofilament functional differences in humans with systemic sclerosis-associated versus idiopathic pulmonary arterial hypertension. *Circulation* **2018**, *137*, 2360–2370. [CrossRef]
65. Tedford, R.J.; Mudd, J.O.; Girgis, R.E.; Mathai, S.C.; Zaiman, A.L.; Housten-Harris, T.; Boyce, D.; Kelemen, B.W.; Bacher, A.C.; Shah, A.A.; et al. Right ventricular dysfunction in systemic sclerosis-associated pulmonary arterial hypertension. *Circ. Heart Fail.* **2013**, *6*, 953–963. [CrossRef]
66. Rommel, K.-P.; von Roeder, M.; Oberueck, C.; Latuscynski, K.; Besler, C.; Blazek, S.; Stiermaier, T.; Fengler, K.; Adams, V.; Sandri, M.; et al. Load-independent systolic and diastolic right ventricular function in heart failure with preserved ejection fraction as assessed by resting and handgrip exercise pressure–volume loops. *Circ. Heart Fail.* **2018**, *11*, e004121. [CrossRef]
67. Guazzi, M.; Dixon, D.; Labate, V.; Beussink-Nelson, L.; Bandera, F.; Cuttica, M.J.; Shah, S.J. RV Contractile Function and its Coupling to Pulmonary Circulation in Heart Failure With Preserved Ejection Fraction. *JACC Cardiovasc. Imaging* **2017**, *10*, 1211–1221. [CrossRef]
68. Tello, K.; Axmann, J.; Ghofrani, H.A.; Naeije, R.; Narcin, N.; Rieth, A.; Seeger, W.; Gall, H.; Richter, M.J. Relevance of the TAPSE/PASP ratio in pulmonary arterial hypertension. *Int. J. Cardiol.* **2018**, *266*, 229–235. [CrossRef]
69. Huston, J.H.; Maron, B.A.; French, J.; Huang, S.; Thayer, T.; Farber-Eger, E.H.; Wells, Q.S.; Choudhary, G.; Hemnes, A.R.; Brittain, E.L. Association of Mild Echocardiographic Pulmonary Hypertension With Mortality and Right Ventricular Function. *JAMA Cardiol.* **2019**, *4*, 1112–1121. [CrossRef]
70. Grünig, E.; Tiede, H.; Enyimayew, E.O.; Ehlken, N.; Seyfarth, H.-J.; Bossone, E.; D'andrea, A.; Naeije, R.; Olschewski, H.; Ulrich, S.; et al. Assessment and prognostic relevance of right ventricular contractile reserve in patients with severe pulmonary hypertension. *Circulation* **2013**, *128*, 2005–2015. [CrossRef] [PubMed]
71. Saito, Y.; Obokata, M.; Harada, T.; Kagami, K.; Sorimachi, H.; Yuasa, N.; Kato, T.; Wada, N.; Okumura, Y.; Ishii, H. Disproportionate exercise-induced pulmonary hypertension in relation to cardiac output in heart failure with preserved ejection fraction: A non-invasive echocardiographic study. *Eur. J. Heart Fail.* 2023, *Online ahead of print*. [CrossRef] [PubMed]
72. Guo, D.-C.; Li, Y.-D.; Yang, Y.-H.; Zhu, W.-W.; Sun, L.-L.; Jiang, W.; Ye, X.; Cai, Q.-Z.; Lu, X.-Z.; Ye, X.-G. Influence of impaired right ventricular contractile reserve on exercise capacity in patients with precapillary pulmonary hypertension: A study with exercise stress echocardiography. *Echocardiography* **2019**, *36*, 671–677. [CrossRef] [PubMed]
73. Almeida, A.R.; Loureiro, M.J.; Lopes, L.; Cotrim, C.; Lopes, L.; Repolho, D.; Pereira, H. Echocardiographic assessment of right ventricular contractile reserve in patients with pulmonary hypertension. *Rev. Port. Cardiol.* **2014**, *33*, 155–163. [CrossRef]
74. Suzuki, K.; Hirano, Y.; Yamada, H.; Murata, M.; Daimon, M.; Takeuchi, M.; Seo, Y.; Izumi, C.; Akaishi, M. Practical guidance for the implementation of stress echocardiography. *J. Echocardiogr.* **2018**, *16*, 105–129. [CrossRef] [PubMed]

75. Rudski, L.G.; Lai, W.W.; Afilalo, J.; Hua, L.; Handschumacher, M.D.; Chandrasekaran, K.; Solomon, S.D.; Louie, E.K.; Schiller, N.B. Guidelines for the Echocardiographic Assessment of the Right Heart in Adults: A Report from the American Society of Echocardiography endorsed by the European Association of Echocardiography, a registered branch of the European Society of Cardiology, and the Canadian Society of Echocardiography. *J. Am. Soc. Echocardiogr.* **2010**, *23*, 685–713. [CrossRef]
76. Borlaug, B.A.; Nishimura, R.A.; Sorajja, P.; Lam, C.S.; Redfield, M.M. Exercise hemodynamics enhance diagnosis of early heart failure with preserved ejection fraction. *Circ. Heart Fail.* **2010**, *3*, 588–595. [CrossRef]

life

Review

A Meta-Analysis of Sampled Maximal Aerobic Capacity Data for Boys Aged 11 Years Old or Less Obtained by Cycle Ergometry

Iva Jurov [1,*], Jure Demšar [2] and Thomas McCurdy [3,†]

1 Sports Medicine Centre, Clinical Institute of Occupational, Traffic and Sports Medicine, University Medical Centre Ljubljana, 1000 Ljubljana, Slovenia
2 Faculty of Computer and Information Science, University of Ljubljana, 1000 Ljubljana, Slovenia
3 Health and Environmental Impacts Division (HEID), Office of Air Quality Planning and Standards (OAQPS), US EPA, Research Triangle Park, 27709 NC, USA
* Correspondence: iva.jurov@kclj.si; Tel.: +386-4-019-18-08
† Retired.

Abstract: The aim of this study was to develop distributions of VO_{2max} based on measured values that exist in the literature in prepubertal boys using cycle ergometry. PRISMA guidelines were followed in conducting this research. One database was searched for peak and maximal VO_2 values in healthy boys with mean age under 11 years old. Data were split into articles reporting absolute and relative VO_{2max} values and analyzed accordingly. Multilevel models grounded in Bayesian principles were used. We investigated associations between VO_{2max} and body mass, year of the study, and country of origin. Differences in "peak" and "maximal" VO_2 were assessed. Absolute VO_{2max} ($Lmin^{-1}$) increases with age (P ~100%) but mean relative VO_{2max} does not change (P ~100%). Absolute VO_{2max} is higher in more recent studies (P = 95.7 ± 0.3%) and mean relative VO_{2max} is lower (P = 99.6 ± 0.1%). Relative VO_{2max} in the USA is lower compared with boys from other countries (P = 98.8 ± 0.2%), but there are no differences in absolute values. Mean aerobic capacity estimates presented as "peak" values are higher than "maximal" values on an absolute basis (P = 97.5 ± 0.3%) but not on a relative basis (P = 99.6 ± 0.1%). Heavier boys have lower cardiorespiratory fitness (P ≈ 100%), and body mass seems to be increasing faster with age in the USA compared with other countries (P = 92.3 ± 0.3%). New reference values for cardiorespiratory fitness are presented for prepubertal boys obtained with cycle ergometry. This is new, as no reference values have been determined so far based on actual measured values in prepubertal boys. Aerobic capacity normalized to body weight does not change with age. Cardiorespiratory fitness in prepubertal boys is declining, which is associated with increasing body mass over the last few decades. Lastly, this study did not find any statistically significant difference in the sample's mean aerobic capacity estimates using the "peak" and "maximum" distinctions identified in the literature.

Keywords: maximal oxygen consumption; children; boys; cycle ergometry; aerobic fitness

Citation: Jurov, I.; Demšar, J.; McCurdy, T. A Meta-Analysis of Sampled Maximal Aerobic Capacity Data for Boys Aged 11 Years Old or Less Obtained by Cycle Ergometry. *Life* **2023**, *13*, 276. https://doi.org/10.3390/life13020276

Academic Editors: Ines Drenjančević and Helena Lenasi

Received: 2 December 2022
Revised: 14 December 2022
Accepted: 22 December 2022
Published: 19 January 2023

1. Introduction

Cardiopulmonary exercise testing is considered the gold standard for cardiorespiratory fitness (CRF) in pediatric medicine [1]. In exercise testing, maximal oxygen uptake (VO_{2max}) is determined by using indirect calorimetry, which requires a skilled clinician and the use of standardized exercise treadmill protocols or cycle ergometry [2]. Reference values for CRF are needed to assess disease progression, for intervention monitoring, or to assess suboptimal aerobic performance [3]. Even though both treadmills and cycle ergometers are considered criterion measures of CRF, the two methods often produce statistically different estimates for the same child. The differences may be as large as 7–15%, with treadmill estimates being higher than cycle estimates [4–6]. Cycle ergometry has an advantage, however, as the test is not easily constrained by the mechanical limitations of the patient, such as deviant walking patterns and soreness in joints. In addition, during cycle ergometry, there is a lower chance of movement artifacts in the ECG and blood pressure recordings [1].

Another issue in exercise physiology is which criteria should be used for determining CRF. For a VO_{2max} determination, a plateau of VO_2 needs to be achieved. Usually, this is hard to attain in children. If relaxed criteria are used, the highest VO_2 value is then called VO_{2peak}. We know that the exercise physiology literature distinguishes between VO_{2max} and VO_{2peak} metrics, and we preserve this terminological difference in the tables but use CRF to designate maximal oxygen consumption in both cases. Differences in the two metrics are another aspect of uncertainty surrounding what is being measured. In fact, some reviewers state that the highly conditional nature of CRF estimates makes their validity and reliability questionable, especially during growth and maturation [7].

The literature has shown that a statistically significant difference in CRF could exist between girls and boys using absolute VO_{2max} values or relative VO_{2max} to body mass, and boys could have higher CRF values [7]. In addition, young girls and boys participate in different activities, and girls are less involved in organized sports and spend less time practicing [8]. These differences in activities cause boys to be outdoors more than girls, on average [9,10]. Thus, there are physiological and behavioral reasons why CRF estimates should be gender specific. Secondly, younger children spend more time outdoors and undertake more moderate and vigorous physical activities than older children, even if that time is modest—not optimal—for almost everyone [8,11,12]. When looking into CRF for children, using only children with mean age under 11 years of age can minimize physiological complications associated with puberty, resulting in significant changes in total body and muscle mass, stroke volume, growth velocity, oxygen uptake kinetics, fat oxidation rates, and blood lactate responses to work [13–15].

Given the lack of age- and gender-specific CRF reference values in prepubertal children, there is a need to develop observed distributions of VO_{2max} based on criterion methods rather than estimated or regression-based predicted values that are currently widely used. To the best of our knowledge, this is the first meta-analysis to critically examine CRF in boys 11 y old or less measured with cycle ergometry and distinguishing between VO_{2max} and VO_{2peak} indicators. The aim of this analysis is to provide researchers, medical experts, and sports practitioners with criterion-based observed values based on sampled studies identified in the literature. In addition, the aim of this paper is to assess whether there are significant differences in VO_{2max} and VO_{2peak} in boys under 11 y old and to compare values in girls of the same age.

2. Materials and Methods

2.1. Design

The Consideration of Population, Intervention, Comparator, Outcomes, and Study design (PICOS) framework was used.

2.2. Population

Included subjects were subjects who: (1) had mean age under 11 years old, (2) were stated to be healthy, (3) were without cardiovascular disease, pulmonary diseases (except asthma), morbid obesity, developmental disabilities, or muscular dystrophies, and (4) were free from injury. Overweight participants were included. Children with asthma were also included, as they seemed to have physical activity levels comparable with those of the normal pediatric population [16].

2.3. Intervention

Regardless of the interventions reported in many of the original articles, only pre-intervention data were used.

2.4. Comparator

VO_{2max} and VO_{2peak} metrics used to denote aerobic capacity were compared. In addition, comparisons were made based on the year and location of the study (USA versus non-USA countries; conducted solely to provide compatible sample sizes).

2.5. Outcomes

The main outcomes were VO_{2max} and VO_{2peak} metrics measured with cycle ergometry.

2.6. Study Design

Articles were considered for the analysis if: (1) they were published in a peer-reviewed journal, (2) they had mean/standard deviation VO_{2max}/VO_{2peak} parameters for each sample, along with mean age data for the subjects, and (3) if maximal effort was achieved during the incremental test.

These measures provided us with children (girls and boys) that used various testing methods for measuring CRF. Consequently, all articles reporting CRF of girls and mixed-gender groups were removed from further analysis. In addition, analysis excluded articles with graphical results only, field studies, treadmill incremental tests, or other nonstandardized protocols and types of incremental tests. The study followed PRISMA guidelines, and the flow diagram presenting the study design can be found in Figure 1.

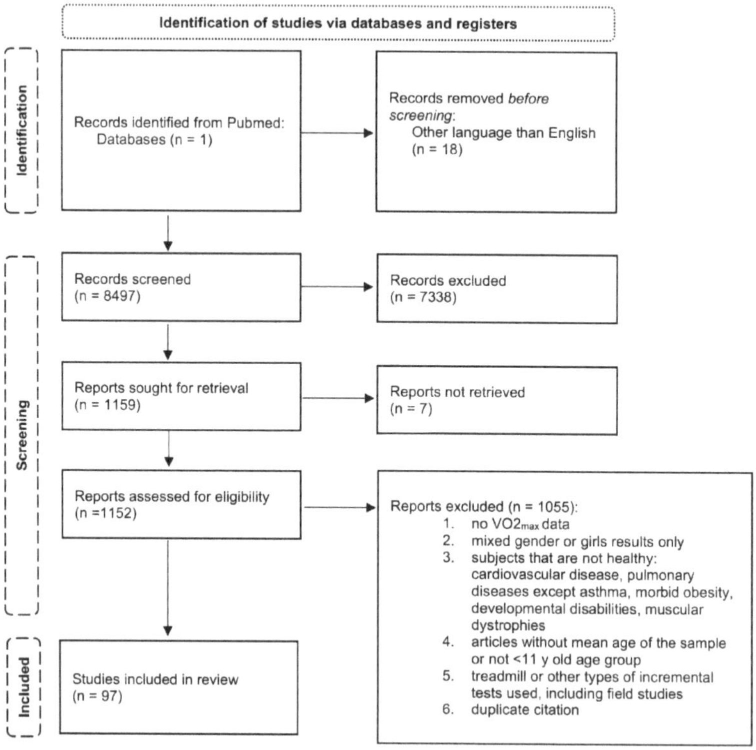

Figure 1. Results of database searches and criteria used in the selection of studies included finding VO_{2max} values in boys under 11 years old.

A systematic electronic literature search was conducted in Pubmed database until 2019 using key search words ((children) AND (oxygen consumption) OR (aerobic power) OR (peak oxygen consumption) OR (VO_2) OR (VO_{2max}) OR (VO_{2peak})). During the first search, potential articles included boys and girls with incremental tests in cycle ergometry and treadmills. All potential articles up to 2019 were hand searched by two researchers. In 2022, the same criteria were used to conduct an additional search from 1 January 2019 to 31 March 2022. Finally, articles that included data from girls and using treadmills were excluded. It is beyond the scope of one manuscript to include boys and girls, so girls were analyzed in a separate paper (in publication). This research includes only the comparison between boys and girls to determine whether there are any differences in cardiorespiratory fitness.

2.7. Statistical Analysis

All statistical conclusions developed in this paper utilize applied Bayesian inferential methodology included in the STAN library for R programming [17]. Among other attributes and capabilities, STAN promotes the use of Bayesian inferential models to allow a researcher to evaluate the likelihood that one distribution—in our case, VO_{2max}—has the same statistical properties as another distribution purported to describe the same phenomenon. In general, random probability Markov chain Monte Carlo (MCMC) algorithms are used to sample from the two distributions to facilitate the comparisons. A wide variety of Bayesian model comparison techniques are available in STAN to facilitate these types of statistical testing.

When comparing two groups, we used a simple normal model (the so-called Bayesian *t*-test):

$$y \sim N(\mu, \sigma),$$

where y is the input data (VO_{2max} or VO_{2peak} measurements), μ is the location parameter, and σ the scale parameter. The default Stan priors (flat improper priors) were used. As observed, y is assumed to be approximately normally distributed.

For the linear regression model, the equation:

$$y \sim N(\alpha_1 + \beta_1 x,\ \alpha_2 + \beta_2 x), \tag{1}$$

was used, where y is the input data (VO_{2max} or VO_{2peak} measurements), x is the dependent variable (e.g., age or year of study), α_1 is the intercept for the location parameter, β_1 is the regression coefficient for the location parameter, α_2 is the intercept for the scale parameter, and β_2 is the regression coefficient for the scale parameter.

In other words, this model can detect both changes in the mean VO_{2max} and the VO_{2max} between-study variance through time or with age. Before making any statistical inferences, we executed all the necessary diagnostics (e.g., trace plots, estimated sample sizes, posterior predictive checks) to ensure the suitability of our models.

With P, we denote the probability that a particular research claim is true. We used a capital P to not confuse the probabilities calculated with Bayesian analyses with P-values from frequentist statistics. Unlike with P-values, with Bayesian statistics, we can directly quantify the probability (P) of a particular research question, which arguably provides us with the most direct, transparent, and intuitive measure of how certain we are about a claim we are making. Note that with Bayesian approaches, we can easily calculate the probability that the opposite of a particular claim is true ($1 - P$). Because of all this, the use of Bayesian statistical analyses has been on the rise over the last couple of years [17–19]. Uncertainty in all our analyses is reported with the Monte Carlo standard error (MCSE) measure.

3. Results

The analyses included 95 study samples that reported absolute values of aerobic capacity (both VO_{2max} and VO_{2peak} metrics) in units of $Lmin^{-1}$ (included articles can be found in Table S1) and 118 study samples that reported relative VO_{2max}/VO_{2peak} measures in units of $mLkg^{-1}min^{-1}$ (included articles can be found in Table S2) [15,20–115]. Observed distributions of $VO_{2max/peak}$ are presented in Table 1.

Table 1. CRF values based on sampled studies for boys under 11 years old obtained with cycle ergometry (mean ± SD).

Age (Years)	4–5	5–6	6–7	7–8	8–9	9–10	10–11
$VO_{2max/peak}$ ($mLkg^{-1}min^{-1}$)	44.29 ± 7.28	44.67 ± 7.20	45.04 ± 7.12	45.41 ± 7.04	45.79 ± 6.96	46.16 ± 6.88	46.54 ± 6.80
Body mass (kg)	16.43 ± 3.58	19.96 ± 4.16	23.50 ± 4.75	27.03 ± 5.34	30.56 ± 5.93	34.09 ± 6.51	37.62 ± 7.10
$VO_{2max/peak}$ ($Lmin^{-1}$)	0.78 ± 0.14	0.93 ± 0.15	1.08 ± 0.16	1.23 ± 0.17	1.38 ± 0.18	1.53 ± 0.19	1.68 ± 0.20
Body mass (kg)	16.37 ± 3.01	19.68 ± 3.24	22.99 ± 3.47	26.31 ± 3.71	29.62 ± 3.94	32.93 ± 4.18	36.24 ± 4.41

$Lmin^{-1}$, absolute cardiorespiratory fitness.

We are as sure as we can be (P ~100%) that absolute CRF increases with age. The probability that the between-study standard deviation increases with age is 87.1 ± 1.4% (Figure 2). Looking into differences between VO_{2max} and VO_{2peak}, this study suggests that mean VO_{2peak} is larger than mean VO_{2max}. We can claim this with a probability of 97.5 ± 0.3% (Figure 3). When checking for any changes across the years, the analysis showed that the mean absolute CRF is higher in more recent studies. We can claim this with a probability of 95.7 ± 0.3%. The between-study variability seems to be dropping, but we can claim this only with less than a 90% certainty (P = 89.9 ± 0.3%) (Figure 4). We also looked for any differences between CRF in the USA and other countries of the world. We cannot claim there are differences here (Figure 5). Lastly, we looked into body mass. In articles reporting absolute values, CRF ($Lmin^{-1}$) is higher in boys with greater body mass ($P \approx 100\%$) (Figure 6). Looking into differences in body mass between the USA and countries in the rest of the world, there are no significant differences between trends in body mass (Figure 7), but in general, USA boys seem to be heavier (P = 96.05 ± 0.4%). What is more, boys in studies using VO_{2peak} seem to be heavier than those with VO_{2max} (P = 98.7 ± 0.2%).

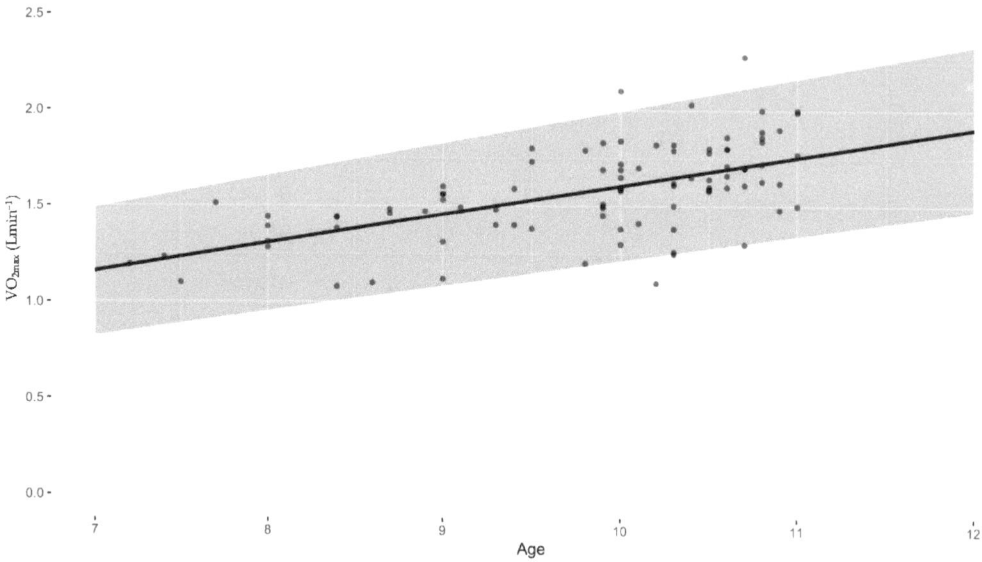

Figure 2. *Cont.*

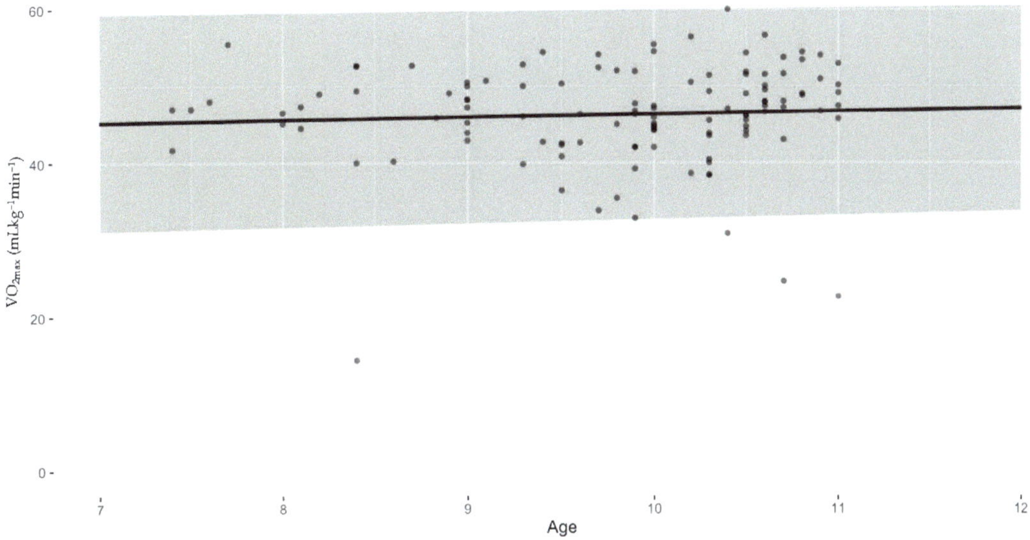

Figure 2. Mean absolute VO_{2max} (Lmin^{-1}) increases with age (**upper** figure) in contrast to mean relative VO_{2max} (mLkg^{-1}min^{-1}) (**lower** figure) and its standard deviation, which do not change with age in boys under 11 years old.

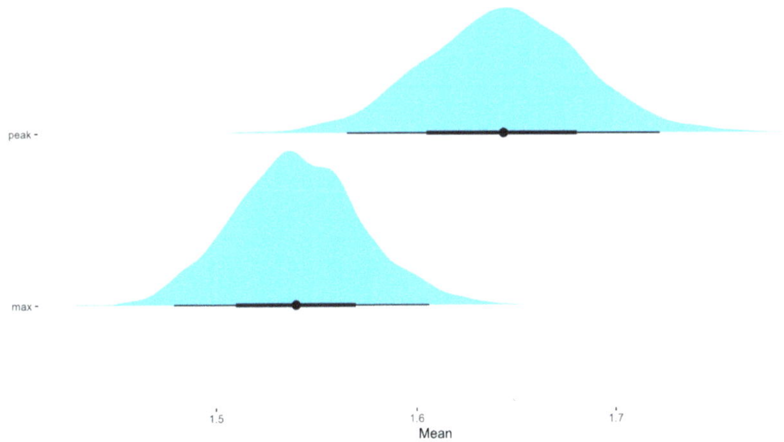

Figure 3. *Cont.*

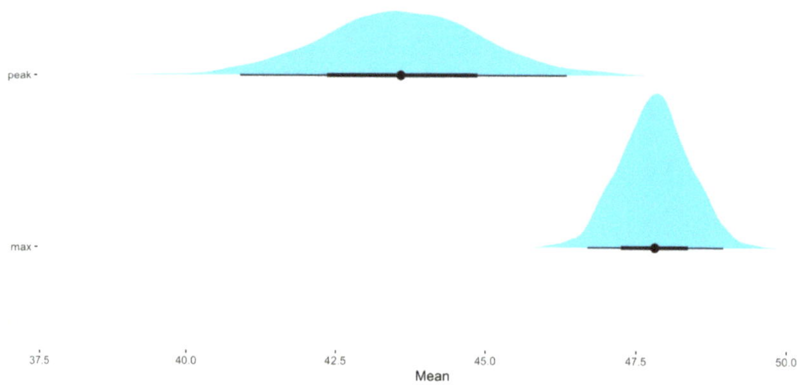

Figure 3. The distribution of mean absolute peak VO_2 values—VO_{2peak} ($Lmin^{-1}$)—and maximal VO_2 values—VO_{2max} ($Lmin^{-1}$)—shows that mean VO_{2peak} is higher (**upper** figure). However, mean relative VO_{2peak} ($mLkg^{-1}min^{-1}$) is lower than mean relative VO_{2max} ($mLkg^{-1}min^{-1}$) (**lower** figure).

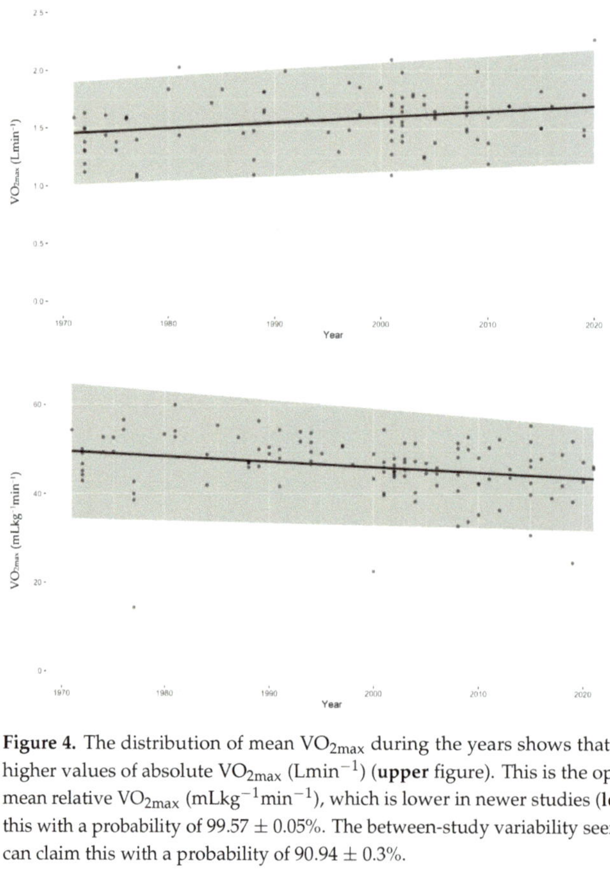

Figure 4. The distribution of mean VO_{2max} during the years shows that more recent studies have higher values of absolute VO_{2max} ($Lmin^{-1}$) (**upper** figure). This is the opposite of the finding with mean relative VO_{2max} ($mLkg^{-1}min^{-1}$), which is lower in newer studies (**lower** figure). We can claim this with a probability of 99.57 ± 0.05%. The between-study variability seems to be dropping, and we can claim this with a probability of 90.94 ± 0.3%.

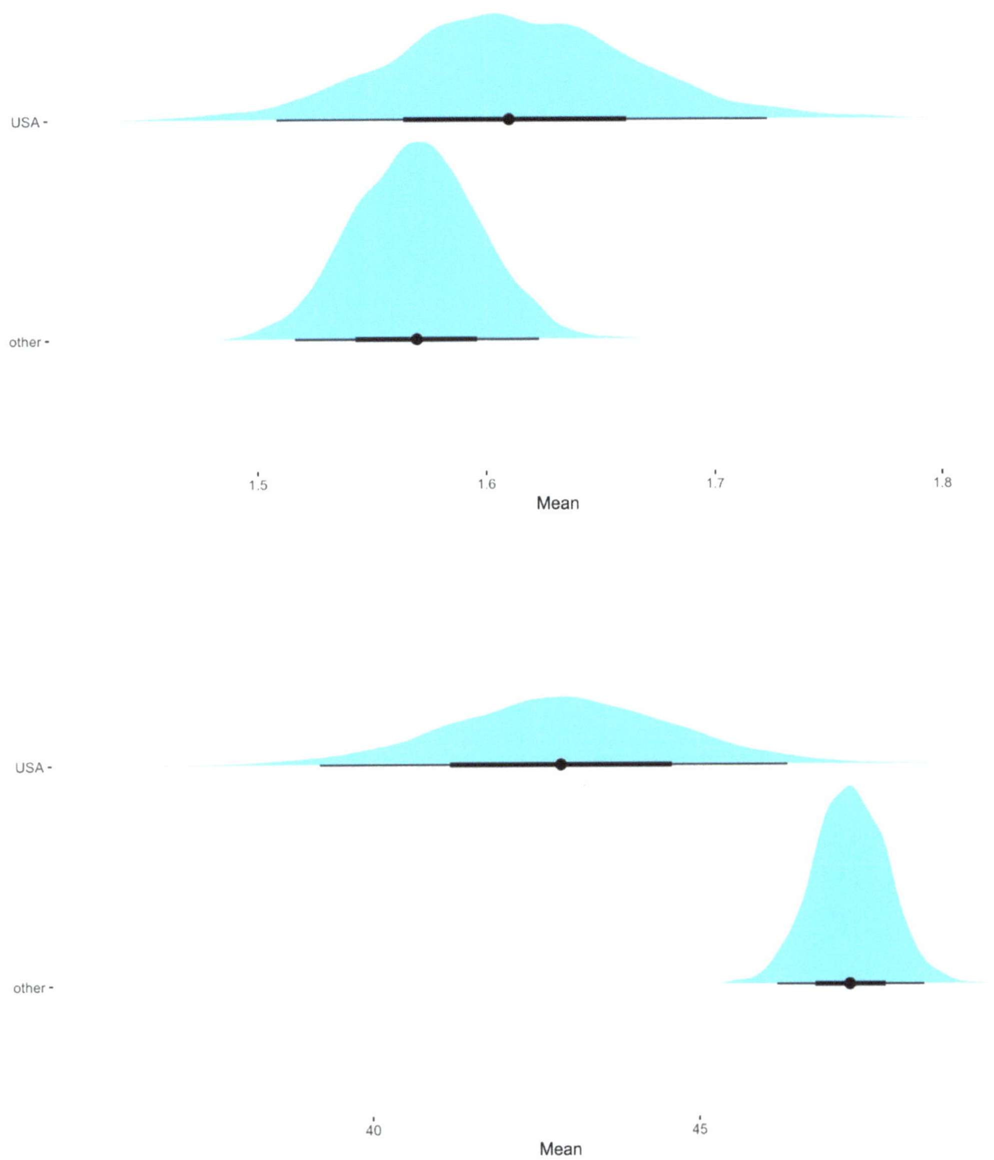

Figure 5. There are no differences in absolute VO_{2max} ($Lmin^{-1}$) between studies with subjects from USA and other countries (**upper** figure), but looking into studies using relative values, higher VO_{2max} ($mLkg^{-1}min^{-1}$) values were reported in subjects from other countries than in USA (**lower** figure). We can claim this with a probability of $98.75 \pm 0.2\%$.

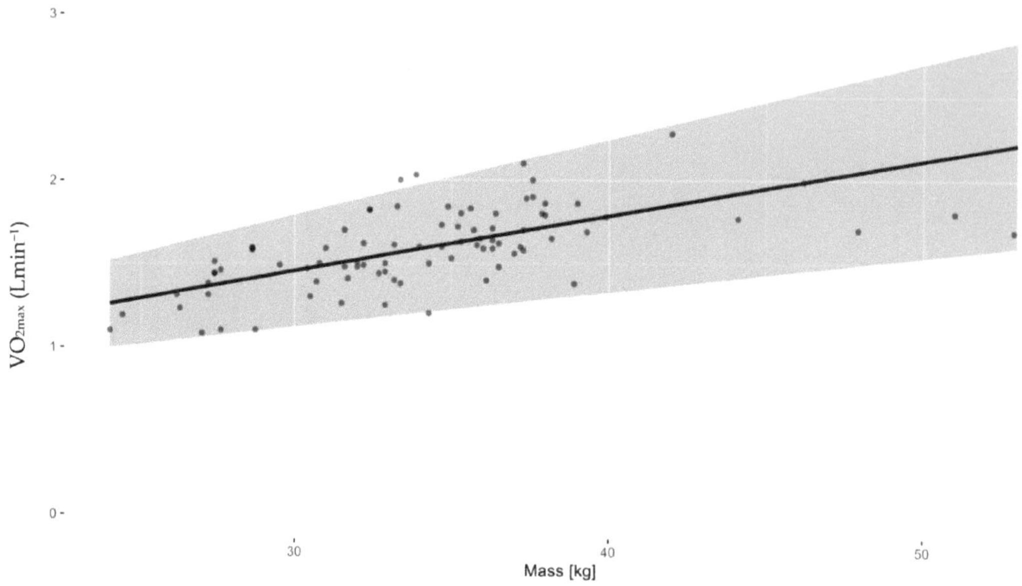

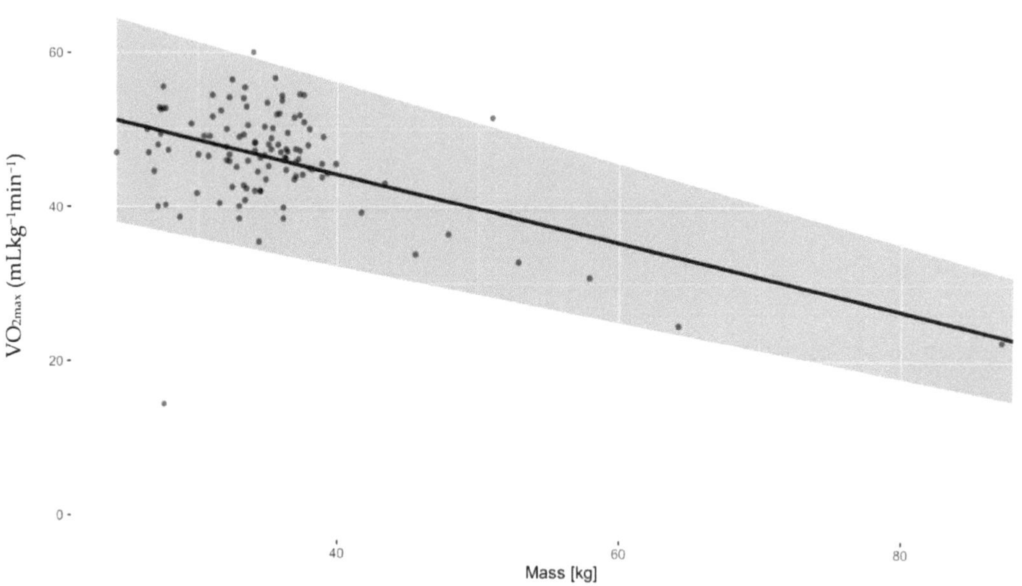

Figure 6. Absolute values of VO_{2max} ($Lmin^{-1}$) are higher in boys with greater body mass ($P \approx 100\%$) (**upper** figure), whereas mean relative VO_{2max} ($mLkg^{-1}min^{-1}$) is lower when participants have higher body mass ($P \approx 100\%$) (**lower** figure).

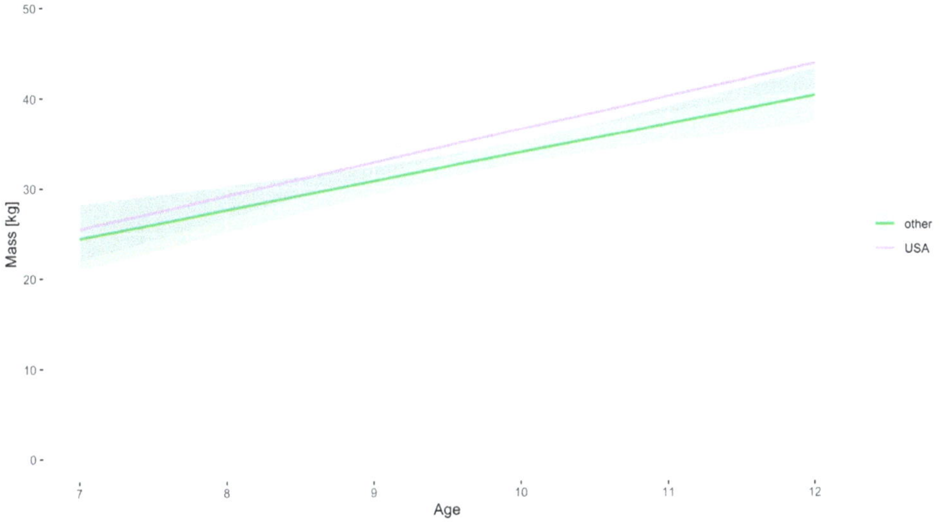

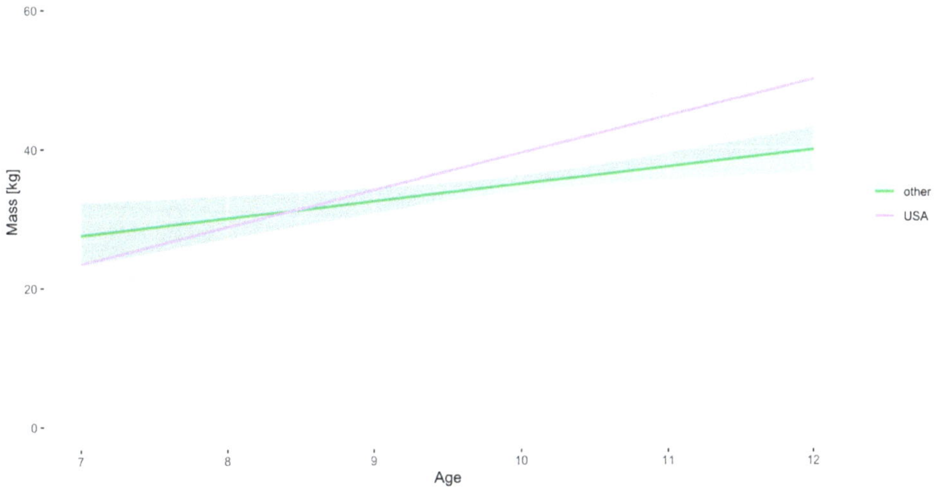

Figure 7. There are no significant differences between trends in articles reporting absolute VO_{2max} values ($Lmin^{-1}$) comparing USA and other countries (**upper** figure) as opposed to articles reporting relative VO_{2max} values ($mLkg^{-1}min^{-1}$) (**lower** figure). Analysis showed that in those articles' bodies: $mLkg^{-1}min^{-1}$, relative cardiorespiratory fitness.

We cannot claim that the mean relative CRF or its standard deviation changes with age (Figure 2). Secondly, we checked for differences between VO_{2peak} and VO_{2max}. The opposite was found in the absolute CRF. Our study suggests that mean VO_{2max} is higher than mean VO_{2peak}. We can claim this with a probability of 99.6 ± 0.1% (Figure 3). Moreover, it seems that the mean relative CRF is lower in more recent studies. We can claim this with a probability of 99.6 ± 0.1%. The between-study variability seems to be dropping, which we can claim with a probability of only 90.9 ± 0.3% (Figure 4). Our study also suggests that the mean relative CRF in other countries is higher than in the USA. We can claim this with a probability of 98.8 ± 0.2% (Figure 5). Investigating body mass in articles reporting relative values, mean relative CRF ($mLkg^{-1}min^{-1}$) is lower when participants have higher body mass ($P \approx 100\%$) (Figure 6). It could be that USA boys are a bit heavier on average, but we cannot claim this with a very high probability (P = 82.9 ± 0.7%). We did, however, observe that in these articles, body mass seems to be increasing faster with age in the USA compared to other countries (P = 92.3 ± 0.3%) (Figure 7). Finally, boys in studies using VO_{2peak} also seem to be heavier in articles reporting relative values compared with studies with VO_{2max} (P = 99.6 ± 0.1%).

3.1. Is There Any Difference between Boys and Girls?

No significant differences exist between relative cardiorespiratory fitness values in prepubertal boys and girls. The probability that boys have lower values than girls is only 73.6 ± 1%.

3.2. Models in Practice

At https://demsarjure.shinyapps.io/vo2max/, (access date: 26 December 2022) a simple app can be found in which the measured VO_{2max} in $Lmin^{-1}$, participant's age, and weight are put in the calculator. The app then uses the fitted Bayesian models to calculate and visualize the percentile for the data that were provided. The app calculates absolute VO_{2max} when VO_{2max} and age are provided and relative VO_{2max} when weight is provided as well. The dashed vertical lines denote the 95% CI.

4. Discussion

To the best of our knowledge, this is a meta-analysis with the largest dataset of CRF measurements in boys with mean age under 11 that performed cycle ergometry. Based on the articles included, normative values for prepubertal boys are presented, and a prediction model based on age has been developed for researchers and clinicians to use.

Children with mean ages 4 to 11 were included in this meta-analysis. Relative CRF (normalized to body mass) or its standard deviation did not change with age, which is in line with norms found in boys from 8 to 18 years old [116]. However, articles reporting CRF not normalized to body mass (mean absolute CRF) showed that CRF and its standard deviation in prepubertal boys are dependent on age. This can be explained by increasing body mass as boys age. Body mass is metabolically active tissue that uses oxygen consumption during exercise. This finding is supported by higher absolute CRF in heavier boys in this meta-analysis. Interestingly, mean relative CRF is lower in heavier boys, which we suggest can be explained by the methodology used. We excluded only morbid obesity, so overweight subjects were included in our analysis. The decision to include them seemed necessary since obesity has become a global epidemic during the last three decades, especially in developed countries. In 2013, 23.8% of boys were overweight or obese [117]. In children with obesity, CRF has declined in the last decades, and it is vital to improve the level of physical activity and to improve their aerobic fitness [118]. Creating normative values for boys with normal body mass index only would not be useful for those who are overweight or obese. Heavier children might be more susceptible to cardiovascular risk later in life and will need clinical evaluation and follow-up. To conclude, heavier boys seem to have lower aerobic capacity, which can be explained by overweight individuals also included in this analysis. Having normative values for boys (both absolute and relative CRF values) is, thus, necessary for understanding an individual's fitness and could gain even greater importance as boys seem to have become heavier in recent years.

Although specifics of determining VO_{2peak} in contrast to VO_{2max} are widely discussed in the literature, there are no recommendations for their use in children based on large studies. This analysis showed that mean VO_{2peak} is higher than mean VO_{2max} in studies reporting absolute values, and the opposite was found in studies with relative values. In theory, lower VO_{2peak} than VO_{2max} could be explained by subjects not reaching their actual maximal oxygen uptake in articles using VO_{2peak} metric, which is a general concern when reporting maximal VO_2 in children and adults. However, there is no clear explanation why children with measured absolute values would show higher VO_{2peak} than mean VO_{2max}. Our study cannot provide clarification of this finding. However, we would like to suggest that higher body mass in boys with VO_{2peak} as compared with boys with VO_{2max} could be a reason for this. We also observed higher body mass in studies using relative VO_{2peak} as compared with boys with relative VO_{2max}, but these values are normalized to body mass.

In more recent studies, absolute CRF values (not normalized to body mass) are higher and mean relative CRF values (normalized to body mass) are lower. We can assume this is the result of increasing body mass in boys involved in the studies analyzed. The between-study variability is dropping, which we suggest can be explained by improved methodological approaches and more articles in recent years (85 groups prior to 1995 vs. 128 subject groups after 1995). Understanding that based on these results, prepubertal boys are becoming less fit, which is not only observed or estimated but can now be supported by actual VO_{2max} measurements as well.

Finally, there is no difference between absolute CRF in the USA and other countries, but relative values are lower in the USA. We found an association between higher body mass and higher absolute values, and we can also say with certainty that USA boys are heavier in the studies included in our analysis. Both absolute and relative values thus indicate that prepubertal boys from the USA have lower endurance capacity than boys from other countries. If we try to interpret that with the data from our analysis that body mass in the USA increases faster than in other countries, we can expect endurance capacity to decrease even further in the future. This is alarming since CRF is the most important marker of health among the health-related physical fitness components in children and adolescents [119–121], and there is an inverse relationship between cardiorespiratory fitness during childhood and cardiovascular disease risk factors in adulthood [122].

Limitations

There are some limitations that should be considered. Firstly, we did not distinguish among the many protocols used for each of the approaches during cycle ergometry. These protocols are quite important in estimating aerobic capacity, but protocol nuances used by individual laboratories make sorting them into logical categories very difficult. The same can be said about criteria used to determine whether a child has attained his personal best CRF for a particular test averaging time [123]. These criteria usually include: RER ($\geq$1.0), no change in VO_2 with increasing workload (i.e., a plateau in VO_2), visible signs of exhaustion, and attainment of age-predicted heart rate or some percentage of it [5,124] but are not identical in all studies involved.

5. Conclusions

New reference values for CRF are presented for prepubertal boys that can be used for physical fitness classification on an individual level for medical experts and sports practitioners. Aerobic capacity normalized to body weight does not change with mean age in boys 4–11 years old, which can be very useful in clinical settings for early diagnosis of reduced cardiorespiratory fitness. It seems that values are not different from those in prepubertal girls.

CRF in prepubertal boys is declining. Our results show that this is associated with increasing body mass, but it also suggests boys with mean age under 11 years old might be less active than in the past. In addition, aerobic capacity is lower in boys in the USA, which is associated with increased body mass. In light of the obesity pandemic, these results indicate more action is needed to improve physical activity in prepubertal boys in order to reduce the likelihood of increased cardiovascular risk later in life. CRF references presented by this analysis can aid in evaluating obesity criteria and physical fitness in prepubertal boys.

Finally, this study did not find any advantage in determining CRF values with the VO_{2peak} or VO_{2max} metrics. Based on our findings, it seems that in prepubertal boys, the differences are not significant enough to be important. This can help researchers and clinicians as they perform cardiopulmonary tests on prepuberal boys.

Supplementary Materials: The following supporting information can be downloaded at: https://www.mdpi.com/article/10.3390/life13020276/s1, Table S1: Included Studies for Absolute $VO_{2max/peak}$ values ($Lmin^{-1}$); Table S2: Included Studies for Relative $VO_{2max/peak}$ values ($Lmin^{-1}$).

Author Contributions: Conceptualization, I.J. and T.M.; methodology, I.J., T.M. and J.D.; software, J.D.; formal analysis, I.J. and J.D.; data curation, I.J. and T.M.; writing—original draft preparation, I.J., T.M. and J.D.; writing—review and editing, I.J. and T.M.; supervision, T.M.; project administration, I.J. All authors have read and agreed to the published version of the manuscript.

Funding: This research received no external funding.

Institutional Review Board Statement: The study was conducted in accordance with the Declaration of Helsinki.

Informed Consent Statement: Not applicable.

Data Availability Statement: All data are available in the Supplementary Materials.

Acknowledgments: Linda McCurdy assisted in the 100% check of data inputs. We thank Alenka Franko for her support in this project.

Conflicts of Interest: The authors declare no conflict of interest.

References

1. Takken, T.; Bongers, B.C.; van Brussel, M.; Haapala, E.A.; Hulzebos, E.H.J. Cardiopulmonary Exercise Testing in Pediatrics. *Ann. Am. Thorac. Soc.* **2017**, *14* (Suppl. S1), S123–S128. [CrossRef] [PubMed]
2. Balady, G.J.; Arena, R.; Sietsema, K.; Myers, J.; Coke, L.; Fletcher, G.F.; Forman, D.; Franklin, B.; Guazzi, M.; Gulati, M.; et al. Clinician's Guide to Cardiopulmonary Exercise Testing in Adults: A Scientific Statement from the American Heart Association. *Circulation* **2010**, *122*, 191–225. [CrossRef] [PubMed]
3. Palange, P.; Laveneziana, P.; Neder, J.A.; Ward, S.A. *Clinical Exercise Testing*; Palange, P., Laveneziana, P., Neder, J.A., Ward, S.A., Eds.; European Respiratory Society: Sheffield, UK, 2018.
4. Forbregd, T.R.; Aloyseus, M.A.; Berg, A.; Greve, G. Cardiopulmonary Capacity in Children During Exercise Testing: The Differences Between Treadmill and Upright and Supine Cycle Ergometry. *Front. Physiol.* **2019**, *10*, 1440. [CrossRef]
5. Bergh, U.; Kanstrup, I.L.; Ekblom, B. Maximal Oxygen Uptake during Exercise with Various Combinations of Arm and Leg Work. *J. Appl. Physiol.* **1976**, *41*, 191–196. [CrossRef] [PubMed]
6. LeMura, L.M.; von Duvillard, S.P.; Cohen, S.L.; Root, C.J.; Chelland, S.A.; Andreacci, J.; Hoover, J.; Weatherford, J. Treadmill and Cycle Ergometry Testing in 5- to 6-Year-Old Children. *Eur. J. Appl. Physiol.* **2001**, *85*, 472–478. [CrossRef]
7. Armstrong, N. Aerobic Fitness and Physical Activity in Children. *Pediatr. Exerc. Sci.* **2013**, *25*, 548–560. [CrossRef]
8. Whiting, S.; Buoncristiano, M.; Gelius, P.; Abu-Omar, K.; Pattison, M.; Hyska, J.; Duleva, V.; Musić Milanović, S.; Zamrazilová, H.; Hejgaard, T.; et al. Physical Activity, Screen Time, and Sleep Duration of Children Aged 6–9 Years in 25 Countries: An Analysis within the WHO European Childhood Obesity Surveillance Initiative (COSI) 2015–2017. *Obes. Facts* **2021**, *14*, 32. [CrossRef] [PubMed]
9. Trost, S.G.; Rosenkranz, R.R.; Dzewaltowski, D. Physical Activity Levels among Children Attending After-School Programs. *Med. Sci. Sports Exerc.* **2008**, *40*, 622–629. [CrossRef]
10. Hinkley, T.; Crawford, D.; Salmon, J.; Okely, A.D.; Hesketh, K. Preschool Children and Physical Activity. A Review of Correlates. *Am. J. Prev. Med.* **2008**, *34*, 435–441. [CrossRef]
11. Bailey, R.C.; Olson, J.; Pepper, S.L.; Porszasz, J.; Barstow, T.J.; Cooper, D.M. The Level and Tempo of Children's Physical Activities: An Observational Study. *Med. Sci. Sports Exerc.* **1995**, *27*, 1033–1041. [CrossRef]
12. Pate, R.R.; O'Neill, J.R.; Liese, A.D.; Janz, K.F.; Granberg, E.M.; Colabianchi, N.; Harsha, D.W.; Condrasky, M.M.; O'Neil, P.M.; Lau, E.Y.; et al. Factors Associated with Development of Excessive Fatness in Children and Adolescents: A Review of Prospective Studies. *Obes. Rev.* **2013**, *14*, 645–658. [CrossRef] [PubMed]
13. Armstrong, N.; Welsman, J.R. Assessment and Interpretation of Aerobic Fitness in Children and Adolescents. *Exerc. Sport Sci. Rev.* **1994**, *22*, 435–476. [CrossRef] [PubMed]
14. Crisp, N.A.; Guelfi, K.J.; Licari, M.K.; Braham, R.; Fournier, P.A. Does Exercise Duration Affect Fat Max in Overweight Boys? *Eur. J. Appl. Physiol.* **2012**, *112*, 2557–2564. [CrossRef] [PubMed]
15. Cunningham, D.A.; Paterson, D.H.; Blimkie, C.J.R.; Donner, A.P. Development of Cardiorespiratory Function in Circumpubertal Boys: A Longitudinal Study. *J. Appl. Physiol. Respir. Environ. Exerc. Physiol.* **1984**, *56*, 302–307. [CrossRef] [PubMed]
16. Welsh, L.; Roberts, R.G.D.; Kemp, J.G. Fitness and Physical Activity in Children with Asthma. *Sport. Med.* **2004**, *34*, 861–870. [CrossRef] [PubMed]

17. Stan Development Team RStan: The R Interface to Stan. Available online: https://mc-stan.org/ (accessed on 5 May 2022).
18. Field, A.P. *Discovering Statistics Using IBM SPSS Statistics*, 5th ed.; SAGE Publications, Inc.: Thousand Oaks, CA, USA, 2018; p. 368.
19. Odor, P.M.; Bampoe, S.; Cecconi, M. Cardiac Output Monitoring: Validation Studies–How Results Should Be Presented. *Curr. Anesthesiol. Rep.* **2017**, *7*, 410–415. [CrossRef] [PubMed]
20. Agbaje, A.O.; Haapala, E.A.; Lintu, N.; Viitasalo, A.; Barker, A.R.; Takken, T.; Tompuri, T.; Lindi, V.; Lakka, T.A. Peak Oxygen Uptake Cut-Points to Identify Children at Increased Cardiometabolic Risk–The PANIC Study. *Scand. J. Med. Sci. Sports* **2019**, *29*, 16–24. [CrossRef]
21. Andersen, K.L.; Ghesquiere, J. Sex Differences in Maximal Oxygen Uptake, Heart Rate and Oxygen Pulse at 10 and 14 Years in Norwegian Children. *Hum. Biol.* **1972**, *44*, 413–431.
22. Arngrímsson, S.Á.; Sveinsson, T.; Jóhannsson, E. Peak Oxygen Uptake in Children: Evaluation of an Older Prediction Method and Development of a New One. *Pediatr. Exerc. Sci.* **2008**, *20*, 62–73. [CrossRef]
23. Crisp, N.A.; Fournier, P.A.; Licari, M.K.; Braham, R.; Guelfi, K.J. Adding Sprints to Continuous Exercise at the Intensity That Maximises Fat Oxidation: Implications for Acute Energy Balance and Enjoyment. *Metabolism* **2012**, *61*, 1280–1288. [CrossRef]
24. Davies, C.T.; Barnes, C.; Godfrey, S. Body Composition and Maximal Exercise Performance in Children. *Hum. Biol.* **1972**, *44*, 195–214. [PubMed]
25. de Jonge, R. Effect of Anthropometric Characteristics and Socio-Economic Status on Physical Performances of Pre-Pubertal Children Living in Bolivia at Low Altitude. *Eur. J. Appl. Physiol. Occup. Physiol.* **1996**, *74*, 367–374. [CrossRef] [PubMed]
26. Del Corral, P.; Mahon, A.D.; Duncan, G.E.; Howe, C.A.; Craig, B.W. The Effect of Exercise on Serum and Salivary Cortisol in Male Children. *Med. Sci. Sports Exerc.* **1994**, *26*, 1297–1301. [CrossRef] [PubMed]
27. Docherty, D.; Gaul, C.A. Relationship of Body Size, Physique, and Composition to Physical Performance in Young Boys and Girls. *Int. J. Sports Med.* **1991**, *12*, 525–532. [CrossRef]
28. Dunstheimer, D.; Hebestreit, H.; Staschen, B.; Starßburg, H.M.; Jeschke, R. Bilateral Deficit during Short-Term, High-Intensity Cycle Ergometry in Girls and Boys. *Eur. J. Appl. Physiol.* **2001**, *84*, 557–561. [CrossRef] [PubMed]
29. Dzurenkova, D.; Marcek, T.; Hajkova, M. Specialities of Assessment of Endurance Capabilities in Sport Active Children. *Bratisl. Lek. Listy* **2001**, *102*, 432–433.
30. Farr, C.; Middlebrooke, A.R.; Armstrong, N.; Barker, A.R.; Fulford, J.; Mawson, D.M.; McManus, A.M. Objectively Measured Aerobic Fitness Is Not Related to Vascular Health Outcomes and Cardiovascular Disease Risk in 9-10 Year Old Children. *J. Sports Sci. Med.* **2019**, *18*, 513–522. [PubMed]
31. Fawkner, S.G.; Armstrong, N. Sex Differences in the Oxygen Uptake Kinetic Response to Heavy-Intensity Exercise in Prepubertal Children. *Eur. J. Appl. Physiol.* **2004**, *93*, 210–216. [CrossRef]
32. Friberg, S.; Bevegård, S.; Graff-Lonnevig, V.; Hallbäck, I. Asthma from Childhood to Adulthood-A Follow-up Study of 20 Subjects with Special Reference to Work Capacity and Pulmonary Gas Exchange. *J. Allergy Clin. Immunol.* **1989**, *84*, 183–190. [CrossRef]
33. Gaisl, G.; Buchberger, J. *Determination of the Aerobic and Anaerobic Thresholds of 10-11-Year-Old Boys Using Blood-Gas Analysis*; Berg, W., Eriksson, B.O., Eds.; University Park Press: Baltimore, MD, USA, 1980.
34. Gilliam, T.B.; Katch, V.L.; Thorland, W.; Weltman, A. Prevalence of Coronary Heart Disease Risk Factors in Active Children, 7 to 12 Years of Age. *Med. Sci. Sports Exerc.* **1977**, *9*, 21–25. [CrossRef]
35. Hansen, H.S.; Froberg, K.; Nielsen, J.R.; Hyldebrandt, N. A New Approach to Assessing Maximal Aerobic Power in Children: The Odense School Child Study. *Eur. J. Appl. Physiol. Occup. Physiol.* **1989**, *58*, 618–624. [CrossRef] [PubMed]
36. Hermansen, L.; Oseid, S. Direct and Indirect Estimation of Maximal Oxygen Uptake in Pre-Pubertal Boys. *Acta Paediatr. Scand. Suppl.* **1971**, *217*, 18–23. [CrossRef] [PubMed]
37. Ikai, M.; Kitagawa, K. Maximum Oxygen Uptake of Japanese Related to Sex and Age. *Med. Sci. Sports Exerc.* **1972**, *4*, 127–131. [CrossRef]
38. Inbar, O.; Bar-Or, O. The Effects of Intermittent Warm-up on 7-9 Year-Old Boys. *Eur. J. Appl. Physiol. Occup. Physiol.* **1975**, *34*, 81–89. [CrossRef]
39. Janz, K.F.; Burns, T.L.; Witt, J.D.; Mahoney, L.T. Longitudinal Analysis of Scaling VO_2 for Differences in Body Size during Puberty: The Muscatine Study. *Med. Sci. Sports Exerc.* **1998**, *30*, 1436–1444. [CrossRef]
40. Janz, K.F.; Dawson, J.D.; Mahoney, L.T. Tracking Physical Fitness and Physical Activity from Childhood to Adolescence: The Muscatine Study. *Med. Sci. Sports Exerc.* **2000**, *32*, 1250–1257. [CrossRef]
41. Kolle, E.; Steene-Johannessen, J.; Andersen, L.B.; Anderssen, S.A. Objectively Assessed Physical Activity and Aerobic Fitness in a Population-Based Sample of Norwegian 9- and 15-Year-Olds. *Scand. J. Med. Sci. Sports* **2010**, *20*, e41–e47. [CrossRef]
42. Kowaluk, A.; Woźniewski, M. Peak Oxygen Uptake and Exercise Capacity of Children Undergoing Leukemia Treatment. *Int. J. Environ. Res. Public Health* **2020**, *17*, 1–18. [CrossRef] [PubMed]
43. Lange Andersen, K.; Seliger, V.; Rutenfranz, J.; Skrobak-Kaczynski, J. Physical Performance Capacity of Children in Norway-Part IV. The Rate of Growth in Maximal Aerobic Power and the Influence of Improved Physical Education of Children in a Rural Community-Population Parameters in a Rural Community. *Eur. J. Appl. Physiol. Occup. Physiol.* **1976**, *35*, 49–58. [CrossRef]
44. Lange Andersen, K.; Seliger, V.; Rutenfranz, J.; Mocellin, R. Physical Performance Capacity of Children in Norway-Part I. Population Parameters in a Rural Inland Community with Regard to Maximal Aerobic Power. *Eur. J. Appl. Physiol. Occup. Physiol.* **1974**, *33*, 177–195. [CrossRef]

45. Lazzer, S.; Molin, M.; Stramare, D.; Facchini, S.; Francescato, M.P. Effects of an Eight-Month Weight-Control Program on Body Composition and Lipid Oxidation Rate during Exercise in Obese Children. *J. Endocrinol. Invest.* **2008**, *31*, 509–514. [CrossRef] [PubMed]
46. Lintu, N.; Viitasalo, A.; Tompuri, T.; Veijalainen, A.; Hakulinen, M.; Laitinen, T.; Savonen, K.; Lakka, T.A. Cardiorespiratory Fitness, Respiratory Function and Hemodynamic Responses to Maximal Cycle Ergometer Exercise Test in Girls and Boys Aged 9–11 Years: The PANIC Study. *Eur. J. Appl. Physiol.* **2015**, *115*, 235–243. [CrossRef] [PubMed]
47. Lu, K.D.; Bar-Yoseph, R.; Radom-Aizik, S.; Cooper, D.M. A New Approach to Estimate Aerobic Fitness Using the NHANES Dataset. *Scand. J. Med. Sci. Sports* **2019**, *29*, 1392–1401. [CrossRef] [PubMed]
48. Mahon, A.D.; Stephens, B.R.; Cole, A.S. Exercise Responses in Boys with Attention Deficit/Hyperactivity Disorder: Effects of Stimulant Medication. *J. Atten. Disord.* **2008**, *12*, 170–176. [CrossRef] [PubMed]
49. Massicotte, D.R.; Gauthier, R.; Markon, P. Prediction of VO_{2max} from the Running Performance in Children Aged 10-17 Years. *J. Sport. Med. Phys. Fit.* **1985**, *25*, 10–17.
50. McManus, A.M.; Cheng, C.H.; Leung, M.P.; Yung, T.C.; Macfarlane, D.J. Improving Aerobic Power in Primary School Boys: A Comparison of Continuous and Interval Training. *Int. J. Sports Med.* **2005**, *26*, 781–786. [CrossRef]
51. McMurray, R.G.; Harrell, J.S.; Bradley, C.; Deng, S.; Bangdiwala, S.I. Predicted Maximal Aerobic Power in Youth Is Related to Gender, Ethnicity, and Body Composition. *Med. Sci. Sports Exerc.* **2001**, *33*, S31. [CrossRef]
52. McNarry, M.A.; Farr, C.; Middlebrooke, A.; Welford, D.; Breese, B.; Armstrong, N.; Barker, A.R. Aerobic Function and Muscle Deoxygenation Dynamics during Ramp Exercise in Children. *Med. Sci. Sports Exerc.* **2015**, *47*, 1877–1884. [CrossRef]
53. Nourry, C.; Fabre, C.; Bart, F.; Grosbois, J.M.; Berthoin, S.; Mucci, P. Evidence of Exercise-Induced Arterial Hypoxemia in Prepubescent Trained Children. *Pediatr. Res.* **2004**, *55*, 674–681. [CrossRef]
54. Ortega, F.B.; Ruiz, J.R.; Hurtig-Wennlöf, A.; Vicente-Rodríguez, G.; Rizzo, N.S.; Castillo, M.J.; Sjöström, M. Cardiovascular Fitness Modifies the Associations between Physical Activity and Abdominal Adiposity in Children and Adolescents: The European Youth Heart Study. *Br. J. Sports Med.* **2010**, *44*, 256–262. [CrossRef]
55. Peralta-Huertas, J.; Livingstone, K.; Banach, A.; Klentrou, P.; O'Leary, D. Differences in Left Ventricular Mass between Overweight and Normal-Weight Preadolescent Children. *Appl. Physiol. Nutr. Metab.* **2008**, *33*, 1172–1180. [CrossRef] [PubMed]
56. Rivera-Brown, A.M.; Alvarez, M.; Rodríguez-Santana, J.R.; Benetti, P.J. Anaerobic Power and Achievement of VO_2 Plateau in Pre-Pubertal Boys. *Int. J. Sports Med.* **2001**, *22*, 111–115. [CrossRef] [PubMed]
57. Robertson, R.J.; Goss, F.L.; Boer, N.; Gallagher, J.D.; Thompkins, T.; Bufalino, K.; Balasekaran, G.; Meckes, C.; Pintar, J.; Williams, A. OMNI Scale Perceived Exertion at Ventilatory Breakpoint in Children: Response Normalized. *Med. Sci. Sports Exerc.* **2001**, *33*, 1946–1952. [CrossRef] [PubMed]
58. Rowland, T.W.; Rambusch, J.M.; Staab, J.S.; Unnithan, V.B.; Siconolfi, S.F. Accuracy of Physical Working Capacity (PWC170) in Estimating Aerobic Fitness in Children. *J. Sport. Med. Phys. Fit.* **1993**, *33*, 184–188.
59. Rowland, T.W. *The Development of Aerobic Fitness in Children*; Armstrong, N., Ed.; E & FN Spon.: London, UK, 1997.
60. Rowland, T.W. Does Peak VO_2 Reflect VO_{2max} in Children?: Evidence from Supramaximal Testing. *Med. Sci. Sports Exerc.* **1993**, *25*, 689–693. [CrossRef]
61. Rutenfranz, J.; Andersen, K.L.; Seliger, V.; Klimmer, F.; Berndt, I.; Ruppel, M. Maximum Aerobic Power and Body Composition during the Puberty Growth Period: Similarities and Differences between Children of Two European Countries. *Eur. J. Pediatr.* **1981**, *136*, 123–133. [CrossRef]
62. Mandigout, S.; Lecoq, A.; Courteix, D.; Guenon, P.; Obert, P. Effect of Gender in Response to an Aerobic Training Programme in Prepubertal Children. *Acta Paediatr.* **2001**, *90*, 9–15. [CrossRef]
63. Sockolov, R.; Irwin, B.; Dressendorfer, R.H.; Bernauer, E.M. Exercise Performance in 6 to 11 Year Old Boys with Duchenne Muscular Dystrophy. *Arch. Phys. Med. Rehabil.* **1977**, *58*, 195–201.
64. Sunnegårdh, J.; Bratteby, L.E. Maximal Oxygen Uptake, Anthropometry and Physical Activity in a Randomly Selected Sample of 8 and 13 Year Old Children in Sweden. *Eur. J. Appl. Physiol. Occup. Physiol.* **1987**, *56*, 266–272. [CrossRef]
65. Swain, K.E.; Rosenkranz, S.K.; Beckman, B.; Harms, C.A. Expiratory Flow Limitation during Exercise in Prepubescent Boys and Girls: Prevalence and Implications. *J. Appl. Physiol.* **2010**, *108*, 1267–1274. [CrossRef]
66. Tolfrey, K.; Campbell, I.G.; Batterham, A.M. Aerobic Trainability of Prepubertal Boys and Girls. *Pediatr. Excercise Sci.* **1998**, *10*, 248–263. [CrossRef]
67. Tompuri, T.; Lintu, N.; Savonen, K.; Laitinen, T.; Laaksonen, D.; Jääskeläinen, J.; Lakka, T.A. Measures of Cardiorespiratory Fitness in Relation to Measures of Body Size and Composition among Children. *Clin. Physiol. Funct. Imaging* **2015**, *35*, 469–477. [CrossRef] [PubMed]
68. Turley, K.R.; Rogers, D.M.; Harper, K.M.; Kujawa, K.I.; Wilmore, J.H. Maximal Treadmill versus Cycle Ergometry Testing in Children: Differences, Reliability, and Variability of Responses. *Pediatr. Exerc. Sci.* **1995**, *7*, 49–60. [CrossRef]
69. Turley, K.R.; Wilmore, J.H. Cardiovascular Responses to Submaximal Exercise in 7- to 9-Yr-Old Boys and Girls. *Med. Sci. Sports Exerc.* **1997**, *29*, 824–832. [CrossRef]
70. Van Praagh, E.; Bedu, M.; Falgairette, G.; Fellmann, N.; Coudert, J. Comparaison Entre over VO_2 Max Direct et Indirect Chez l'enfant de 7 et 12 Ans. Validation d'une Épreuve de Terrain. *Sci. Sports* **1988**, *3*, 327–332. [CrossRef]

71. Vandekerckhove, K.; Coomans, I.; Moerman, A.; De Wolf, D.; Boone, J. Characterizing Cerebral and Locomotor Muscle Oxygenation to Incremental Ramp Exercise in Healthy Children: Relationship with Pulmonary Gas Exchange. *Eur. J. Appl. Physiol.* **2016**, *116*, 2345–2355. [CrossRef] [PubMed]
72. Vinet, A.; Mandigout, S.; Nottin, S.; Nguyen, L.D.; Lecoq, A.M.; Courteix, D.; Obert, P. Influence of Body Composition, Hemoglobin Concentration, and Cardiac Size and Function of Gender Differences in Maximal Oxygen Uptake in Prepubertal Children. *Chest* **2003**, *124*, 1494–1499. [CrossRef] [PubMed]
73. Washington, R.L.; van Gundy, J.C.; Cohen, C.; Sondheimer, H.M.; Wolfe, R.R. Normal Aerobic and Anaerobic Exercise Data for North American School-Age Children. *J. Pediatr.* **1988**, *112*, 223–233. [CrossRef]
74. Weber, G.; Kartodihardjo, W.; Klissouras, V. Growth and Physical Training with Reference to Heredity. *J. Appl. Physiol.* **1976**, *40*, 211–215. [CrossRef]
75. Weinstein, Y.; Kamerman, T.; Berry, E.; Falk, B. Mechanical Efficiency of Normal-Weight Prepubertal Boys Predisposed to Obesity. *Med. Sci. Sports Exerc.* **2004**, *36*, 567–573. [CrossRef]
76. Welsman, J.; Bywater, K.; Farr, C.; Welford, D.; Armstrong, N. Reliability of Peak VO_2 and Maximal Cardiac Output Assessed Using Thoracic Bioimpedance in Children. *Eur. J. Appl. Physiol.* **2005**, *94*, 228–234. [CrossRef]
77. Winsley, R.J.; Fulford, J.; Roberts, A.C.; Welsman, J.R.; Armstrong, N. Sex Difference in Peak Oxygen Uptake in Prepubertal Children. *J. Sci. Med. Sport* **2009**, *12*, 647–651. [CrossRef] [PubMed]
78. Yamaji, K.; Miyashita, M. Oxygen Transport System during Exhaustive Exercise in Japanese Boys. *Eur. J. Appl. Physiol. Occup. Physiol.* **1977**, *36*, 93–99. [CrossRef] [PubMed]
79. Zebrowska, A.; Zwierzchowska, A. Spirometric Values and Aerobic Efficency of Children and Adolescents with Hearing Loss. *J. Physiol. Pharmacol.* **2006**, *57*, 443–447. [PubMed]
80. Aandstad, A.; Berntsen, S.; Hageberg, R.; Klasson-Heggebø, L.; Anderssen, S.A. A Comparison of Estimated Maximal Oxygen Uptake in 9 and 10 Year Old Schoolchildren in Tanzania and Norway. *Br. J. Sports Med.* **2006**, *40*, 287–292. [CrossRef]
81. Adegboye, A.R.A.; Anderssen, S.A.; Froberg, K.; Sardinha, L.B.; Heitmann, B.L.; Steene-Johannessen, J.; Kolle, E.; Andersen, L.B. Recommended Aerobic Fitness Level for Metabolic Health in Children and Adolescents: A Study of Diagnostic Accuracy. *Br. J. Sports Med.* **2011**, *45*, 722–728. [CrossRef]
82. Barkley, J.E.; Roemmich, J.N. Validity of the CALER and OMNI-Bike Ratings of Perceived Exertion. *Med. Sci. Sports Exerc.* **2008**, *40*, 760–766. [CrossRef]
83. Birat, A.; Bourdier, P.; Piponnier, E.; Blazevich, A.J.; Maciejewski, H.; Duché, P.; Ratel, S. Metabolic and Fatigue Profiles Are Comparable between Prepubertal Children and Well-Trained Adult Endurance Athletes. *Front. Physiol.* **2018**, 9. [CrossRef]
84. Boissière, J.; Maufrais, C.; Baquet, G.; Schuster, I.; Dauzat, M.; Doucende, G.; Obert, P.; Berthoin, S.; Nottin, S. Specific Left Ventricular Twist-Untwist Mechanics during Exercise in Children. *J. Am. Soc. Echocardiogr.* **2013**, *26*, 1298–1305. [CrossRef]
85. Cooper, D.M.; Weiler-Ravell, D.; Whipp, B.J.; Wasserman, K. Aerobic Parameters of Exercise as a Function of Body Size during Growth in Children. *J. Appl. Physiol. Respir. Environ. Exerc. Physiol.* **1984**, *56*, 628–634. [CrossRef]
86. Cunningham, D.A.; Telford, P.; Swart, G.T. The Cardiopulmonary Capacities of Young Hockey Players: Age 10. *Med. Sci. Sports Exerc.* **1976**, *8*, 23–25. [CrossRef]
87. Danielson, A.; Thorsson, O.; Karlsson, M.K.; Wollmer, P.; Andersen, L.B.; Dencker, M. School-Based Study Found That Physical Activity and Aerobic Fitness Predicted Increases in Total Body Fat and Abdominal Fat at a Mean Age of 9.8 Years. *Acta Paediatr. Int. J. Paediatr.* **2018**, *107*, 1810–1817. [CrossRef] [PubMed]
88. Dencker, M.; Thorsson, O.; Karlsson, M.K.; Lindén, C.; Svensson, J.; Wollmer, P.; Andersen, L.B. Daily Physical Activity and Its Relation to Aerobic Fitness in Children Aged 8-11 Years. *Eur. J. Appl. Physiol.* **2006**, *96*, 587–592. [CrossRef]
89. DeStefano, R.A.; Caprio, S.; Fahey, J.T.; Tamborlane, W.V.; Goldberg, B. Changes in Body Composition after a 12-Wk Aerobic Exercise Program in Obese Boys. *Pediatr. Diabetes* **2000**, *1*, 61–65. [CrossRef]
90. Dotan, R.; Falk, B.; Raz, A. Intensity Effect of Active Recovery from Glycolytic Exercise on Decreasing Blood Lactate Concentration in Prepubertal Children. *Med. Sci. Sports Exerc.* **2000**, *32*, 564–570. [CrossRef]
91. Falgairette, G.; Bedu, M.; Fellmann, N.; Van-Praagh, E.; Coudert, J. Bio-Energetic Profile in 144 Boys Aged from 6 to 15 Years with Special Reference to Sexual Maturation. *Eur. J. Appl. Physiol. Occup. Physiol.* **1991**, *62*, 151–156. [CrossRef]
92. Falk, B.; Bar-Or, O. Longitudinal Changes in Peak Aerobic and Anaerobic Mechanical Power of Circumpubertal Boys. *Pediatr. Exerc. Sci.* **1993**, *5*, 318–331. [CrossRef]
93. Fellmann, N.; Beaune, B.; Coudert, J. Blood Lactate after Maximal and Supramaximal Exercise in 10- to 12-Year-Old Bolivian Boys Effects of Altitude and Socioeconomic Status. *Int. J. Sports Med.* **1994**, *15*, S90–S95. [CrossRef]
94. Foraita, R.; Brandes, M.; Günther, F.; Bammann, K.; Pigeot, I.; Ahrens, W. The Influence of Aerobic Fitness on Obesity and Its Parent-Offspring Correlations in a Cross-Sectional Study among German Families. *BMC Public Health* **2015**, *15*, 638. [CrossRef]
95. García-Prieto, J.C.; Martinez-Vizcaino, V.; García-Hermoso, A.; Sánchez-López, M.; Arias-Palencia, N.; Fonseca, J.F.O.; Mora-Rodriguez, R. Energy Expenditure in Playground Games in Primary School Children Measured by Accelerometer and Heart Rate Monitors. *Int. J. Sport Nutr. Exerc. Metab.* **2017**, *27*, 467–474. [CrossRef]
96. Guerrero, L.; Naranjo, J.; Carranza, M.D. Influence of Gender on Ventilatory Efficiency during Exercise in Young Children. *J. Sports Sci.* **2008**, *26*, 1455–1457. [CrossRef] [PubMed]
97. Haapala, E.A.; Gao, Y.; Hartikainen, J.; Rantalainen, T.; Finni, T. Associations of Fitness, Motor Competence, and Adiposity with the Indicators of Physical Activity Intensity during Different Physical Activities in Children. *Sci. Rep.* **2021**, *11*, 12521. [CrossRef]

98. Herman, K.M.; Chaput, J.P.; Sabiston, C.M.; Mathieu, M.E.; Tremblay, A.; Paradis, G. Combined Physical Activity/Sedentary Behavior Associations with Indices of Adiposity in 8- to 10-Year-Old Children. *J. Phys. Act. Health* **2015**, *12*, 20–29. [CrossRef]
99. Inbar, O.; Bar-Or, O.; Dotan, R.; Gutin, B. Conditioning versus Exercise in Heat as Methods for Acclimatizing 8- to 10-Yr-Old Boys to Dry Heat. *J. Appl. Physiol. Respir. Environ. Exerc. Physiol.* **1981**, *50*, 406–411. [CrossRef]
100. Petzl, D.H.; Haber, P.; Schuster, E.; Popow, C.; Haschke, F. Reliability of Estimation of Maximum Performance Capacity on the Basis of Submaximum Ergometric Stress Tests in Children 10-14 Years Old. *Eur. J. Pediatr.* **1988**, *147*, 174–178. [CrossRef]
101. Rivas, E.; Huynh, H.; Galassetti, P.R. Obesity Affects Submaximal Oxygen Uptake-Heart Rate Relationship and Exercise Economy Differently in Pre- and Post-Pubescent Boys and Girls. *Int. J. Exerc. Sci.* **2019**, *12*, 748–763.
102. Rutenfranz, J.; Máček, M.; Lange Andersen, K.; Bell, R.D.; Vávra, J.; Radvanský, J.; Klimmer, F.; Kylian, H. The Relationship between Changing Body Height and Growth Related Changes in Maximal Aerobic Power. *Eur. J. Appl. Physiol. Occup. Physiol.* **1990**, *60*, 282–287. [CrossRef] [PubMed]
103. Schöffl, I.; Ehrlich, B.; Stanger, S.; Rottermann, K.; Dittrich, S.; Schöffl, V. Exercise Field Testing in Children: A New Approach for Age-Appropriate Evaluation of Cardiopulmonary Function. *Pediatr. Cardiol.* **2020**, *41*, 1099–1106. [CrossRef] [PubMed]
104. Springer, C.; Barstow, T.J.; Wasserman, K.; Cooper, D.M. Oxygen Uptake and Heart Rate Responses during Hypoxic Exercise in Children and Adults. *Med. Sci. Sports Exerc.* **1991**, *23*, 71–79. [CrossRef] [PubMed]
105. Steene-Johannessen, J.; Anderssen, S.A.; Kolle, E.; Andersen, L.B. Low Muscle Fitness Is Associated with Metabolic Risk in Youth. *Med. Sci. Sports Exerc.* **2009**, *41*, 1361–1367. [CrossRef]
106. van der Steeg, G.E.; Takken, T. Reference Values for Maximum Oxygen Uptake Relative to Body Mass in Dutch/Flemish Subjects Aged 6–65 Years: The LowLands Fitness Registry. *Eur. J. Appl. Physiol.* **2021**, *121*, 1189–1196. [CrossRef] [PubMed]
107. Wedderkopp, N.; Froberg, K.; Hansen, H.S.; Riddoch, C.; Andersen, L.B. Cardiovascular Risk Factors Cluster in Children and Adolescents with Low Physical Fitness: The European Youth Heart Study (EYHS). *Pediatr. Exerc. Sci.* **2003**, *15*, 419–427. [CrossRef]
108. Yoshizawa, S. A Comparative Study of Aerobic Work Capacity in Urban and Rural Adolescents. *J. Hum. Ergol.* **1972**, *1*, 45–65.
109. Zunquin, G.; Theunynck, D.; Sesboüé, B.; Arhan, P.; Bouglé, D. Evolution of Fat Oxidation during Exercise in Obese Pubertal Boys: Clinical Implications. *J. Sports Sci.* **2009**, *27*, 315–318. [CrossRef]
110. Langer, R.D.; da Costa, K.G.; Bortolotti, H.; Fernandes, G.A.; de Jesus, R.S.; Gonçalves, E.M. Phase Angle Is Associated with Cardiorespiratory Fitness and Body Composition in Children Aged between 9 and 11 Years. *Physiol. Behav.* **2020**, 215. [CrossRef]
111. Leclair, E.; Berthoin, S.; Borel, B.; Thevenet, D.; Carter, H.; Baquet, G.; Mucci, P. Faster Pulmonary Oxygen Uptake Kinetics in Children vs Adults Due to Enhancements in Oxygen Delivery and Extraction. *Scand. J. Med. Sci. Sports* **2013**, *23*, 705–712. [CrossRef]
112. Leclair, E.; Mucci, P.; Borel, B.; Baquet, G.; Carter, H.; Berthoin, S. Time to Exhaustion and Time Spent at a High Percentage of Vo_{2max} in Severe Intensity Domain in Children and Adults. *J. Strength Cond. Res.* **2011**, *25*, 1151–1158. [CrossRef]
113. Mahon, A.D.; Anderson, C.S.; Hipp, M.J.; Hunt, K.A. Heart Rate Recovery from Submaximal Exercise in Boys and Girls. *Med. Sci. Sports Exerc.* **2003**, *35*, 2093–2097. [CrossRef]
114. Martin, R.; Hautier, C.; Bedu, M. Effect of Age and Pedalling Rate on Cycling Efficiency and Internal Power in Humans. *Eur. J. Appl. Physiol.* **2002**, *86*, 245–250. [CrossRef]
115. Obert, P.; Mandigouts, S.; Nottin, S.; Vinet, A.; N'Guyen, L.D.; Lecoq, A.M. Cardiovascular Responses to Endurance Training in Children: Effect of Gender. *Eur. J. Clin. Invest.* **2003**, *33*, 199–208. [CrossRef]
116. Bongers, B.C.; Hulzebos, E.H.J.; van Brussel, M.; Takken, T. *Pediatric Norms for Cardiopulmonary Exercise Testing: In Relation to Gender and Age*; Uitgeverij BOXPRESS: 's-Hertogenbosch, The Netherlands, 2012; p. 129.
117. Ng, M.; Fleming, T.; Robinson, M.; Thomson, B.; Graetz, N.; Margono, C.; Mullany, E.C.; Biryukov, S.; Abbafati, C.; Abera, S.F.; et al. Global, Regional, and National Prevalence of Overweight and Obesity in Children and Adults during 1980–2013: A Systematic Analysis for the Global Burden of Disease Study 2013. *Lancet* **2014**, *384*, 766–781. [CrossRef]
118. Johansson, L.; Brissman, M.; Morinder, G.; Westerståhl, M.; Marcus, C. Reference Values and Secular Trends for Cardiorespiratory Fitness in Children and Adolescents with Obesity. *Acta Paediatr. Int. J. Paediatr.* **2020**, *109*, 1665–1671. [CrossRef] [PubMed]
119. Ortega, F.B.; Ruiz, J.R.; Castillo, M.J.; Sjöström, M. Physical Fitness in Childhood and Adolescence: A Powerful Marker of Health. *Int. J. Obes.* **2008**, *32*, 1–11. [CrossRef] [PubMed]
120. Raghuveer, G.; Hartz, J.; Lubans, D.R.; Takken, T.; Wiltz, J.L.; Mietus-Snyder, M.; Perak, A.M.; Baker-Smith, C.; Pietris, N.; Edwards, N.M. Cardiorespiratory Fitness in Youth: An Important Marker of Health: A Scientific Statement From the American Heart Association. *Circulation* **2020**, *142*, E101–E118. [CrossRef]
121. Aadland, E.; Kvalheim, O.M.; Rajalahti, T.; Skrede, T.; Resaland, G.K. Aerobic Fitness and Metabolic Health in Children: A Clinical Validation of Directly Measured Maximal Oxygen Consumption versus Performance Measures as Markers of Health. *Prev. Med. Rep.* **2017**, *7*, 74–76. [CrossRef]
122. Högström, G.; Nordström, A.; Nordström, P. High Aerobic Fitness in Late Adolescence Is Associated with a Reduced Risk of Myocardial Infarction Later in Life: A Nationwide Cohort Study in Men. *Eur. Heart J.* **2014**, *35*, 3133–3140. [CrossRef] [PubMed]

123. Martin-Rincon, M.; Calbet, J.A.L. Progress Update and Challenges on VO_{2max} Testing and Interpretation. *Front. Physiol.* **2020**, *11*, 1070. [CrossRef]
124. Howley, E.T.; Bassett, D.R.; Welch, H.G. Criteria for Maximal Oxygen Uptake: Review and Commentary. *Med. Sci. Sports Exerc.* **1995**, *27*, 1292–1301. [CrossRef]

life

MDPI

Case Report

Recovery after Running an "Everesting" Mountain Ultramarathon

Anton Ušaj [1,*], Jon Lihteneger Vidmajer [1] and Sonja Lojen [2]

1 Laboratory of Biodynamics, Faculty of Sport, University of Ljubljana, 1000 Ljubljana, Slovenia; jon.lihteneger@gmail.com

2 Department of Environmental Sciences, Institute Jožef Stefan, Jamova 39, 1000 Ljubljana, Slovenia; sonja.lojen@ijs.si

* Correspondence: anton.usaj@siol.net or anton.usaj@fsp.uni-lj.si

Abstract: Blood markers of muscle microdamage and systemic inflammation do not adequately explain the reduced performance observed over a prolonged recovery after running a mountain ultramarathon. This case study aimed to determine whether the reduced performance after the Everesting mountain ultramarathon can be further assessed by considering cardiorespiratory and metabolic alterations determined via repeated incremental and continuous running tests. A single runner (age: 24 years, BM: 70 kg, BMI: 22, $\text{Vo}_{2\text{peak}}$: 74 mL·min^{-1}·kg^{-1}) was observed over a preparatory period of two months with a one-month recovery period. The Everesting consisted of nine ascents and descents of 9349 vertical metres completed in 18:22 (h:min). During the first phase of the recovery, enhanced peak creatine kinase (800%) and C-reactive protein (44%) levels explained the decreased performance. In contrast, decreased performance during the second, longer phase was associated with a decreased lactate threshold and Vo_2 (21% and 17%, respectively), as well as an increased energetic cost of running (15%) and higher endogenous carbohydrate oxidation rates (87%), lactate concentrations (170%) and respiratory muscle fatigue sensations that remained elevated for up to one month. These alterations may represent characteristics that can explain the second phase of the recovery process after Everesting.

Keywords: relative running intensity; energetic cost; lactate threshold; carbohydrate oxidation rate; respiratory fatigue

Citation: Ušaj, A.; Lihteneger Vidmajer, J.; Lojen, S. Recovery after Running an "Everesting" Mountain Ultramarathon. *Life* **2023**, *13*, 1946. https://doi.org/10.3390/life13101946

Academic Editors: Helena Lenasi and Ines Drenjančević

Received: 5 September 2023
Revised: 18 September 2023
Accepted: 20 September 2023
Published: 22 September 2023

1. Introduction

Everesting ultramarathon running events evolved from the general category of mountain ultramarathons (MUM) and ultramarathons. This running event consists of several uphill and downhill runs that reach or even exceed the cumulative terrestrial altitude of Mount Everest (8864 m). The altitude, the steepness, the nature of the routes (dry land, ice and rocks) and thus a high proportion of eccentric work and different, even extreme environmental conditions are the main differences compared to other ultramarathons. The decreased performance during these runs is a result of accumulated fatigue. Ultramarathons last several hours to days and are interrupted by rest breaks that are as short as possible. Decreased performance has been associated with an increased energy cost (Cr) during this event [1,2], muscle microdamage due to mechanical stress [3,4], oxidative stress [5,6], neuromuscular fatigue [7,8], systemic inflammation [9,10], respiratory muscle fatigue [11] and an imbalance between the effects of the anabolic hormone testosterone and the catabolic hormone cortisol [12]. The changes during an ultramarathon persist after these events in the recovery phase but eventually revert to their resting values. This phenomenon could be related to the return of an athlete's ultra-endurance performance to pre-competition levels. The most dramatic changes occur in the first few days after an ultra-endurance competition. The typical blood markers for muscle microdamage, represented by creatine phosphokinase (CPK), and for systemic inflammation, represented by C-reactive protein (CRP), have been associated with a significant decrease in performance. Notably, the aforementioned changes reach peak blood concentration about 1–2 days after the ultramarathon and then decrease toward the resting values within the next 3–5 days [13].

In contrast, performance tests used to assess recovery after ultramarathons showed a sustained decline in performance even after a longer period. Sherman et al. [14] reported decreased isokinetic torque during knee extension after as many as seven days. Similarly, Chambers et al. [15] found that vertical jump height decreased for 18 days. Warhol et al. [4] reported that muscle biopsies showed incomplete regeneration 12 weeks after an ultramarathon. Although these parameters and the corresponding performance tests do not represent extreme endurance performance, they suggest that a return to pre-competition performance may take longer than ten days [4,13,16].

The aforementioned dramatic decrease in performance could be associated with an early dramatic increase in the blood markers of muscle damage and systemic inflammation. Despite these phenomena disappearing within days, the performance remained reduced. Hypothetically, the recuperation of performance during the recovery period after an ultramarathon could represent two phases reflecting the typical behaviour of a dynamic system after the termination of an impulse [17]. We hypothesised that the same two phases may also occur after Everesting. The first phase could be related to the early dramatic decrease in performance. The second phase follows the first phase and could be a longer process associated with slow recovery from microdamage in exercising muscles [4]. These phenomena cannot be detected by blood markers measured during resting because they already reach their resting values typical for a period before the ultramarathon, with the exception of the values obtained via muscle biopsy [4]. Therefore, they may be associated with other factors. One of these could be the decrease in running efficiency, as may be evidenced by increased energetic costs. Other markers should also be identified and selected to potentially assess changes in cardiorespiratory function through an increase in heart rate and ventilation and changes in metabolism through an altered selection between carbohydrates and fats as fuels. However, these alterations require adequate exercise, perhaps in an ultramarathon, where they manifest themselves in an organism that has not yet fully recovered. Therefore, we hypothesise that fluctuations in the potential parameters to be selected and tested in advance will be greater after Everesting than during the preparatory period and will disappear in conjunction with the increased performance during the second recovery phase, lasting one month or longer. Testing this hypothesis requires an incremental testing protocol and a continuous running test instead of resting measurements. A continuous running test needs initial, runner-specific calibration according to intensity and duration to ascertain possible differences influenced by training and those caused by Everesting and recovery. These tests should be repeated several times during the experimental procedure before Everesting (preparatory period) and after Everesting (recovery period). Managing a complex experiment needs careful planning, which can enable standardization of testing conditions, but also needs to be flexible to disable interferences with training in the preparatory period and with recovery time course during the recovery period. To achieve these standards effectively, a case study research design was selected.

2. Case Report

2.1. Participant and Ethical Approval

A single male subject was used in this study, an amateur ultramarathon runner (age: 24 years; BM: 70 kg; BMI: 22; Vo_{2peak} = 74 (ml·min^{-1}·kg^{-1})). The runner had five years of experience in endurance running. In the last two years, the subject specialised in Everesting. Preparation for the competition season started in November, but special preparation for Everesting started only two months before the event. At the time of the experiment, the runner was healthy and had no injuries. The subject gave his written informed consent, and the experimental procedure was approved by the Faculty of Sport Ethics Committee (FS8:2021).

2.2. Experimental Design and Protocol

The preparatory period for the Everesting trial consisted of two months of regular training. The additional month after Everesting represented a recovery period. The runner participated in the following tests: resting haematological venous blood measurements, a body composition test, an incremental test and a submaximal continuous running test (CRT), all on the treadmill. These tests were repeated according to the schedule shown in Table 1.

Table 1. Testing schedule during the experimental period.

Testing Day	−58	−54	−47	−27	−25	−18	−2	0	1	3	5	8	10	12	19	25	30
Incremental test	X			X			X						X		X		
Continuous running test		Xw	Xc		Xc	Xw						Xc		Xw		Xc	Xw
Hematological analyses					X	X	X	TRIAL	X	X	X						
Body composition			X			X	X		X	X	X	X			X	X	

X—realised test; Xw—water beverage; Xc—carbohydrate beverage.

Training in the last two months before Everesting consisted of two one-month mesocycles. The training characteristics were designed by the runner. The first of the two mesocycles consisted of five 6-day microcycles of equal length. The first day consisted of a continuous 30–40 km run at 3.5–3.8 $m{\cdot}s^{-1}$ (LSD) (Figure 1). The fourth day consisted of a similar but shorter continuous 20–30 km run at about 4.6–4.8 $m{\cdot}s^{-1}$ (LSD) (Figure 1). The second day of the microcycle consisted of 2 km runs repeated five times at a speed of about 5.1–5.3 $m{\cdot}s^{-1}$ (close to Vo_{2peak} intensity), with five minutes of recovery in between (repeated distances) (Figure 1). The third and sixth days were recovery days.

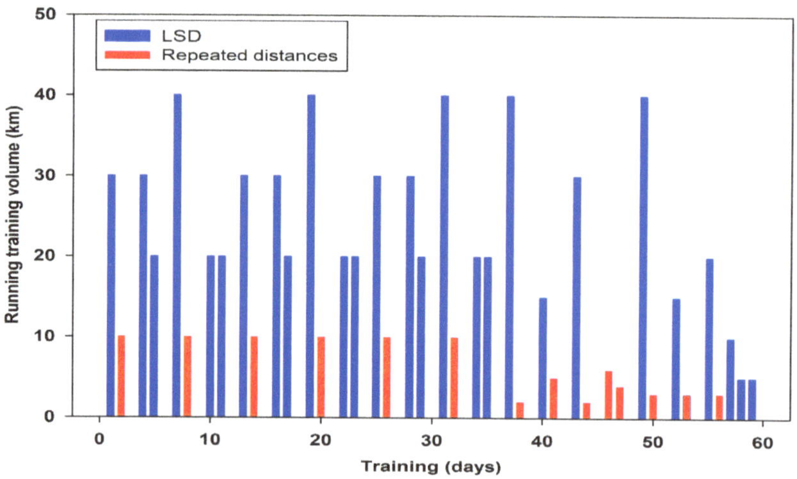

Figure 1. Training characteristics during the preparatory period for Evereting consisted of LSD—long, slow distance and repeated distances.

The first day consisted additionally of light stretching and general strength exercises. The fifth training day consisted of about 20 km of fartlek and Nordic walking in the mountains. The fartlek training started with a steady run of light to moderate intensity, followed by two repetitions of about one kilometre of high–intensity running on slightly sloping terrain, and continued with a two–kilometre run with very steep descents. The first half of the training ended with about three-kilometres of very steep Nordic walking. The second half of the training starts with a fast run of four-kilometres downhill, followed by two repetitions of runs of one kilometre uphill. The fartlek ends with a four–kilometre steep descent to the starting place. The first two microcycles of the second mesocycle were a continuation of the microcycles in the first mesocycle. The four other microcycles represented tapering in which the running volume was linearly reduced to the last 60% of the training volume in the first mesocycle (Figure 1). The total training volume in both mesocycles was 750 km.

2.3. Measurements

2.3.1. Haematological Measurements

Haematological tests were performed in the morning (fasted). Two venous blood samples of 5 and 3.5 mL were collected in SST II Advance BD Vacutainers and centrifuged at 1700 RPM for 10 min. Analyses were performed using Beckmann Coulter AU 680 (Beckmann, Boston, MA, USA) and Abbott Architect i1000 (Abbott, Abbott Park, IL, USA) devices. The third sample of 3 mL was collected in a BD Vacutainer K2E (Becton Dickinson, Franklin Lakes, NJ, USA) with EDTA. The haemogram was obtained using a Sysmex XN-1000 analyser. The tests were repeated six times according to a specific scheme (Table 1).

2.3.2. Body Composition

Body composition was analysed using an InBody720 Body Composition Analyser (Seoul, Republic of Korea) in the morning (fasted). The test was repeated nine times according to a specific scheme (Table 1).

2.3.3. Incremental Testing Protocol

The incremental test was repeated five times according to a specific schedule (Table 1). This test consisted of 4-min runs on a Pulsar treadmill (HP Cosmos, Nußdorf, Germany). The running speed was increased by 2 $km \cdot h^{-1}$ from an initial 2.2 $m \cdot s^{-1}$ (8 $km \cdot h^{-1}$). This exercise was interrupted for 0.5 min for blood sampling until the subject could no longer maintain his running speed due to fatigue. Continued breath-by-breath gas exchange was analysed using a Vmax Spectra (SensorMedics, Orange City, FL, USA). At every exercise stage, the blood lactate concentration [LA] was measured from a hyperaemic earlobe capillary blood sample (5 µL) using the Lactate Pro 2 analyser (Arkray, Kyoto, Japan).

2.3.4. Continuous Running Test (CRT)

A continuous 120-min submaximal running test (CRT) at 3.9 $m \cdot s^{-1}$ (14 $km \cdot h^{-1}$) was performed eight times: four times with a carbohydrate drink (CHO, Xc) and four times with water (WAT, Xw), according to a specific schedule (Table 1). The velocity was selected to correspond with the Lactate Threshold [18,19] determined during the first test, which was classified as "somewhat heavy" according to the Borg scale [20]. A 25-min rest period before the test was followed by a 10-min warm-up run of 2.2 $m \cdot s^{-1}$ (8 $km \cdot h^{-1}$), followed by a two-hour run briefly interrupted for measurement purposes (Figure 2). The runner arrived at the laboratory having consumed a diet of foods naturally low in ^{13}C for three days. The runner also avoided strenuous endurance training during this time. About 25 min before the start, Capsolin cream (Laboratorio Farmaceutico SIT, Mede, Italy) was applied to the skin of the earlobe to influence hyperaemia. Respiratory gases were also collected at rest in a comfortable sitting position using a Vmax Spectre metabolic cart (Sensor Medics, Yorba Linda, CA, USA). In addition, two vacutainers (20 mL each) were filled with exhaled air to determine the $^{13}C/^{12}C$ ratio of exhaled air at rest. A micro-sample of 95 µL capillary blood from a hyperaemic earlobe was collected to measure blood gas, acid–base, electrolyte, lactate [LA], and glucose [GLU] concentrations using an ABL800FLEX analyser (Radiometer, København, Denmark). The runner ingested a 6 $mL \cdot kg^{-1}$ bolus of a 15% sugarcane solution (CHO) (Mascavo, Brusque, Brazil) 20 min before the start of the test. The sugarcane consisted of 87% sucrose, 2% glucose and 2% fructose with a high natural abundance of ^{13}C ($\delta^{13}C_{VPDB} = -9.8$ ‰). Breath analysis to determine the $^{13}C/^{12}C$ ratio was performed using a Europa Scientific 20–20 isotope ratio mass spectrometer with an ANCA-TG trace gas separation module (Europa Scientific, Cheshire, UK). The test run began with a 10-min warm-up at 2.8 $m \cdot s^{-1}$ (Figure 2), which continued at 3.9 $m \cdot s^{-1}$ (14 $km \cdot h^{-1}$) until the first interruption for approximately 20–30 s at 25 min to remove the breathing mask and put on a mask adapted for isotope breath sampling. The runner continued to run until the 27th minute. Breath sampling was performed in the last three minutes. Then, the run was interrupted for about 1 min to collect blood samples, consume the drink (this time, 2 $mL \cdot kg^{-1}$) and put the breathing mask back on (Figure 2). The run continued until the second interruption, this time between the 45th and 50th minutes (and again in the interval between the 55th and 60th minutes) to remove and put on the breathing mask to collect breath for the isotope measurement (Figure 2). The third repetition of the procedure occurred between the 75th and 80th minutes and again in the interval between the 85th and 90th minutes. The fourth and final repetition was performed between the 105th and 110th minutes and again in the interval between the 115th and 120th minutes (Figure 2). The last blood samples were taken immediately after the end of the run.

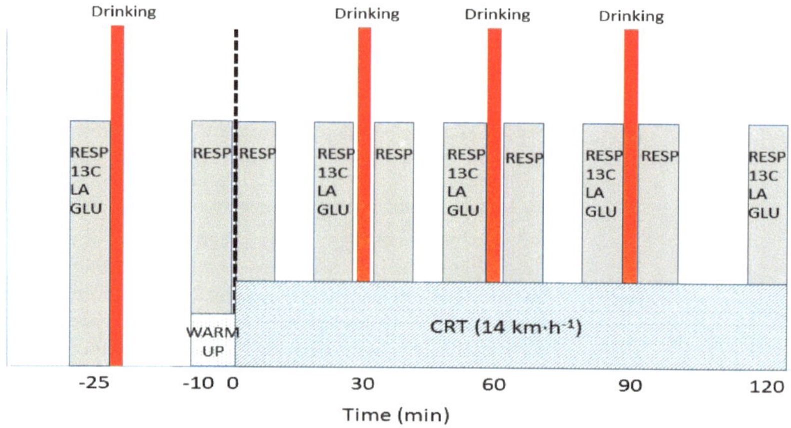

Figure 2. Test setup during CRT. Testing consisted of intervals for collecting breath and isotope samples and for collecting blood micro samples, represented by grey rectangles, and drinking breaks, represented by thick red lines interrupting the continuous run. Testing activities began about 25 min before the start of the CRT with measurements of gas exchange at rest, blood samples and bolus drinking. About 10 min before the test, the second measurements of respiration and gas exchange were taken, followed by a warm-up run. The CRT started at 0 min and continued with interruptions for measurements and drinking until 120 min.

2.3.5. Calculations

CHO and FAT oxidation rates were calculated using the standard indirect calorimetry equations [21,22]. From the measured Vo_2 and Vco_2, the total CHO and FAT oxidation rates were calculated as follows:

$$\text{CHOox} = 4.55 \cdot Vco_2 - 3.21 \cdot Vo_2$$

$$\text{FATox} = 1.67 \cdot Vo_2 - 1.67 \cdot Vco_2.$$

We applied the assumption that protein oxidation during exercise was negligible.

The isotope composition is expressed as the ‰ difference between the $^{13}C/^{12}C$ of the sample and a known laboratory reference standard [23,24] according to the following:

$$\delta = \left(\frac{(^{13}C/^{12}C)\text{sample}}{(^{13}C/^{12}C)\text{standard}} - 1\right) \cdot 10^3\ [‰]$$

where δ is the relative deviation of the heavy-to-light C isotope ratio of the sample from that of the standard (VPDB), expressed in per mil (‰). The EXOox ($g \cdot min^{-1}$) was then calculated [25] as

$$EXOox = \frac{V_{CO2} \cdot (\delta exp - \delta b)}{(\delta ing - \delta b)} \cdot \frac{1}{k}$$

where δexp is the isotope composition expressed as δ value of expired air during exercise, δing is the ^{13}C enrichment of ingested CHO (sucrose), and δb is the ^{13}C enrichment of expired air during resting (background) before exercise began. The amount of CO_2 (in litres) produced by the oxidation of 1 g of glucose (0.7467) was k. When water was drunk, the δ^{13}C values remained similar to those at rest. EXOox remained close to 0 g/min.

Changes in endurance performance were estimated from two points of view. The use of the incremental test data allowed us to calculate the Lactate Threshold (LT) [18,19], the Onset of Blood Accumulation (OBLA) [26], and the peak running velocity reached (V_{peak}). Time to fatigue was measured via CRT. It was assumed that the inability to complete a 2 h run due to fatigue would be a clear sign of the expected reduced performance after Everesting. In addition, variations in the measured parameters between the preparation and recovery periods were compared to identify any differences due to Everesting.

2.4. Results

The subject started the event at 00:00:00 (midnight) and completed the Everesting trial at 18:22:30 (h:min:s) (Figure 3). The distance covered was 81.850 km, and the cumulative terrestrial altitude was 9349 m. The temperature at the start (600 m) was 12 °C. The wind was calm, and fog somewhere on the route. The temperature at the peak of the mountain (1609 m) was 6 °C, and the wind was calm and moderately cloudy. The temperature at the end of the trial was 18 °C, wind calm, moderately cloudy. The average HR during the first four ascent intervals was 170 ± 1 min^{-1} and decreased to 156 ± 4 min^{-1} during the last three ascents (Figure 3). The heart rate was 137 ± 3 min^{-1} during the first six descent intervals and decreased to 125 ± 6 min^{-1} during the last three descent intervals (Figure 3). The runner experienced increased fatigue of his respiratory muscles during the last two ascents. These sensations disappeared during descents.

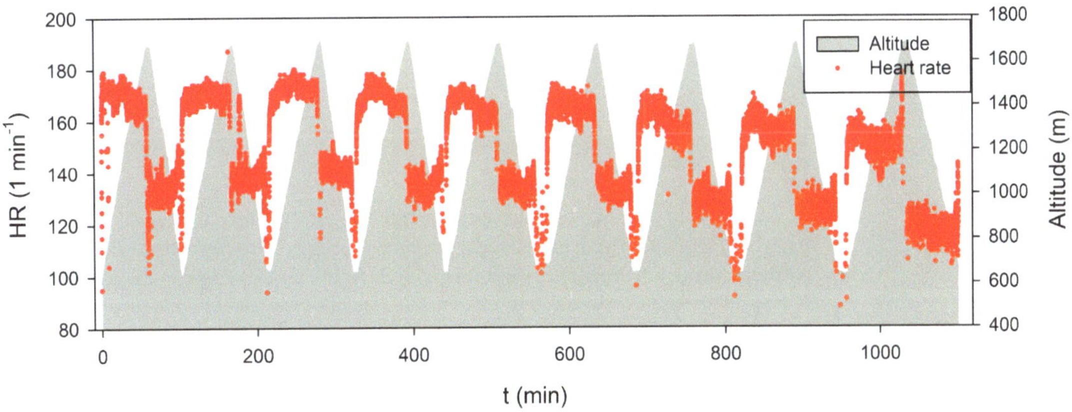

Figure 3. The Everesting event consisted of nine ascents and descents (grey), each beginning and ending at the same location. The HR time course is shown with red dots. Towards the end of Everesting, the time for ascents increased (shown by the more gentle slopes of the grey structures), and the HR increased less significantly.

The subject consumed food with an energy equivalent of about 5830 kcal while Everesting. The runner's food consisted of energy gels (10 × 400 g), energy bars (6 × 240 g), bananas (6 × 600 g), pasta with tuna (155 g), an isotonic drink (4.5 L), water (2 L) and an energy drink (0.5 L). The consumed drinks and food enabled the runner to maintain body mass within −3% of the pre–event BM.

The runner's body composition characteristics remained stable during the preparation period (Table 2) but changed during the recovery period after the trial (Table 2). All observed parameters were minimally decreased in a similar interval between 5 and 19 days after the trial. Small changes in the body composition enable comparison between different parameters without corrections.

Table 2. Fluctuations in body composition parameters throughout the preparatory and recovery periods.

	Preparatory Period				Recovery Period					
Testing (Days)	**−47**	**−18**	**−2**	**0**	**+1**	**+3**	**+5**	**+8**	**+19**	**+25**
Body mass (kg)	69	69	69		69	68	68	67	69	70
BMI (kg·m^2)	21.6	21.7	21.7		21.6	21.3	21.3	21	21.2	21.8
Fat mass (kg)	4.6	4.2	4.9		3.5	4.2	4.6	4.7	5.2	5.9
Visceral fat (%)	18.9	17.5	19.2		11.1	11.4	14.1	11.1	14.7	21.8
Muscle mass (kg)	37.3	36.9	37.7		37.7	36.7	36.6	36	36.8	37.4

The haematological resting status was determined using three tests during the preparation period before Everesting and four tests within five days after the trial (Table 3). All markers were stable during the preparatory period (Table 3). During the recovery period, typical changes were seen in the markers for muscle microdamage (CK and Mb), which increased dramatically and peaked on the first day (Table 3, Figure 4).

Table 3. Fluctuations in resting blood markers of systemic inflammation and muscle damage throughout the preparatory and recovery periods.

	Preparatory Period				Recovery Period		
Testing (Days)	−25	−18	−2	0	+1	+3	+5
AST (μkat·L^{-1})	0.86	0.51	0.58		2.13	1.12	0.73
CRP (mg·L^{-1})	5	5	5		27	11	5
LDH (μkat·L^{-1})	3.86	3.63	3.37		5.47	4.60	4.00
CK (μkat·L^{-1})	8.34	3.55	4.64		43.96	10.01	5.55
Mb (μg·L^{-1})	36	33.3	34.6		134.6	65.9	32.8
Testosterone (μg·L^{-1})	2.81	2.76	2.84		1.72	1.90	2.77
Cortisol (nmol·L^{-1})	506	502	519		273	430	490

Abbreviations: AST—aspartate aminotransferase, CRP—C-reactive protein, LDH—lactate dehydrogenase, CK—creatine kinase, Mb—muscle myoglobin.

Markers of systemic inflammation (AST, CRP and LDH) peaked on the first day after the trial (Figure 4). These markers decreased within five days after the trial to the values reached during the preparatory period (Table 3, Figure 4). Decreases in the concentrations of testosterone and cortisol peaked on the first day after the trial but returned to the values measured during the preparatory period after five days (Table 3).

Performance characteristics typical of submaximal intensity (LT and OBLA) were stable during the preparatory period (Table 4) but remained at 90% of the pre-trial values 19 days after Everesting (Table 4, Figure 5). The peak performance equivalent to running velocity (V_{peak}) at maximal intensity during the incremental test (Table 4) was stable at 5.5 m·s^{-1} during the preparation period but decreased to 5.0 m·s^{-1} (a 9% decrease) for 5–15 days after the trial (Table 4, Figure 5). On day 19, during the recovery period, the V_{peak} again presented values similar to those in the preparation phase. The CRT tests showed decreased performance due to the shorter running duration. Instead of 120 min, as in the preparation period, the runner stopped running after 60 min due to fatigue. The CRT velocity of 3.9 m·s^{-1} corresponded to 71% of the V_{peak} during the preparatory period. After Everesting, this value increased to 78% of V_{peak} at 0.5 months. Based on a comparison of Cr within runs during the preparatory period, Cr increased steadily (by 0.02 $mL \cdot kg^{-1} \cdot m^{-1}$ on average). In each run, Cr reached values that were 0.03 $mL \cdot min^{-1} \cdot m^{-1}$ (9%) higher than those during the preparatory period after 0.5 months and 0.05 $mL \cdot min^{-1} \cdot m^{-1}$ (19%) higher than those during the preparatory period about one month after Everesting (Table 5).

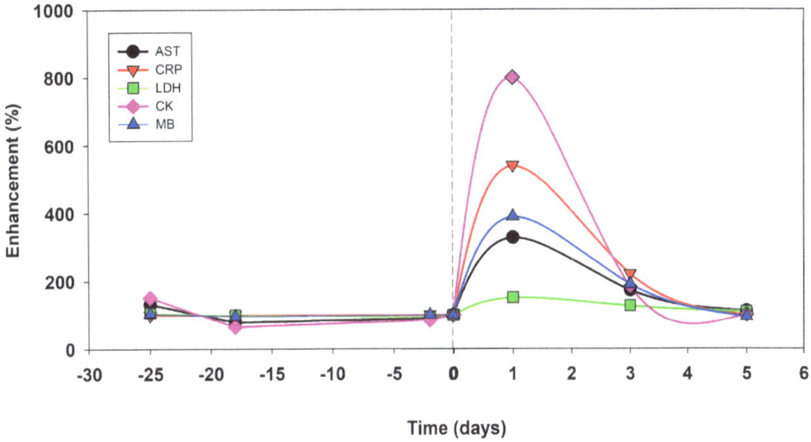

Figure 4. The time course of the fluctuations in blood markers for muscle damage and systemic inflammation shows a synchronous peak on the first day after Everesting and a decrease in this phenomenon on the fifth day after Everesting. The differences between the values during the preparation and recovery phases are obvious.

Table 4. Fluctuations of submaximal and maximal performance characteristics are determined via the incremental testing protocol throughout the preparatory and recovery periods.

	Preparatory Period				Recovery Period	
Testing (Days)	−58	−27	−3	0	+10	+19
LT ($m \cdot s^{-1}$)	3.6	3.8	3.8		3.4	3.5
OBLA ($m \cdot s^{-1}$)	4.7	4.7	4.7		4.4	4.5
V_{peak} ($m \cdot s^{-1}$)	5.5	5.5	5.5		5.0	5.5
Vo_{2peak} ($ml \cdot min^{-1} \cdot kg^{-1}$)	73	74	74		65	67
V_{Epeak} ($L \cdot min^{-1}$)	160	161	162		137	150
fR_{peak} (min^{-1})	62	62	58		53	56
VT_{peak} (L)	2.58	2.60	2.80		2.58	2.68
HR_{peak} (min^{-1})	201	197	194		196	200
LA_{peak} ($mmol \cdot L^{-1}$)	11.9	14.9	13.2		8.1	11.0

LT—Lactate Threshold; OBLA—Onset of Blood Lactate Accumulation; V_{peak}—peak velocity; Vo_{2peak}—peak oxygen consumption; V_{Epeak}—peak ventilation; fR_{peak}—peak respiratory frequency; VT_{peak}—peak of tidal volume; HR_{peak}—peak heart rate; LA_{peak}—peak blood lactate concentration; all values were achieved during the incremental test.

Vo_2 did not change during the incremental test in the preparatory or recovery periods. Peak Vo_2 values followed the pattern of peak performance (Figure 5) but did not return to pre-trial values. These values remained at 90% after 19 days of recovery (Figure 5). During CRT, Vo_2 increased about 0.5 months after Everesting and remained elevated one month after this event (Table 5).

V_E did not change after Everesting during the incremental test and remained similar to that in the preparatory period. Nevertheless, fR typically remained elevated until ten days after Everesting and decreased one month after the event to the values measured during the preparatory period (Figure 6). The runner experienced respiratory muscle fatigue and reported chest muscle pain during the CRT on day eight after the trial, so he had to stop the test after 60 min. Increased V_E and fR accompanied this phenomenon (Table 5) during the CRT test about 12 days after the trial.

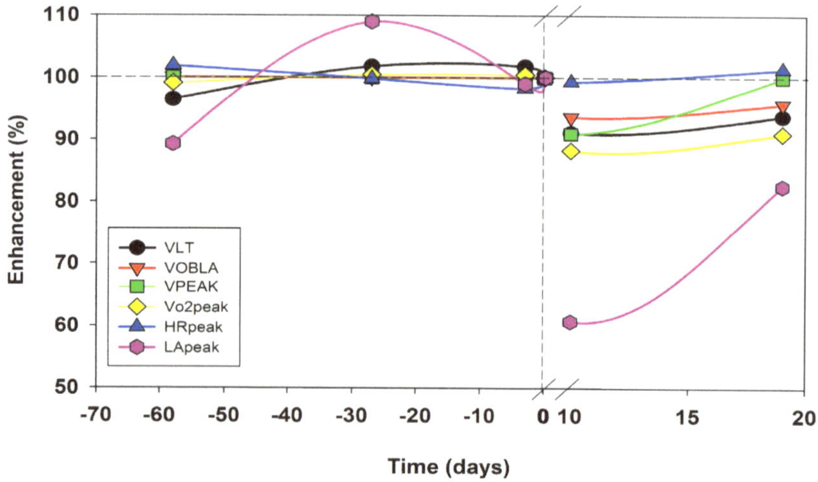

Figure 5. The time course of the relative (%) values of performance characteristics estimated during the incremental test shows a significant decrease in all parameters except HR_{peak} on the 10th day after Everesting. The other variations continued to decrease or remained at constant values. Nevertheless, even on day 20 during the recovery period, these values did not reach the range observed in the preparation period, with the exception of V_{peak}.

Table 5. Responses during the 2 h continuous run at 3.9 m·s^{-1}.

	Preparatory Period								**Recovery Period**							
	−2 Months				**−1 Month**				**+0.5 Month**				**+1 Month**			
Time (Min)	**30**	**60**	**90**	**120**	**30**	**60**	**90**	**120**	**30**	**60**	**90**	**120**	**30**	**60**	**90**	**120**
Vo_2 (L·min^{-1})	3.422	3.627	3.662	3.634	3.400	3.695	3.588	3.451	3.344	3.712			3.314	3.636	3.893	4.034
Vo_2 (ml·min^{-1}·kg^{-1})	50	53	53	53	49	53	54	50	49	55			48	53	56	58
Cr (ml·min^{-1}·m^{-1})	0.19	0.21	0.21	0.21	0.19	0.21	0.20	0.19	0.19	0.23			0.18	0.22	0.22	0.25
V_E (L· min^{-1})	80	83	83	82	82	91	87	82	84	95			81	89	96	102
fR (min^{-1})	30	35	37	38	34	40	38	40	35	43			33	40	41	42
HR (min^{-1})	168	174	177	180	163	167	170	175	180	188			170	175	180	185

Vo_2—absolute and relative oxygen consumption, Cr—energy cost of running; V_E—ventilation, fR—respiratory frequency, HR—heart rate.

The HR showed a tendency to increase during the incremental test about ten days after Everesting but returned to the values reached during the preparatory period after about one month (Figure 5). HR_{peak} did not change and remained at a similar level (Table 4). During the CRT test, HR increased by 5–8 min^{-1}, similar to the results of the tests performed around 0.5 to one month after Everesting (Table 5).

Carbohydrate and fat metabolism were monitored during CRT. First, the test running velocity of 3.0 m·s^{-1} presented a 7% higher relative intensity (about 0.5 m·s^{-1}) because V_{peak} and LT decreased during the recovery period, just as they did throughout the preparatory period. Therefore, the presentation of all other results must be interpreted in the context of this phenomenon. CHOox values stabilised or decreased throughout each run and differed little depending on the CHO or WAT beverages consumed during the preparation period (Figure 7, upper figures). The Everesting trial increased CHOox values, which remained unexpectedly high (3.8 g·min^{-1}) about eight days after the Everesting trial (Figure 7, top row of figure). However, the subsequent recovery phase decreased these values, bringing them closer to those of the preparation period (Figure 7, top row of figure). FATox (Figure 7, top row of figure) was lower than CHOox in both the preparatory and recovery periods. FATox achieved its lowest values about eight days after Everesting (Figure 7, top row of figure), which practically reflects a reversal of CHOox changes. EXOox, the oxidation rate from the beverage, presented an expected difference between CHO and WAT (Figure 7, middle row), with no effect of the trial on this parameter. The difference between CHOox and EXOox reflects the oxidation of endogenous carbohydrate sources (ENDOox) (Figure 7, middle row of figure), which showed an unexpectedly large increase to 3.6 g·min^{-1} 0.5 months after Everesting. About one month after Everesting, this value decreased to 2–2.5 g·min^{-1} (Figure 7, middle row of figure). The ENDOox time course was associated with [LA] fluctuations during the preparatory and recovery periods (Figure 7, bottom row).

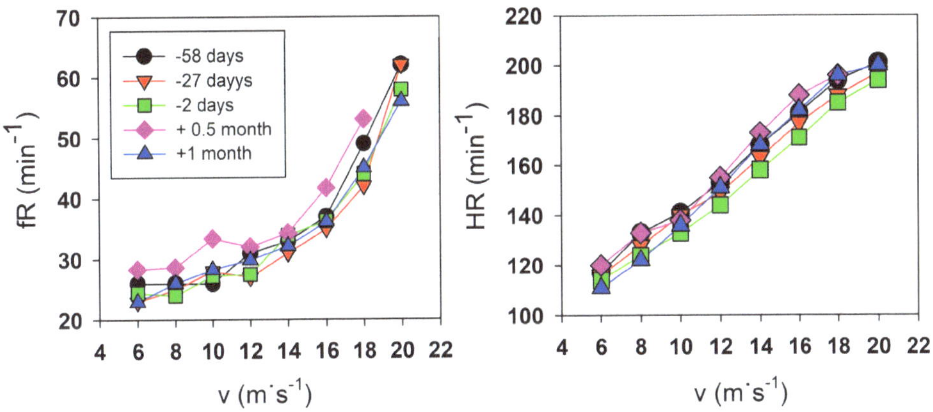

Figure 6. Breathing frequency (fR, **left figure**) and heart rate (HR, **right figure**) during incremental tests repeated throughout the experimental period. The images show relatively stable responses during the preparatory period in March (58 days before Everesting) and April (27 and 2 days before Everesting). The values of fR and HR in May showed higher values about 0.5 months after Everesting, while both fR and HR decreased to values similar to those in the preparatory period about one month after Everesting.

During the incremental test, [LA] showed no difference due to Everesting. The LA_{peak}, which dropped to 60% of the pre-trial value around day ten after the Everesting trial (Figure 5), did not return to the pre-trial values. On day 19 of the recovery period, LA_{peak} was 80% (Figure 5). During CRT, [LA] initially fluctuated around 2–3 mmol·L^{-1} throughout the preparatory period but then increased to 5 mmol·L^{-1} after about 0.5 months in the recovery period and returned to pre-trial values about one month after Everesting (Figure 7, bottom row of figure).

During CRT, similar to other parameters representing CHO metabolism, [GLU] fluctuated in a steady state at about 5–6 mmol·L^{-1} during the preparatory period and increased to 6 mmol·L^{-1} at about 0.5 months during the recovery period (Figure 7, bottom row of figure). Finally, about one month after Everesting, [GLU] decreased to values similar to those in the preparation period (Figure 7, bottom row of figure).

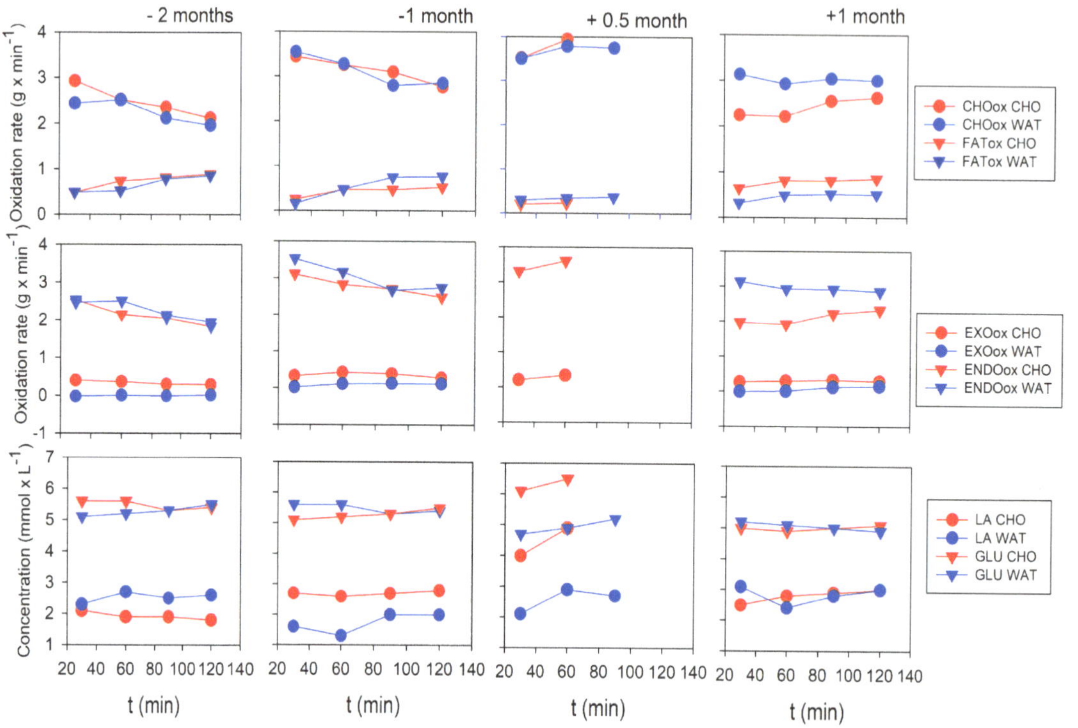

Figure 7. Changes in CHOox and FATox during a two-hour continuous running test at 3.9 $m \cdot s^{-1}$ (14 $km \cdot h^{-1}$) (**top row figures**), changes in EXOox (**middle row figures**) and changes in [LA] and [GLU] (**bottom row figures**) when CHO or WAT (water) was consumed throughout the experiment. The figure shows an increase in CHOox over the period of 0.5 months during the recovery period, accompanied by increased [LA] and [GLU], but not EXOox, values.

3. Discussion

A single runner was observed during a two-month preparatory period for an Everesting mountain ultramarathon and one month during recovery. Decreased running performance during recovery persisted in two phases over one month after the Everesting trial. This confirms the hypothesis of two recovery phases after Everesting. The first phase was characterised by a dramatic decrease in performance that prevented running during the first week after Everesting. This phenomenon was associated with increased blood markers of muscle microdamage and systemic inflammation at rest, which lasted only a few days. This phase disappeared when the blood markers returned to their fluctuation ranges observed during the preparatory period.

Nevertheless, the reduced performance continued into the second phase of the recovery period, with a slow but incomplete return to the values of the preparatory period. This supports the hypothesis that the second recovery phase is extended by a month or longer. A decreased Lactate Threshold, the Onset of Blood Lactate Accumulation, the Vo_{2peak} and peak running velocity determined during the incremental test showed decreased running performance. Consequently, the increased energetic costs of running, heart rate, the increased total carbohydrate oxidation rate (CHOox), endogenous carbohydrate oxidation rate (ENDOox) and blood levels of [LA] and [GLU] were elevated during the 0.5-month recovery period. They typically showed increased relative running intensity at the same absolute running velocity of 14 $km \cdot h^{-1}$ during constant running tests. Increased ventilation and respiratory rate during the constant running test were typical characteristics of respiratory muscle fatigue and increased relative running intensity. The hypothesis that the characteristics of the second phase can be determined by a continuous running test was confirmed. The changes in the presented parameters throughout the recovery period were partially associated with a return of performance toward the values typical of the preparatory period.

The runner's performance, as determined by a repeated incremental running test and continuous running test (CRT), did not change during the preparatory period but significantly decreased during the first phase (week) of the recovery period after Everesting. The dramatic increase in blood markers for muscle microdamage (CK and MB) and systemic inflammation (AST and CRP) peaked on the first day after Everesting. It disappeared within one week, in agreement with previous findings [7,10,12,26]. It is expected that depleted or reduced glycogen stores in the legs were also replenished during this phase due to reduced exercise and a normal diet [16,27]. In contrast, the runner's performance recovered more slowly and became similar to that in the preparation period after about one month during the second phase of the recovery period. The most likely explanation for this difference is an increase in relative running intensity (effort) at a similar absolute running velocity due to decreased performance and slow recovery of the damaged muscle fibres [4,28,29]. This phenomenon was indirectly observed in our study during the incremental test based on a reduced running velocity according to LT and OBLA during the second phase after the first week of the recovery period, with CRT as a shortened time to fatigue. In the subsequent part of the second phase in the recovery period, the values of LT and OBLA remained similarly reduced and showed incomplete recovery. During the one-month recovery period, the time to fatigue for CRT returned to the targeted two hours observed before Everesting. However, fatigue was still present. Notably, a feeling of severe respiratory muscle fatigue in addition to leg fatigue seems to be why the runner stopped CRT at 60 min, 0.5 months after Everesting, and fatigue was still observable one month later in spite of the runner reached the targeted two hours. Decreases in performance after ultramarathons have been reported previously but using different, non-specific, short-term and maximal tests [4,14,15].

The energy cost of running (Cr) in CRT increased during recovery. It was detected after 60 min of running 0.5 months after Everesting and after 120 min of running one month after Everesting. This result primarily reflects increased relative running intensity due to a decrease in V_{peak} by about 0.5 $m{\cdot}s^{-1}$ and Vo_{2peak} by 9 $mL{\cdot}kg^{-1}{\cdot}min^{-1}$, both representing decreases of about 10%, compared with the pre–Everesting values, as well as a decrease in submaximal running velocity LT by 0.4 $m{\cdot}s^{-1}$ (10%). This result also likely reflects persistent microdamage to the muscle fibres [4,14,15] due to mechanical and oxidative stresses [5,6,9]. The aforementioned changes and microdamage to muscle fibres can be responsible for the high increase in ENDOox and CHOox 0.5 months after Everesting during CRT at running intensities of 80–85% Vo_{2max}. As mentioned above, these changes increased relative running intensity and, consequently, influenced metabolic changes. To our knowledge, this is the first mention of this phenomenon during recovery after Everesting.

However, explaining this phenomenon is difficult due to a single case study. Moreover, a typical reference study by Derman et al. [30] shows a similar CHOox of 4 $g{\cdot}min^{-1}$ as in ours. The increase in CHOox was not associated with EXOox but may be explained by increased ENDOox, i.e., increased utilisation of glycogen from the limbs and respiratory muscles. The fatigue of the limb muscles was the result of pre-existing microdamage to muscle fibres due to mechanical stress, oxidative stress and glycogen depletion of skeletal muscles during Everesting, whereas the fatigue of respiratory muscles was more likely due to glycogen depletion of the diaphragm [29] and oxidative stress [6]. Although EXOox did not change, the increase in [GLU] may be explained by the greater selection of intramuscular glycogen as fuel. This phenomenon may be due to increased relative running intensity and, consequently, additional recruitment of motor units, resulting in the activation of IIa fibres and/or less well-trained motor units that are less efficient at using fat as fuel [16,30]. Another consequence of the increased activation of muscle fibres was an increased demand for oxygen supply due to increased Cr. This increase in oxygen supply can be achieved by increasing HR, which consistently increases cardiac output and oxygen delivery [16]. Increased fR could be a compensatory response to increased respiratory muscle fatigue [31]. However, whether the respiratory muscles play an important role in this process can only be speculated. On the other hand, both leg and respiratory muscles compete for blood flow when the respiratory muscles are already more fatigued at moderate intensities, especially when the relative running intensity (effort) increases [31,32]. This factor can increase the progression of leg muscle fatigue.

The results of this study should be used with caution due to the limitations of a single case-study design. Generally, lack of rigour, weak possibility for generalization of results, difficult for replication and bias of researcher feeling in interpretations and conclusions are typical sources of limitations. We have tried as much as possible to reduce the effects of the first and fourth limitations by using a large spectre of variables determined in different tests. This research provides novel insights into the recovery period after an extreme endurance run, focusing specifically on Everesting. However, the analysis of the recovery period using three tests and more than twenty characteristics needs to be revised in the future. Instead of testing running performance with the same absolute running velocity, another test should be performed with the same relative running intensity, especially during recovery. However, these changes would further complicate the study design, as the additional testing may strongly interfere with recovery characteristics.

During the CRT tests at 1-month intervals (Figure 6, right column), there was an unexpected increase in CHOox WAT and ENDOox WAT during water consumption. Both values were higher than those of the corresponding CHO drink consumed five days earlier during the previous CRT. A possible explanation for this phenomenon is related to the particular training conditions during this five-day interval. Namely, the first test was performed by consuming CHO. The following five-day training consisted of two long-distance training sessions: a continuous 35 km run of about 3.7 $km \cdot h^{-1}$ and, three days later, a 25 km Nordic walk in the mountains. Regardless of whether both training sessions were performed for the first time after Everesting, they could influence the renewal of partial muscle microdamage and increase CHOox and ENDOox with the WAT drink.

4. Conclusions

Our observations of a single ultramarathon runner during a two-month preparatory period before Everesting and a one-month recovery period after Everesting indicate that a temporary decrease in performance can be detected by tracking various parameters over a relatively long period through repeated observations. Comparing pre-and post-Everesting values allowed us to analyse the extent and time course of post-Everesting enhancements based on calibrated reference values for pre-Everesting fluctuations. This study showed that recovery after Everesting took longer than previously observed from blood markers of muscle damage and systemic inflammation and consisted of two phases. The first phase lasted less than one week, although the decrease in performance was the greatest. The second phase lasted longer than one month and could be recognised by typical changes in various characteristics measured by incremental tests, especially the continuous running test. Candidate characteristics could be employed in constructing a future biological network to describe the recovery period.

Author Contributions: Conceptualization, A.U. and J.L.V.; methodology, A.U. and S.L.; software, A.U.; validation, A.U. and S.L.; formal analysis, A.U.; investigation, A.U.; resources, A.U. and S.L.; data curation, A.U. and J.L.V.; writing—original draft preparation, A.U.; writing—review and editing, A.U. and S.L.; visualization, A.U.; supervision, A.U. and S.L.; project administration, A.U. All authors have read and agreed to the published version of the manuscript.

Funding: This research received no external funding.

Institutional Review Board Statement: The study was conducted in accordance with the Declaration of Helsinki and approved by the Faculty of Sport Ethics Committee (FS8:2021).

Informed Consent Statement: The participant (runner) was verbally informed regarding the aim of the study and the potential risks and challenges in the experimental procedures. He has provided written informed consent before the study.

Data Availability Statement: Due to ethical and privacy issues, the additional data, except those already presented in the manuscript, is not publicly available.

Conflicts of Interest: The authors declare no conflict of interest.

References

1. Lazzer, S.; Salvadego, D.; Rejc, E.; Buglione, A.; Antonutto, G.; di Prampero, P.E. The energetics of ultra-endurance running. *Eur. J. Appl. Physiol.* **2012**, *112*, 1709–1715. [CrossRef]
2. Gimenez, P.; Kerherve, H.; Messonnier, L.A.; Feasson, L.; Millet, G.Y. Changes in the energy cost of running during a 24 h treadmill exercise. *Med. Sci. Sports Exerc.* **2013**, *45*, 1807–1813. [CrossRef]
3. Hikida, R.S.; Staron, R.S.; Hagerman, F.C.; Sherman, W.M.; Costill, D.L. Muscle fiber necrosis associated with human marathon-runners. *J. Neuro. Sci.* **1983**, *59*, 185–203. [CrossRef] [PubMed]
4. Warhol, M.J.; Siegel, A.J.; Evans, W.J.; Silverman, L.M. Skeletal muscle injury and repair in marathon runners after competition. *Am. J. Pathol.* **1985**, *118*, 331–339. [PubMed]
5. Mrakic-Sposta, S.; Gussoni, M.; Moretti, S.; Pratali, L.; Giardini, G.; Tacchini, P.; Vezzoli, A. Effects of mountain ultra-marathon running on ROS production and oxidative damage by micro-invasive analytical techniques. *PLoS ONE* **2015**, *10*, e0141780. [CrossRef] [PubMed]
6. Thirupathi, A.; Pinho, R.A.; Ugbolue, U.C.; He, Y.; Meng, Y.; Gu, Y. Effect of running exercise on oxidative stress biomarkers: A systematic review. *Front. Physiol.* **2021**, *11*, 1789. [CrossRef]
7. Rubio-Arias, J.A.; Avila-Gandia, V.; Romain, J.A.; Soto-Mendez, F.; Alcaraz, P.E.; Ramos-Campo, D.J. Muscle damage and inflammation biomerkers after two ultra-endurance mountain races of different distances: 54 km vs 111 km. *Physiol. Behav.* **2018**, *205*, 51–57. [CrossRef]
8. Saugy, J.; Place, N.; Millet, G.Y.; Degache, F.; Schena, F.; Millet, G.P. Alterations of neuromuscular function after the world's most challenging mountain ultra-marathon. *PLoS ONE* **2013**, *8*, e65596. [CrossRef]
9. Khodaee, M.; Spittler, J.; VanBaak, K.; Changstrom, B.G.; Hill, J.C. Effects of running an ultramarathon on cardiac, hematologic, and metabolic biomarkers. *Int. J. Sports Med.* **2015**, *36*, 867–871. [CrossRef]
10. Kim, H.J.; Lee, Y.H.; Kim, C.K. Biomarkers of muscle and cartilage damage and inflammation during 200 km run. *Eur. J. Appl. Physiol.* **2007**, *99*, 443–447. [CrossRef]
11. Romer, L.M.; Polkey, M.I. Exercise-induced respiratory muscle fatigue: Implications for performance. *J. Appl. Physiol.* **2008**, *104*, 879–888. [CrossRef] [PubMed]
12. Kupchak, B.R.; Kraemer, W.J.; Hoffman, M.D.; Phinney, S.D.; Volek, J.S. The impact of an ultramarathon on hormonal and biochemical parameters in men. *Wilderness Environ. Med.* **2014**, *25*, 278–288. [CrossRef] [PubMed]
13. Hoffman, M.D.; Badowski, N.; Chin, J.; Stuempfle, K.J.; Parise, C.A. Determinants of recovery from a 161-km ultramarathon. *J. Sports Sci.* **2016**, *35*, 669–677. [CrossRef] [PubMed]
14. Sherman, W.M.; Costill, D.L.; Fink, W.J.; Hagerman, F.C.; Armstrong, L.E.; Murray, T.F. Effects of 42.2 km footrace and subsequent rest or exercise on muscle glycogen and enzymes. *J. Appl. Physiol.* **1983**, *55*, 1219–1224. [CrossRef]
15. Chambers, C.; Noaks, T.D.; Lambert, E.V.; Lambert, M.I. Time course of recovery of vertical jump height and heart rate vs running speed after a 90 km foot race. *J. Sport. Sci.* **1998**, *16*, 645–651. [CrossRef]
16. Noaks, T.D. *Lore of Running*; Human Kinetics: Champaign, IL, USA, 2003.
17. Milsum, J.H. *Biological Control Systems Analysis*; Mc Graw-Hill: New York, NY, USA, 1966.
18. Beaver, W.L.; Wasserman, K.; Whipp, B. A new method for detecting anaerobic threshold by gas exchange. *J. Appl. Physiol.* **1986**, *60*, 2020–2027. [CrossRef]
19. Usaj, A.; Starc, V. Blood pH and Lactate Kinetics in the Assessment of Running Endurance. *Int. J. Sports Med.* **1996**, *17*, 34–40. [CrossRef]
20. Borg, G. Psychophysical bases of perceived exertion. *Scand. J. Work. Environ. Health* **1990**, *14*, 55–58. [CrossRef]
21. Frayn, K.N. Calculation of substrate oxidation rates in vivo from gaseous exchange. *J. Appl. Physiol.* **1983**, *55*, 628–634. [CrossRef]
22. Jentjens, R.; Venables, M.C.; Jeukendrup, A.E. Oxidation of exogenous glucose, sucrose, and maltose during prolonged cycling exercise. *J. Appl. Physiol.* **2003**, *96*, 1285–1291. [CrossRef]
23. Mosora, F.; Lefebvre, P.; Pirnay, F.; Lacroix, M.; Luyckx, A.; Duchesne, J. Quantitative evaluation of the oxidation of an exogenous glucose load using naturally labeled 13C-Glucose. *Metabolism* **1976**, *25*, 1575–1582. [CrossRef]
24. Pirnay, F.; Crielaard, J.M.; Pallikarakis, N.; Lacroix, M.; Mosora, F.; Krzentowski, G.; Lefebvre, P.J. Fate of exogenous glucose during exercise of different intensities in humans. *J. Appl. Physiol.* **1982**, *53*, 1620–1624. [CrossRef] [PubMed]
25. Peronnet, F.; Massicotte, D.; Brisson, G.; Hillaire-Marcel, C. Use of 13C substrates for metabolic studies in exercise: Methodological considerations. *J. Appl. Physiol.* **1990**, *69*, 1047–1052. [CrossRef]
26. Sjodin, B.; Schele, R.; Karlsson, J.; Linarsson, D.; Wallensten, R. The physiological background of onset of blood lactate accumulation (OBLA). In *Exercise and Sport Biology*; Komi, P.V., Ed.; Human Kinetics: Champaign, IL, USA, 1982; pp. 43–56.

27. Panizo González, N.; Reque Santivañez, J.E.; Hernando Fuster, B.; Collado Boira, E.J.; Martinez-Navarro, I.; Chiva Bartoll, Ó.; Hernando Domingo, C. Quick recovery of renal alterations and inflammatory activation after a marathon. *Kidney Dis.* **2019**, *5*, 259–265. [CrossRef] [PubMed]
28. Martinez-Navarro, I.; Sanchez-Gomez, J.M.; Aparicio, I.; Priego-Quesada, J.I.; Perez-Soriano, P.; Collado, E.; Hernando, C. Effect of mountain ultramarathon distance competition on biochemical variables, respiratory and lower-limb fatigue. *PLoS ONE* **2020**, *15*, e0238846. [CrossRef] [PubMed]
29. Lambert, M.I.; Gibson, A.S.C.; Derman, W.; Noaks, T.D. Regeneration after ultra-endurance exercise. In *Overload, Performance Incompetence, and Regeneration in Sport*; Lehman, M., Foster, C., Gastmann, U., Keizer, H., Steinacker, J.M., Eds.; Kluwer Academic/Plenum Publishers: New York, NY, USA, 1999; pp. 163–172.
30. Derman, K.D.; Hawley, J.A.; Noaks, T.D.; Dennis, S.C. Fuel kinetics during intense running and cycling when fed carbohydrate. *Europ. J Appl Physiol.* **1996**, *74*, 36–43. [CrossRef]
31. Vernillo, G.; Rinaldo, N.; Giorgi, N.; Esposito, F.; Trabucci, P.; Millet, G.; Schena, F. Changes in lung function during an extreme mountain ultramarathon. *Scand. J. Med. Sci. Sports* **2014**, *25*, 374–380. [CrossRef]
32. Wuthrich, T.; Marty, J.; Kerherve, H.; Millet, G.Y.; Verges, S.; Spengler, C.M. Aspects of respiratory muscle fatigue in a mountain ultramarathon race. *Med. Sci. Sports Exerc.* **2015**, *47*, 519–527. [CrossRef]

life

MDPI

Case Report

Takotsubo Cardiomyopathy Occurring Simultaneously with Acute Myocardial Infarction

Ilija Srdanović [1,2], Dragana Dabović [1,2,*], Vladimir Ivanović [1,2], Milenko Čanković [1,2], Teodora Pantić [1,2], Maja Stefanović [1,2], Sonja Dimić [1,2], Branislav Crnomarković [1,2], Marija Bjelobrk [1,2], Miljana Govedarica [1,3] and Marija Zdravković [4,5]

1 Faculty of Medicine, University of Novi Sad, 21000 Novi Sad, Serbia; ilija.srdanovic@mf.uns.ac.rs (I.S.); vladimir.ivanovic@mf.uns.ac.rs (V.I.); milenko.cankovic@mf.uns.ac.rs (M.Č.); teodora.pantic@ikvbv.ns.ac.rs (T.P.); maja.stefanovic@mf.uns.ac.rs (M.S.); sonja.dimic@ikvbv.ns.ac.rs (S.D.); branislav.crnomarkovic@ikvbv.ns.ac.rs (B.C.); marija.bjelobrk@mf.uns.ac.rs (M.B.); miljana.govedarica.94@gmail.com (M.G.)
2 Clinic of Cardiology, Institute of Cardiovascular Diseases of Vojvodina, 21204 Sremska Kamenica, Serbia
3 Department of Obstetrics and Gynaecology, Clinical Centre of Vojvodina, 21000 Novi Sad, Serbia
4 University Clinical Hospital Center Bezanijska Kosa, 11000 Belgrade, Serbia; sekcija.kardioloska@gmail.com
5 Faculty of Medicine, University of Belgrade, 11000 Belgrade, Serbia
* Correspondence: dragana.dabovic@mf.uns.ac.rs; Tel.: +381-638358888

Abstract: Introduction: Takotsubo cardiomyopathy (TCM) is a reversible form of cardiomyopathy characterized by transient regional systolic dysfunction of the left ventricle. Case outline: A 78-year-old woman was admitted to the general hospital due to acute inferior STEMI late presentation. Two days after admission, the patient reported intense chest pain and an ECG registered diffuse ST-segment elevation in all leads with ST-segment denivelation in aVR. The patient also showed clinical signs of cardiogenic shock and was referred to a reference institution for further evaluation. Echocardiography revealed akinesia of all medioapical segments, dynamic obstruction of the left ventricular outflow tract (LVOT), moderate mitral regurgitation, and pericardial effusion. Coronary angiography showed the suboccluded right coronary artery, and a primary percutaneous coronary intervention was performed, which involved implanting a drug-eluting stent. The patient's condition worsened as pericardial effusion increased and led to tamponade. Pericardiocentesis was performed, resulting in the patient's stabilization. At this point, significant gradients at the LVOT and pericardial effusion were not registered. After eight days without symptoms and stable status, the patient was discharged. Conclusions: The simultaneous presence of AMI and TCM increases the risk of developing cardiogenic shock. The cardio-circulatory profile of these patients is different from those with AMI.

Keywords: Takotsubo cardiomyopathy; acute myocardial infarction; cardiogenic shock

Citation: Srdanović, I.; Dabović, D.; Ivanović, V.; Čanković, M.; Pantić, T.; Stefanović, M.; Dimić, S.; Crnomarković, B.; Bjelobrk, M.; Govedarica, M.; et al. Takotsubo Cardiomyopathy Occurring Simultaneously with Acute Myocardial Infarction. *Life* **2023**, *13*, 1770. https://doi.org/10.3390/life13081770

Academic Editors: Helena Lenasi and Ines Drenjančević

Received: 19 July 2023
Revised: 7 August 2023
Accepted: 10 August 2023
Published: 18 August 2023

1. Introduction

Takotsubo cardiomyopathy (TCM) is a reversible cardiomyopathy characterized by transient regional systolic dysfunction of the left ventricle [1]. It is estimated that 1–3% of all patients with acute coronary syndrome have TCM [2]. TCM is registered in 5–6% of women with acute myocardial infarction with ST elevation (STEMI) [3]. The "Fourth Definition of Myocardial Infarction" does not consider TCM as myocardial infarction [4]. Although TCM can clinically mimic myocardial infarction without obstruction of coronary arteries, it appears to be a distinctly different syndrome and should be considered separately. Today, in the era of the COVID-19 pandemic, the incidence of TCM has increased significantly and is about 7.8% [5]. The TCM mechanism is still unclear, but it is believed to be triggered by emotional or physical stress. TCM is most often described as a nonsignificant coronary artery disease (CAD). However, according to the InterTak and Mayo diagnostic criteria, significant CAD is not an exclusive factor for the development of TCM [6]. The literature describes rare cases in which acute myocardial infarction (AMI) and TCM occurred simultaneously. Acute coronary syndrome is thought to cause somatic stress and thus triggers the development of TCM [7].

In this article, we will discuss the case of a 78-year-old woman who experienced acute STEMI of the inferior region and TCM simultaneously.

2. Case Report

A 78-year-old woman was admitted to the general hospital due to chest pain and ECG signs of acute inferior STEMI. She had experienced chest pain 36 h prior to the admission, following a period of intense emotional stress. The patient also had anxiety as a comorbidity. Upon admission, the patient was alert and orientated, hemodynamically and rhythmically stable, and without signs of heart failure. It was concluded that the patient was a late presenter of myocardial infarction, since 36 h had passed from the beginning of the symptoms. As a result, she did not receive primary percutaneous coronary intervention, nor was she given fibrinolytic therapy. Two days after admission, the patient reported intense chest pain. The ECG registered diffuse ST-segment elevation with ST-segment denivelation in aVR and QTc interval (406 ms) (Figure 1).

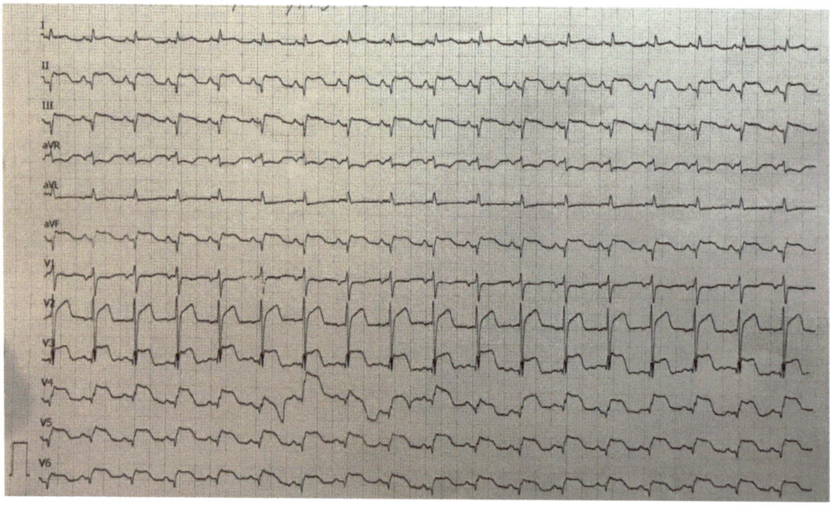

Figure 1. ECG registered diffuse ST-segment elevation with ST-segment denivelation in aVR and QTc interval (406 ms).

As a result, the patient was referred to a specialized medical institution. Upon admission, the patient was conscious, confused, hypotensive (TA 80/50 mmHg), with a heart rate of about 110/min, and showing clinical signs of hypoperfusion and cardiogenic shock, Killip IV. The medical staff administered sedation, inserted an endotracheal tube, and placed her on invasive mechanical ventilation. The patient was given crystalline solutions, inotrope, and vasopressor medication. An urgent echocardiographic examination was performed due to a rough systolic murmur over the precordium. It revealed akinesia of all medioapical segments of the left ventricle and akinesia basally inferior, where the myocardium was fibrously altered. Other hyperkinetic basal segments formed a dynamic obstruction of the left ventricular outflow tract (LVOTO) with turbulent flow and moderate mitral regurgitation (Figures 2 and 3). The maximum gradient above the LVOT was 160 mmHg (Figure 4). The aortic valve area was 1.8 cm^2 (Figure 5). The ejection fraction of the left ventricle (LVEF) was estimated to be 25%. The examination also revealed pericardial effusion with separation between pericardial layers along the right ventricle and atrium of up to 1.2 cm, but without any signs of tamponade.

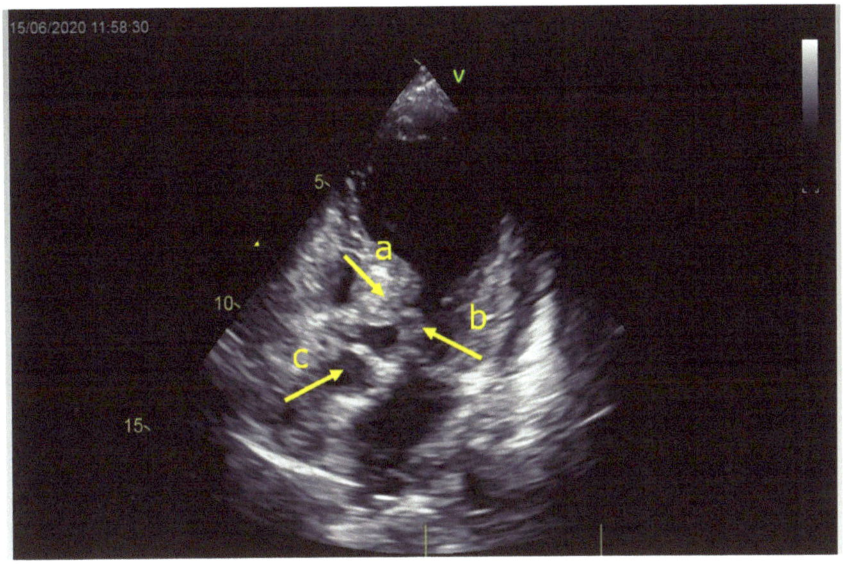

Figure 2. Echocardiography registered basal hyperkinetic segments of the left ventricle and formed a dynamic obstruction of the left ventricular outflow tract (a—LVOT, b—cuspi anterioris valvulae mitralis, c—aortic valve).

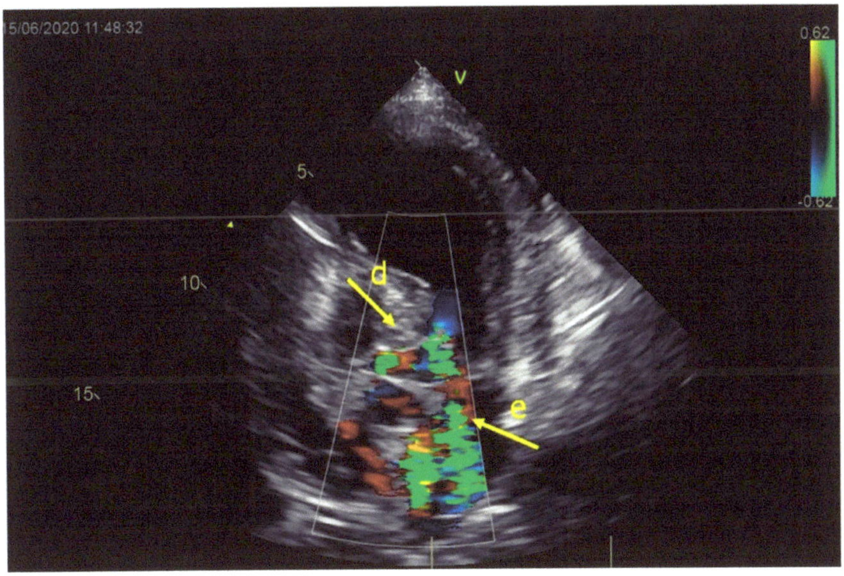

Figure 3. Echocardiography registered moderate mitral regurgitation (d—LVOT, e—mitral regurgitation).

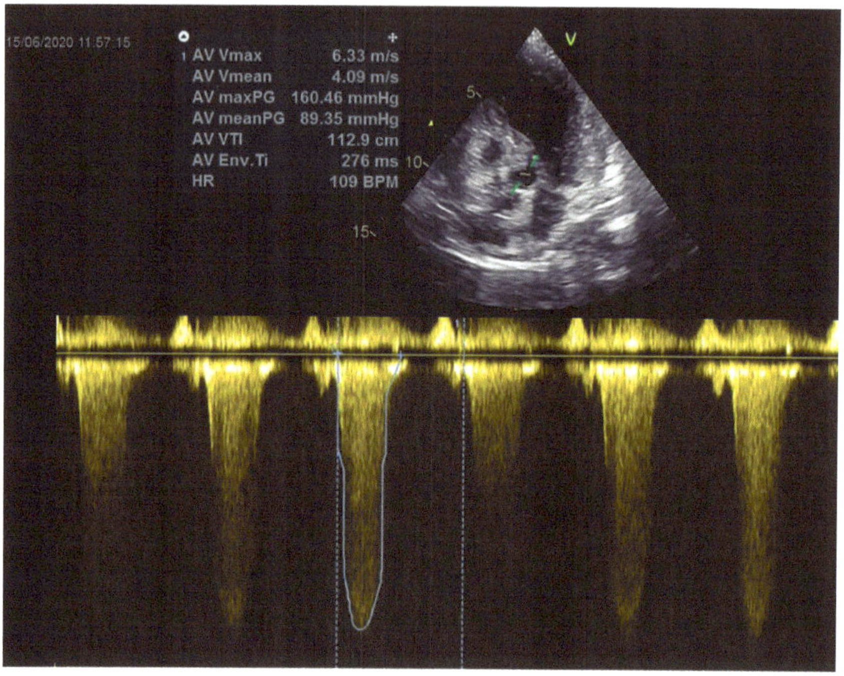

Figure 4. The maximum gradient above the LVOT was 160 mmHg.

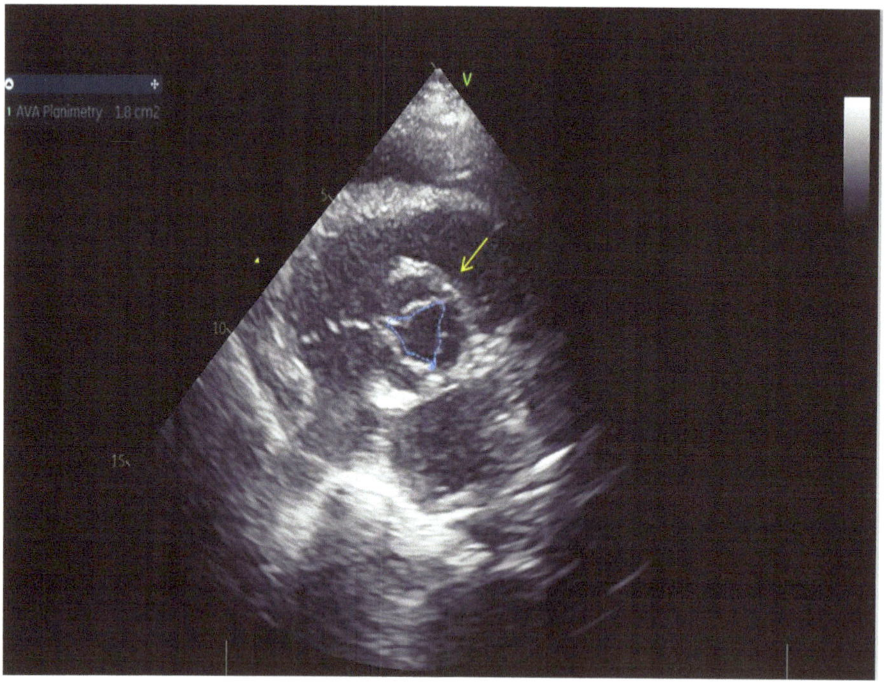

Figure 5. Aortic valve area was 1.8 cm^2.

Coronary angiography was performed, showing the left coronary artery (LCA) was without significant lesions (Figure 6). On the right coronary artery (RCA), a subocclusive lesion was registered in the distal segment (Figure 7). Initially, it was thought to be a spasm, but the lesion persisted even after administering nitroglycerin intracoronary. Therefore, a primary percutaneous coronary intervention (pPCI) was performed with the implantation of a drug-eluting stent 16 × 25 mm (Boston Scientific, Marlborough, MA, USA) in the RCA, achieving the optimal result of the intervention (Figure 8).

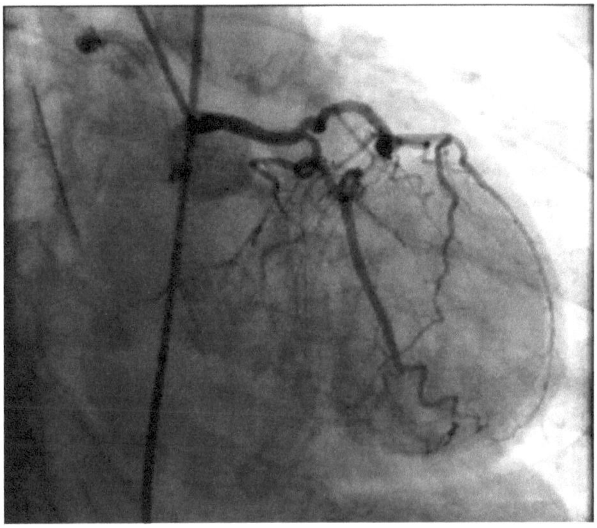

Figure 6. Coronary angiography registered left coronary artery without significant stenosis.

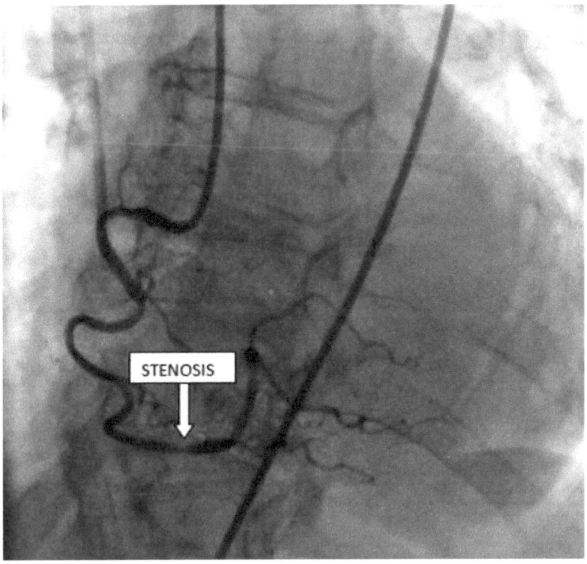

Figure 7. Coronary angiography registered a subocclusive lesion in the distal segment of the right coronary artery.

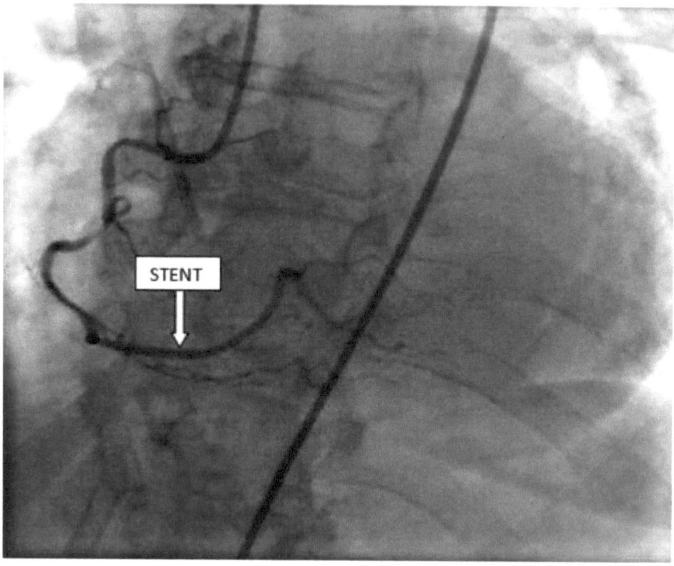

Figure 8. Percutaneous coronary intervention was performed, and a drug-eluting stent was implanted in the RCA, with the optimal result.

In the following course of treatment, the patient became hemodynamically unstable, despite high doses of inotropes and vasopressors. Echocardiography registered a more significant amount of effusion around the heart compared to the previous exam, with signs of cardiac tamponade (Figure 9). Pericardiocentesis was performed, and 260 mL of hemorrhagic fluid was drained (Figure 10).

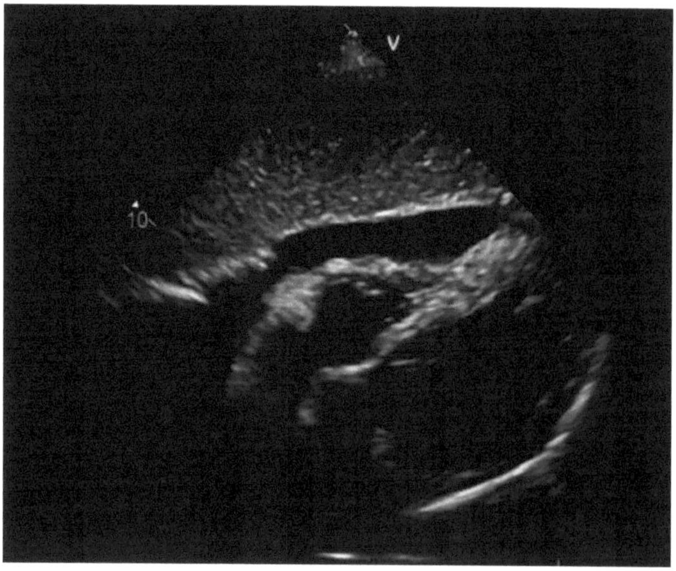

Figure 9. Echocardiography registered a larger amount of effusion around the heart compared to the first exam, with signs of cardiac tamponade.

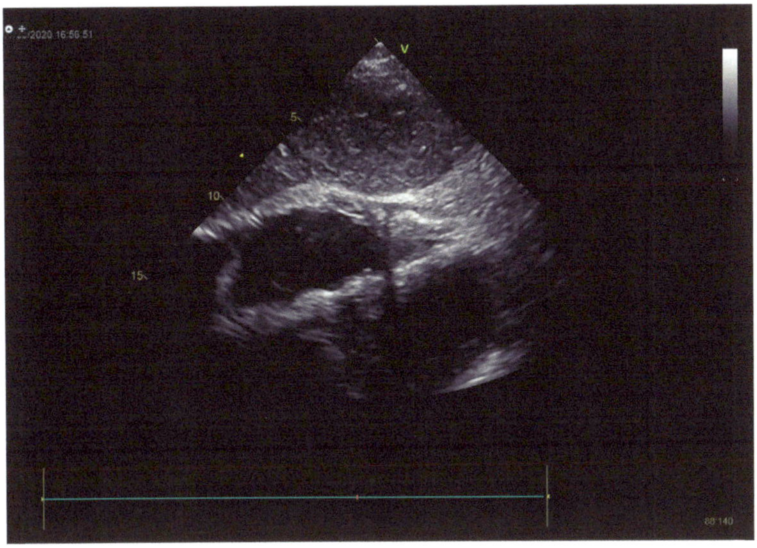

Figure 10. On echocardiography after pericardiocentesis, pericardial effusion was not registered.

Stabilization was achieved gradually, and vasopressor and inotrope support were excluded. By the second day of hospitalization, sedation was stopped, and the patient was alert and responsive. Invasive mechanical ventilation was no longer necessary, and the patient was successfully extubated. On the seventh day of hospitalization, a control echocardiographic examination registered inferior wall akinesia and hypokinesia apically, anteroseptally, inferoseptally, anteriorly, and inferiorly with an estimated LVEF of 52%. No significant gradients were found above the LVOT and there was no pericardial effusion (Figure 11). Mild mitral regurgitation was also registered (Figure 12), but there was no systolic murmur over the precordium.

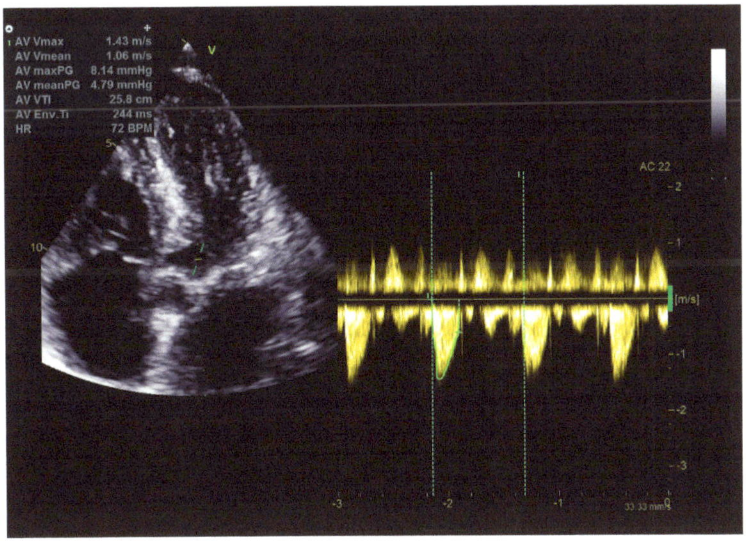

Figure 11. No significant gradients were registered above the LVOT, measured by CW Doppler echocardiography.

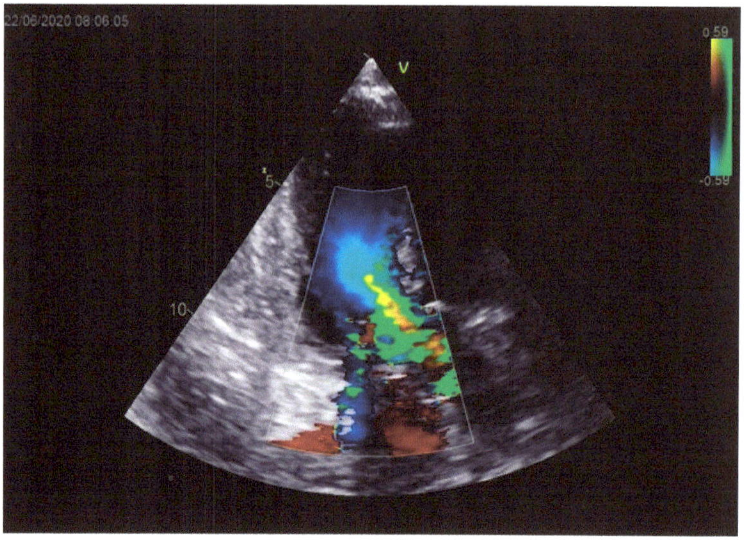

Figure 12. Mild mitral regurgitation was registered with color Doppler echocardiography.

On the eighth hospital day, the patient was asymptomatic, hemodynamically and rhythmically stable, and without signs of heart failure. The ECG showed a negative T-wave in the anterior leads (Figure 13). She was discharged with acetylsalicylic acid, Ticagrelor, Bisoprolol, Ramipril, a statin, and a proton pump inhibitor.

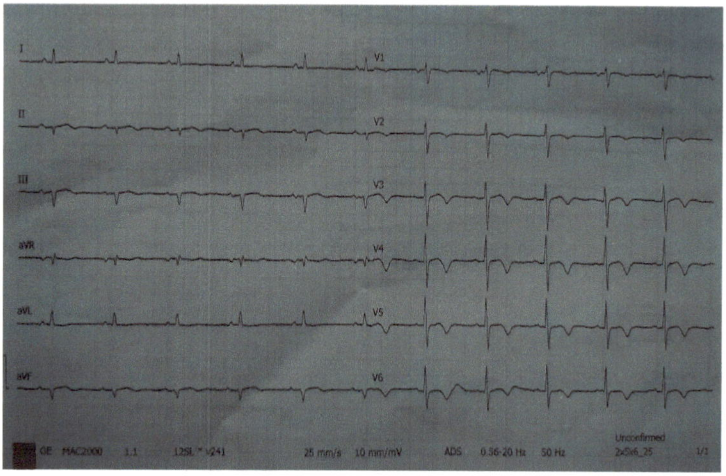

Figure 13. ECG registered negative T-wave in anterior leads.

At the one-year follow-up examination, the patient was asymptomatic.

Cardiac magnetic resonance (CMR) imaging performed per protocol showed no visible zones of late pericardial enhancement (LGE) phenomenon (Figure 14).

The native T1 mapping sequence was without areas of prolonged native T1 time (edema/fibrosis). The post-contrast T1 mapping sequence was without areas of shortened post-contrast T1 time (fibrosis) as well (Figure 15).

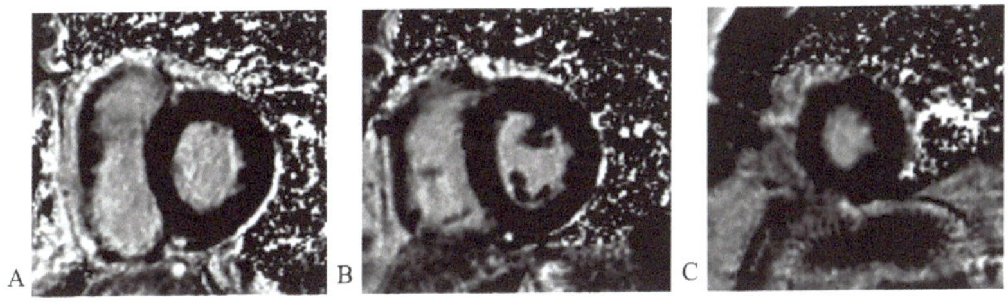

Figure 14. Cardiac magnetic resonance imaging: (**A**) LGE PSIR sequence, short axis, basal view; without visible zones of LGE phenomenon; (**B**) LGE PSIR sequence, short axis, mid chamber view; without visible zones of LGE phenomenon; (**C**) LGE PSIR sequence, short axis, apical view; without visible zones of LGE phenomenon.

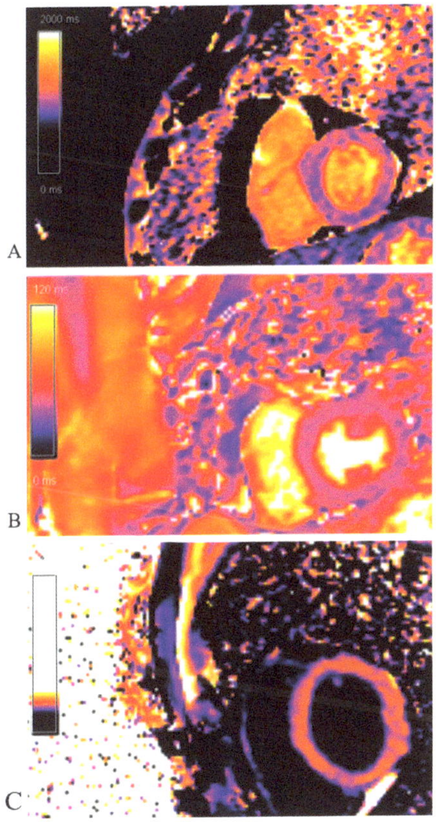

Figure 15. Cardiac magnetic resonance imaging: (**A**)Native T1 mapping sequence; without areas of prolonged native T1 time (edema/fibrosis); (**B**) T2 mapping sequence; without areas of prolonged T2 time (edema); (**C**) Post-contrast T1 mapping sequence; without areas of shortened post-contrast T1 time (fibrosis).

A written consent to publish this report was obtained from the patient.

3. Discussion

The mechanism of TCM is not yet fully understood, but it is believed to be triggered by emotional or physical stress. Emotional stress is the cause in 20–39% of cases, while somatic stress is responsible in 35–55% [8]. There are two important aspects of physiology to take into account. The first is the cognitive centers of the brain and hypothalamic–pituitary–adrenal axis, which plays a role in how stress is perceived and how much epinephrine and norepinephrine are released in response to stress [9]. The second is the response of the cardiovascular and sympathetic nervous systems to the sudden sympathetic activation and a surge in circulating catecholamines [9]. It is stated in the literature that the increased concentration of catecholamines caused by emotional or physical stress causes coronary vasospasm and microcirculation abnormalities, which may be one of the explanations for the development of TCM [10,11]. However, the SMINC-2 trial, which included patients with Takostubo cardiomyopathy, showed no evidence of massive catecholamine elevations [12]. Somatic stressors include intracranial events, severe infections, and surgical trauma [13]. The literature has also reported cases of elderly patients experiencing both AMI and TCM simultaneously [1]. AMI, as an intense somatic stressor, is thought to contribute to increased catecholamine concentrations and thus is a trigger for the development of TCM. Also, there is a significantly elevated concentration of catecholamines in the peri-infarction zone, and simultaneous occurrence of AMI and TCM can be expected.

TCM is most often described with nonsignificant CAD. However, according to the Mayo and InterTak diagnostic criteria, significant CAD is not an exclusive criterion for TCM [14,15]. In a study involving 413 patients admitted to the intensive care unit due to acute coronary syndrome, 5 patients also had CAD and TCM simultaneously [16]. This conclusion was reached after reviewing echocardiographic findings retrospectively. The simultaneous presence of CAD and TCM is associated with a higher risk of developing cardiogenic shock, the need for invasive mechanical ventilation, and the occurrence of death of any etiology. Our patient also developed cardiogenic shock, followed by the necessary involvement of invasive mechanical ventilation.

Diagnosing TCM is often a challenge. Echocardiographic examination with coronary angiography and ventriculography is the gold standard in the diagnosis. Careful examination of the coronary anatomy from several angiographic sections is necessary. Transthoracic echocardiographic examination is the first line in the diagnosis. Echocardiographic parameters indicating a high risk of TCM are low minute volume, LVEF below 35%, diastolic dysfunction, LVOTO, mitral regurgitation, right ventricular involvement, left ventricular thrombus, pericardial effusion, and rupture of the free wall [17]. Most of the listed parameters were registered in our patient upon admission.

CMR is useful for differential diagnoses. Typically, patients with stress cardiomyopathy do not present significant late enhancement, while subendocardial late enhancement is common in myocardial infarction, and focal or subepicardial late enhancement is frequent in myocarditis [18]. Our patient did not exhibit any late enhancement.

According to our findings, the evidence of Takotsubo cardiomyopathy (TCM) was transient akinesia of all medioapical segments of the left ventricle and hyperkinetic basal segments that were not registered on control echocardiography. Per the consensus document, the apical ballooning type was known as the typical TCM form, which occurs in most cases [14].

Pericardial effusion in our patient was most likely a consequence of bleeding per diapedesis since the effusion was hemorrhagic and the patient was a late presenter of myocardial infarction.

An InterTac score has been developed to assess the likelihood of TCM. It consists of seven variables: female sex (25 points), emotional stress (24 points), physical stress (13 points), absence of ST-segment depression (12 points), psychiatric disorders (11 points), neurological disorders (9 points), and prolonged QTc interval (6 points) [19]. When the patient's score is more than 70, the probability of TCM is over 95%. In our patient, the value of InterTak was almost 72 (female sex, emotional stress, no ST segment depression, psychiatric disorders).

There are still no results from a randomized clinical study on the prognostic significance of any medication. Given the possibility that the toxic effect of catecholamines may cause TCM, the use of beta-blockers should be considered. ACE inhibitors also play a significant role in long-term therapy. On the other hand, about 20% of patients with TCM present with LVOTO and cardiogenic shock. The cardio-circulatory profile of these patients is different from those with AMI and cardiogenic shock. Low peripheral resistance and low blood pressure with poor tissue perfusion are registered. In these patients, levosimendan and mechanical circulatory support should be preferred, while inotropes and vasopressors should be avoided [20]. In our case, stabilization was rapidly achieved after the pPCI and pericardiocentesis, and inotropes and vasopressors were quickly stopped. Although levosimendan was not available at the time, mechanical circulatory support was considered, but ultimately not necessary due to the fast stabilization. It is important to note that while a beta-blocker may help reduce obstruction of LVOT, it should not be used in severe acute heart failure and hypotension.

Today, in the era of the COVID-19 pandemic, a large percentage of patients with COVID-19 infection have complications in the cardiovascular system. The most common are acute myocarditis, myocardial infarction, arrhythmias, and pulmonary thromboembolism. However, cases of COVID-19 infection complicated by TCM have also been described in the literature [21]. On the other hand, being afraid of COVID-19 infection, the patients contact the medical service later, and there is a significantly higher percentage of patients with AMI who appear as late presenters of the disease, as in the case of our patient [22]. Patients must be educated about the importance of immediately contacting an emergency department if they experience chest pain. It is crucial because the success of pPCI in late presenters is limited.

The prognosis of patients with TCM significantly depends on the presence of CAD. According to data from the Swedish Registry for Coronary Angiography and Angioplasty from the period from 2009 to2013, mortality is substantially higher in patients with TCM and CAD [3]. After one year, our patient is asymptomatic, hemodynamically and rhythmically stable, and without signs of heart failure.

In conclusion, the simultaneous presence of AMI and TCM is associated with a higher risk of developing cardiogenic shock. Patients with TCM and AMI have a unique cardio-circulatory profile compared to those with only AMI. They often present with low blood pressure, poor tissue perfusion, and low peripheral resistance. Treating these patients is challenging, as limited research is available, and each case must be approached individually. Treatment options such as levosimendan and mechanical circulatory support are recommended, while inotropes and vasopressors should be avoided.

Author Contributions: Conceptualization: I.S., D.D., Methodology: V.I., M.Č., Writing—original draft prepearation: T.P., M.S., supervision: S.D., B.C., writing review and editing: M.B., M.G., visualization: M.Z. All authors have read and agreed to the published version of the manuscript.

Funding: This research received no external funding.

Institutional Review Board Statement: The study was conducted in accordance with the Declaration of Helsinki, and approved by the Institutional Review Board (or Ethics Committee) of Instite of cardiovascular diseases of Vojvodina (protocol code 2256-1/1 and date of approval 15 August 2023).

Informed Consent Statement: Written informed consent has been obtained from the patient(s) to publish this paper.

Data Availability Statement: No new data were created or analyzed in this study. Data sharing is not applicable to this article.

Conflicts of Interest: The authors declare no conflict of interest.

References

1. Koeth, O.; Zeymer, U.; Schiele, R.; Zahn, R. Inferior ST elevation myocardial infarction associated with takotsubo cardiomyopathy. *Case Rep. Med.* **2010**, *2*, 467867. [CrossRef] [PubMed]
2. Prasad, A.; Dangas, G.; Srinivasan, M.; Yu, J.; Gersh, B.J.; Mehran, R.; Stone, G.W. Incidence and angiographic characteristics of patients with apical ballooning syndrome (takotsubo/stress cardiomyopathy) in the HORIZONS-AMI trial: An analysis from a

multicenter, international study of ST-elevation myocardial infarction. *Catheter. Cardiovasc. Interv.* **2014**, *83*, 343–348. [CrossRef] [PubMed]
3. Redfors, B.; Vedad, R.; Angerås, O.; Råmunddal, T.; Petursson, P.; Haraldsson, I.; Omerovic, E. Mortality in Takotsubo syndrome is similar to mortality in myocardial infarction—A report from the SWEDEHEART registry. *Int. J. Cardiol.* **2015**, *185*, 282–289. [CrossRef]
4. Thygesen, K.; Alpert, J.S.; Jaffe, A.S.; Chaitman, B.R.; Bax, J.J.; Morrow, D.A.; White, H.D.; Executive Group on behalf of the Joint European Society of Cardiology (ESC)/American College of Cardiology (ACC)/American Heart Association (AHA)/World Heart Federation (WHF) Task Force for the Universal Definition of Myocardial Infarction. Fourth Universal Definition of Myocardial Infarction (2018). *J. Am. Coll. Cardiol.* **2018**, *72*, 2231–2264. [PubMed]
5. Dullabh, D.; Zachariah, D. Takotsubo cardiomyopathy—An unusual manifestation of the COVID-19 pandemic. *Wits J. Clin. Med.* **2020**, *2*, 255–256. [CrossRef]
6. Kurisu, S.; Inoue, I.; Kawagoe, T.; Ishihara, M.; Shimatani, Y.; Nakama, Y. Prevalence of incidental coronary artery disease in Takotsubo cardiomyopathy. *Coron. Artery Dis.* **2009**, *20*, 214–218. [CrossRef]
7. Redfors, B.; Ramunddal, T.; Shao, Y.; Omerovic, E. Takotsubo triggered by acute myocardial infarction: A common but overlooked syndrome? *J. Geriatr. Cardiol.* **2014**, *11*, 171–173.
8. Okura, H. Echocardiographic assessment of takotsubo cardiomyopathy: Beyond apical ballooning. *J. Echocardiogr.* **2016**, *14*, 13–20. [CrossRef] [PubMed]
9. Lyon, A.R.; Citro, R.; Schneider, B.; Morel, O.; Ghadri, J.R.; Templin, C.; Omerovic, E. Pathophysiology of Takotsubo Syndrome: JACC State-of-the-Art Review. *J. Am. Coll. Cardiol.* **2021**, *77*, 902–921. [CrossRef]
10. Gianni, M.; Dentali, F.; Grandi, A.M.; Sumner, G.; Hiralal, R.; Lonn, E. Apical ballooning syndrome or takotsubo cardiomyopathy: A systematic review. *Eur. Heart J.* **2006**, *27*, 1523–1529. [CrossRef] [PubMed]
11. Wittstein, I.S.; Thiemann, D.R.; Lima, J.A.C.; Baughman, K.L.; Schulman, S.P.; Gerstenblith, G.; Wu, K.C.; Rade, J.J.; Bivalacqua, T.J.; Champion, H.C. Neurohumoral features of myocardial stunning due to sudden emotional stress. *N. Engl. J. Med.* **2005**, *352*, 539–548. [CrossRef] [PubMed]
12. Y-Hassan, S.; Sörensson, P.; Ekenbäck, C.; Lundin, M.; Agewall, S.; Brolin, E.B.; Caidahl, K.; Cederlund, K.; Collste, O.; Daniel, M.; et al. Plasma catecholamine levels in the acute and subacute stages of takotsubo syndrome: Results from the Stockholm myocardial infarction with normal coronaries 2 study. *Clin. Cardiol.* **2021**, *44*, 1567–1574. [CrossRef]
13. Redfors, B.; Shao, Y.; Omerovic, E. Stress-induced cardiomyopathy in the critically ill. Why inotropes fail to improve outcome. *Int. J. Cardiol.* **2013**, *168*, 4489–4490. [CrossRef] [PubMed]
14. Prasad, A.; Lerman, A.; Rihal, C.S. Apical ballooning syndrome (tako-tsubo or stress cardiomyopathy): A mimic of acute myocardial infarction. *Am. Heart J.* **2008**, *155*, 408–417. [CrossRef]
15. Ghadri, J.R.; Wittstein, I.S.; Prasad, A.; Sharkey, S.; Dote, K.; Akashi, Y.J. International Expert Consensus Document on Takotsubo Syndrome (Part I): Clinical characteristics, diagnostic criteria, and pathophysiology. *Eur. Heart J.* **2018**, *39*, 2032–2046. [CrossRef] [PubMed]
16. Okura, H.; Nakada, Y.; Ishihara, S.; Nogi, M.; Onoue, K.; Soeda, T. "Hidden & quot; takotsubo cardiomyopathy in cardiac care unit. *J. Echocardiogr.* **2020**, *18*, 113–116.
17. Citro, R.; Okura, H.; Ghadri, J.R.; Izumi, C.; Meimoun, P.; Izumo, M. Multimodality imaging in takotsubo syndrome: A joint consensus document of the European Association of Cardiovascular Imaging (EACVI) and the Japanese Society of Echocardiography (JSE). *J. Echocardiogr.* **2020**, *18*, 199–224. [CrossRef]
18. Friedrich, M.G.; Sechtem, U.; Schulz-Menger, J.; Fausto, J.P.; Brito, D. Cardio-vascular magnetic resonance in myocarditis: A JACC White Paper. *J. Am. Coll. Cardiol.* **2009**, *53*, 1475–1487. [CrossRef]
19. Ghadri, J.R.; Cammann, V.L.; Jurisic, S.; Seifert, B.; Napp, L.C.; Diekmann, J. A novel clinical score (InterTAK Diagnostic Score) to differentiate takotsubo syndrome from acute coronary syndrome: Results from the International Takotsubo Registry. *Eur. J. Heart Fail.* **2017**, *19*, 1036–1042. [CrossRef]
20. Santoro, F.; Ieva, R.; Ferraretti, A.; Ienco, V.; Carpagnano, G.; Lodispoto, M.; Di Biase, L.; Di Biase, M.; Brunetti, N.D. Safety and feasibility of levosimendan administration in Takotsubo cardiomyopathy: A case series. *Cardiovasc. Ther.* **2013**, *31*, e133–e137. [CrossRef]
21. Meyer, P.; Degrauwe, S.; Van Delden, C.; Ghadri, J.R.; Templin, C. Typical takotsubo syndrome triggered by SARS-CoV-2 infection. *Eur. Heart J.* **2020**, *41*, 1860. [CrossRef] [PubMed]
22. Aldujeli, A.; Hamadeh, A.; Briedis, K.; Tecson, K.M.; Rutland, J.; Krivickas, Z.; Stiklioraitis, S.; Briede, K.; Aldujeili, M.; Unikas, R.; et al. Delays in presentation in patients with acute myocardial infarction during the COVID-19 pandemic. *Cardiol. Res.* **2020**, *11*, 386–391. [CrossRef] [PubMed]

MDPI AG
Grosspeteranlage 5
4052 Basel
Switzerland
Tel.: +41 61 683 77 34

Life Editorial Office
E-mail: life@mdpi.com
www.mdpi.com/journal/life